Vickery

Ecology of
Soil Seed Banks

Ecology of Soil Seed Banks

Edited by

Mary Allessio Leck
Department of Biology
Rider College
Lawrenceville, New Jersey

V. Thomas Parker
Department of Biology
San Francisco State University
San Francisco, California

Robert L. Simpson
School of Science and Mathematics
William Paterson College
Wayne, New Jersey

ACADEMIC PRESS, INC.
Harcourt Brace Jovanovich, Publishers
San Diego   New York   Berkeley   Boston
London   Sydney   Tokyo   Toronto

Academic Press, Inc.
San Diego, California 92101

United Kingdom Edition published by
Academic Press Limited
24–28 Oval Road, London NW1 7DX

Library of Congress Cataloging-in-Publication Data

Ecology of soil seed banks / edited by Mary Allessio Leck, V. Thomas
    Parker, Robert L. Simpson.
        p.   cm.
    Bibliography: p.
    Includes index.
    ISBN   0-12-440405-7   (alk. paper)
    1. Soil seed banks. 2. Seeds--Ecology. 3. Vegetation dynamics.
I. Leck, Mary Allessio. II. Parker, V. Thomas. III. Simpson,
Robert L. (Robert Lee), Date.
QK910.E27   1989
582'.05'6--ec19   88-28815
                                                                CIP

Printed in the United States of America
89 90 91 92   9 8 7 6 5 4 3 2 1

Contents

Contributors

O. W. Archibold
Department of Geography
University of Saskatchewan
Saskatoon, Saskatchewan
Canada S7N 0W0

Herbert G. Baker
Department of Botany
University of California
Berkeley, California 94720

Carol C. Baskin
School of Biological Sciences
University of Kentucky
Lexington, Kentucky 40506

Jerry M. Baskin
School of Biological Sciences
University of Kentucky
Lexington, Kentucky 40506

Diane L. Benoit
Agriculture Canada
Research Station, C.P. 457
St.-Jean-sur-Richelieu
Quebec, Canada J3B 6Z8

Paul B. Cavers
Department of Plant Sciences
University of Western Ontario
London, Ontario
Canada N6A 5B7

Nancy C. Garwood
Department of Botany
Field Museum of
   Natural History
Chicago, Illinois 60605, and
Smithsonian Tropical Research
   Institute
Balboa, Panama

Connie Gaudet
Department of Biology
University of Ottawa
Ottawa, Ontario
Canada K1N 6N5

J. P. Grime
Unit of Comparative Plant
   Ecology (NERC)
Department of Plant Sciences
University of Sheffield
Sheffield, England S10 2T

Paul A. Keddy
Department of Biology
University of Ottawa
Ottawa, Ontario
Canada K1N 6N5

Victoria R. Kelly
Institute of Ecosystem Studies
Mary Flagler Cary Arboretum
Millbrook, New York 12545

Paul R. Kemp
Systems Ecology Research Group
College of Sciences
San Diego State University
San Diego, California 92182

Mary Allessio Leck
Department of Biology
Rider College
Lawrenceville, New Jersey 08648

Svaťa M. Louda
School of Biological Sciences
University of Nebraska
Lincoln, Nebraska 68588

M. J. McDonnell
Institute of Ecosystem Studies
The New York Botanical Garden
Mary Flagler Cary Arboretum
Millbrook, New York 12545

James B. McGraw
Department of Biology
West Virginia University
Morgantown, West Virginia 26506

V. Thomas Parker
Department of Biology
San Francisco State University
San Francisco, California 94132

Roger L. Pederson
Ducks Unlimited, Inc.
Western Regional Office
9823 Old Winery Place
Suite 16
Sacramento, California 95827

S. T. A. Pickett
Institute of Ecosystem Studies
The New York Botanical Garden
Mary Flagler Cary Arboretum
Millbrook, New York 12545

Kevin J. Rice
Department of Agronomy and
    Range Science
University of California
Davis, California 95616

Bill Shipley
Department of Biology
University of Ottawa
Ottawa, Ontario
Canada K1N 6N5

Robert L. Simpson
School of Science and
    Mathematics
William Paterson College
Wayne, New Jersey 07470

A. G. van der Valk
Department of Botany
Iowa State University
Ames, Iowa 50011

Milan C. Vavrek
Department of Biology
West Virginia University
Morgantown, West Virginia 26506

D. Lawrence Venable
Department of Ecology and
    Evolutionary Biology
University of Arizona
Tucson, Arizona 85721

Irene C. Wisheu
Department of Biology
University of Ottawa
Ottawa, Ontario
Canada K1N 6N5

Preface

The old adage, "One year's seeding—seven years' weeding," indicates that gardeners have long understood the potential of seeds to remain viable in soil. In recent years many ecologists interested in plant population and community dynamics have also come to realize the importance of seeds in the soil. The soil seed bank refers to seeds and fruits, such as achenes and caryopses, on or in the soil.

How seeds enter and leave the soil and their persistence are determined by a variety of factors. These include the seed rain, dispersal, predation, longevity, and factors controlling germination and recruitment. The importance of these varies with species, and in turn, species vary in their dependence on persistent seed banks. Some have transitory seed banks, lasting less than a year, whereas others persist in the soil for decades or perhaps even a millenium or more. Knowledge of the seed bank dynamics of species that make up a community can provide an understanding of important limiting factors or processes that occur within that community. Ultimately, the dynamics of the seed bank guarantee the ability of a community to maintain itself and to respond to change. The reservoir of seeds in soil, the seed bank, may contain diverse species, genotypes, and phenotypes that provide substantial flexibility for potential community response.

The ecology of seeds in soils has been considered by Heydecker (1973), Mayer and Poljakoff-Mayber (1975), Harper (1977), Roberts (1981), Fenner (1985), and Priestly (1986). A number of basic aspects of seed ecology have been widely studied, such as seed physiology and germination (Murray, 1984; Bewley and Black, 1985), dispersal (Murray, 1986), and variations in life history in relationship to seed size or number (Harper 1977). While this volume provides an update and brings together much scattered information, it approaches the topic from a different perspective. We feel that understanding the ecology of seeds is critical for developing theory on community structure and function.

Our primary objectives are to examine factors influencing seed bank dynamics and the variety of patterns found among different species. In addition, we present seed banks in a community context to explore the ecological implications of different patterns and thus begin the development of a synthesis by comparing various communities.

The history of seed bank studies illustrates a disparate and fragmented genesis, the result of a multitude of approaches, goals, and methodological shortcomings. Life history studies of individual species have provided the basis for understanding the underlying patterns.

In this volume we first examine the general processes that influence inputs or losses from the seed bank: predation, dormancy/germination mechanisms, and their evolutionary importance. While we recognize the importance of dispersal, a manuscript on that topic was not forthcoming. Fortunately, most chapters touch on dispersal in the context of seed bank dynamics. Interested readers should consult recent reviews (e.g., Howe and Smallwood, 1982; Murray, 1986) for more details on dispersal.

Second, we examine seed banks in a community context. Only eight vegetation types are included, but the range in diversity of life form, length of growing season, and dominant environmental conditions allow comparisons of seed bank patterns. Finally, the role of seed banks in vegetation management is examined. In some managed habitats the goal is elimination of species, e.g., agricultural weeds, whereas in others the goal is to increase populations, e.g., rare species or those required for vegetation restoration. The diverse approaches to the study of seed banks presented by the contributors provide a framework on which to base generalizations, and provide direction for future studies.

Because we examine processes, specific vegetation types, and management practices as they relate to the role of seed banks in vegetation dynamics, this book should be of interest to population and community ecologists and managers. Moreover, evolutionary consequences of seed banks should be of interest to population and theoretical biologists.

We thank the contributors for insights regarding methodological deficiencies on which part of Chapter 1 is based. All indicated the importance of standardization of techniques; toward this end a working group should explore the means to bring guidelines to seed bank methods.

We wish to express our gratitude to those who reviewed manuscripts, including Susan H. Bicknell, Dwight Billings, Gregory P. Cheplick, Laurel R. Fox, Susan Kaliz, Cees M. Karssen, Mark J. McDonnell, Robert W. Patterson, Steward T. A. Pickett, Francis E. Putz, O. James Reichman, Kevin J. Rice, J. Stan Rowe, Loren M. Smith, Robert D. Sutter, Irwin A. Ungar, Arnold G. van der Valk, and Dennis F. Whigham. We are indebted to Florence Sackett for cheerfully keeping up with typing; Elizabeth Faulkenstein, Linda Feeney, and Richard Deni for computer assistance; and James H. Carlson. Rider College, William Paterson College, and San Francisco State University supported many aspects of the preparation of this volume. We express our gratitude to Academic Press, especially Jean Thomson Black and Kerry Pinchbeck. We especially thank our spouses, Charles F. Leck, Judith D. Parker, and Penelope Simpson, for their considerable tolerance and understanding.

Mary Allessio Leck
V. Thomas Parker
Robert L. Simpson

Bewley, J. R., and Black, M. (1985). "Seeds: Physiology of Development and Germination." Plenum Press, New York.

Fenner, M. (1985). "Seed Ecology." Chapman and Hall, London.

Harper, J. L. (1977). "Population Biology of Plants." Academic Press, London.

Heydecker, W., ed. (1973). "Seed Ecology." Pennsylvania State University Press, University Park and London.

Howe, H. F., and Smallwood, J. (1982). Ecology of Seed Dispersal. *Annu. Rev. Ecol. Syst.* **13,** 201–228.

Mayer, A. M., and Poljakoff-Mayber, A. (1975). "The Germination of Seeds." Macmillan, New York.

Murray, D. R., ed. (1984). "Seed Physiology." Vols. 1 and 2. Academic Press, Sydney.

Murray, D. R., ed. (1986). "Seed Dispersal." Academic Press, Sydney.

Priestly, D. A. (1986). "Seed Aging: Implications for Seed Storage and Persistence in the Soil." Cornell University Press, Ithaca, New York.

Roberts, H. A. (1981). Seed banks in soil. *Adv. Appl. Biol.* **6,** 1–55.

Seed Banks in Ecological Perspective

J.P. Grime

Unit of Comparative Plant Ecology (NERC)
Department of Plant Sciences
University of Sheffield
Sheffield, England

I. Introduction

The appearance of this volume signifies that, at last, investigations of seed and spore banks have become a recognized and indispensable part of plant ecology. It is salutory to reflect that 59 years lie between this first synthesis and Ridley's classical work on the dispersal of propagules (Ridley, 1930). This delay is a measure of the priority which, until recently, spatial patterns have exerted over temporal dynamics in the minds of plant ecologists.

The recent promotion of seed banks into the foreground of ecological theory and practice is therefore one symptom of a major shift in emphasis toward the analysis of events over time, rather than the detection of correlations through space, as an approach to understanding the functioning of populations and communities. This transition has brought to fruition the work of the pioneers of plant population dynamics (e.g., Watt, 1947; Tamm, 1956) and has been interpreted by some authors (e.g., Harper, 1977, 1982) as a formula for ultimate success in the attempt to develop generalizing principles in ecology. However, the ordination of data along temporal, as opposed to spatial, base lines does not by itself guarantee a sure path to general ecological theory. Demography is as capable as cartography of descending into a fragmentary descriptive exercise when pursued without reference to underlying principle. Hence, in the specific case of seed bank information, it seems imperative that demographic analysis is allied to complementary forms of study. Of particular importance are those which relate the emerging typology of seed banks to the fundamental design constraints which restrict both ranges and amplitudes in seed form and function. Equally important, however, is the need to place seed banks securely in relation to the forms of natural selection that over evolutionary time and in the generation times of contemporary populations influence the fate of juveniles in natural habitats.

II. Origins

The seed banks of arable weeds were among the first to receive intensive study (Brenchley, 1918), and their role in the rapid exploitation of disturbed ground achieved public recognition when the Flanders poppy (*Papaver dubium*) was adopted as a memorial symbol for those who fell on the cratered battlefields of World War I. The presence of substantial numbers of dormant seeds in the soil has been detected in a wide range of other habitats, including some occupied by a closed cover of perennial vegetation. Especially notable contributions include those of Darwin (1859), Milton (1939), Gómez-Pompa (1967), and Marks (1974), all of whom recognized that the presence of a reserve of dormant seeds conferred the potential for population recovery following disturbance of the established vegetation.

Until recently the characterization of seed banks mainly consisted of attempts to determine whether particular species developed persistent seed banks in the soil. This usually involved a census of buried dormant seeds conducted at one occasion; in the majority of studies viable seeds were detected through the appearance of seedlings during laboratory incubation of soil samples, but several investigations (e.g., Kropáč, 1966; Major and Pyott, 1966) assumed heroic proportions through the commitment of the authors to the task of recovering and identifying all living, ungerminated seeds by extremely tedious fractionating procedures. Seed persistence has been demonstrated also by examination of the viable seed content of soil samples from beneath buildings of known antiquity (Ødum, 1965) and by experimental burials of seeds, followed by germination tests on samples exhumed at various intervals (Darlington and Steinbaur, 1961).

The decade 1970–1980 coincided with a growing appreciation of the critical importance of regeneration mechanisms in the functioning of plant populations and the structuring of plant communities. It was soon apparent that the objectives in both of these fields of inquiry could not be met by a simple dichotomy between plants which were capable and those which were incapable of long-term seed persistence.

III. Seed Bank Classification

A critical step in the development of a more sophisticated typology of seed banks was the completion of studies in which the fate of seeds was monitored by programs of surface soil sampling at frequent intervals throughout the year. In the investigation of Sarukhán (1974), for example, this approach allowed quantitative assessments of seed longevity and turnover in the seed banks of three coexisting pasture species of *Ranunculus,* and it was possible to incorporate the seed bank data into models representing the population dynamics of each species. In a more comprehensive but less detailed study, Thompson and Grime (1979) measured seasonal variation in the densities of germinable seeds in 10 contrasting habitats. As in many earlier investigations, few consistent relationships were found between the density of buried seeds exhibited by a species and its abundance in the community of established plants. More significant, however, was the recognition that seasonal patterns in the density of germinable seeds observed in various communities fell into four basic types that were consistent with the morphology and germination characteristics of seeds examined in the laboratory (Fig. 1).

This study, in attempting a general linkage between seed bank type and germination physiology, built upon a number of pioneering studies, such as those of Went (1949), Ratcliffe (1961), Vegis (1964), and Wesson and Wareing (1969a). In retrospect, the lack of interaction between ecologists and germination physiologists, which persisted well into the 1970s, can now be viewed as one of the most puzzling examples of missed opportunities in the checkered history of biological science; all of the physiological phenomena that are now crucial to a mechanistic understanding of seed banks (e.g., afterripening, chilling requirements, light requirements, responses to fluctuating temperatures, canopy-induced dormancy) had been the subject of intensive study over many years prior to the development of a seed bank classification.

The classification in Fig. 1 is lacking in subtlety, but has the merit that logical connections can be established with major factors influencing the fate of juveniles. It also enables seed bank types to be placed in relation to other regenerative strategies (Table 1), such as vegetative expansion and the capacity to develop a bank of persistent seedlings. At this relatively crude level of classification, it is also possible to explore the parallels between seed banks and comparable phenomena in various heterotrophs, including fungi, insects, and fish (Wourms, 1972). This is not to suggest, however, that seed bank classification should not be further elaborated; several attempts have been made already (e.g., Roberts, 1981; Grime, 1981), and there is opportunity for further refine-

A

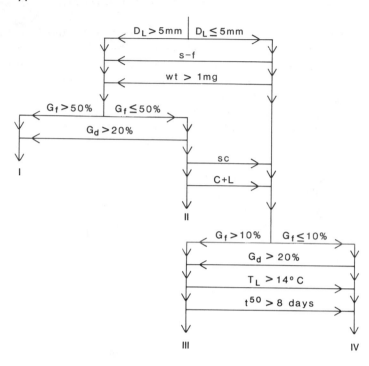

B

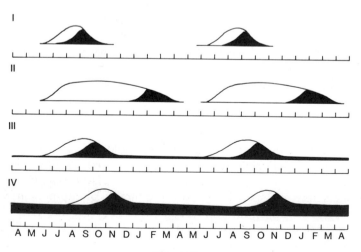

Figure 1.

ments that take account of factors such as seed polymorphism, dispersal vectors, and more complex patterns of secondary dormancy.

The rapid progress which has been achieved recently in our understanding of seed banks must not obscure the fact that we remain largely ignorant of some important aspects of their functioning and ecology. In particular, in certain habitats there is a lack of precise information on the role of seed banks that remain attached to aerial parts of the parent plant. These have been recognized in closed cone pines (Vogl, 1973) and desert ephemerals (Dr. Y. Gutterman, personal communication), where protection of the seeds against predation appears to be of crucial importance. These rather exotic examples provide a reminder that we are still far from a complete understanding of the role of herbivory in the evolution and functioning of the seed banks of many of the most universally common plants. As Thompson (1987) has suggested, there is need to test the hypothesis that the minute size of the seeds in the majority of species forming persistent seed banks in perennial vegetation is the result of the preference for large food items exercised by many herbivores. The extreme vulnerability of seedlings originating from tiny seeds with associated development of "depth sensing" and "gap detection" mechanisms *sensu* Thompson *et al.* (1977) suggests that there must be extremely strong counterveiling selection forces operating to resist natural selection for larger seeds.

Figure 1. (A) A key using the laboratory characteristics of the seed to predict four seed bank types developed in the field (after Grime and Hillier, 1981). Guide to symbols: D_L, length of dispersule; s–f, seed not readily detached from fruit; wt, weight of seed; G_f, maximum percentage germination achieved by fresh seed; G_d, maximum percentage germination achieved by seed stored dry at 20°C for 1 month; SC, seed requires scarification; C + L, seed requires chilling and light to break dormancy; T_L, lowest temperature at which 50% germination is achieved; t^{50}, time taken by seed stored dry at 20°C for 1 month to reach 50% germination. (B) Scheme describing four types of seed banks of common occurrence in temperate regions (after Thompson and Grime, 1979). Shaded area: seeds capable of germinating immediately after removal to suitable laboratory conditions. Unshaded area: seeds viable but not capable of immediate germination. (I) Annual and perennial grasses of dry or disturbed habitats (e.g., *Hordeum murinum, Lolium perenne,* and *Catapodium rigidum*) capable of immediate germination. (II) Annual and perennial herbs, colonizing vegetation gaps in early spring (e.g., *Impatiens glandulifera, Anthriscus sylvestris,* and *Heracleum sphondylium*). (III) Annual and perennial herbs, mainly germinating in the autumn but maintaining a small seed bank (e.g., *Arenaria serpyllifolia, Holcus lanatus,* and *Agrostis tenuis*). (IV) Annual and perennial herbs and shrubs with large, persistent seed banks (e.g., *Stellaria media, Chenopodium rubrum,* and *Calluna vulgaris*).

Table 1
Five Regenerative Strategies of Widespread Occurrence in Terrestrial Vegetation

Strategy	Functional characteristics	Conditions under which strategy appears to enjoy a selective advantage
Vegetative expansion	New shoots vegetative in origin and remaining attached to parent plant until well established	Productive or unproductive habitats subject to low intensities of disturbance
Seasonal regeneration (includes seed banks Types I and II)[a]	Independent offspring (seeds or vegetative propagules) produced in a single cohort	Habitats subjected to seasonally predictable disturbance by climate or biotic factors
Persistent seed or spore bank (includes seed banks Types III and IV)[a]	Viable but dormant seeds or spores present throughout the year; some persisting more than 12 months	Habitats subjected to temporally unpredictable disturbance
Numerous widely dispersed seeds or spores	Offspring numerous and exceedingly buoyant in air; widely dispersed and often of limited persistence	Habitats subjected to spatially unpredictable disturbance or relatively inaccessible (cliffs, walls, tree trunks, etc.)
Persistent juveniles	Offspring derived from an independent propagule, but seedling or sporeling capable of long-term persistence in a juvenile state	Unproductive habitats subjected to low intensities of disturbance

[a]Seed bank classification explained in Fig. 1B.

IV. Seed Banks in a Wider Ecological Context

As will become abundantly clear from later chapters in this book, the presence of different types of seed banks provides a vital clue to some of the mechanisms which permit species to coexist in perennial plant communities. This diversity in seed banks not only suggests how variation

Table 2
The Proportion of Increasing and Decreasing Species
Associated with Different Types of Seed Banks
in the British Flora[a,b]

Seed bank type	Number of species	Percentage increasing	Percentage decreasing	Increasing/ decreasing
I	33	27	52	0.53
II	39	18	66	0.32
III	48	50	33	1.50
IV	151	39	41	0.95

[a]After Grime et al., 1988.
[b]Seed bank classification the same as that explained in Fig. 1B.

in the form, intensity, and seasonal distribution of habitat disturbance facilitates complementary forms of regeneration, but it also provides a basis for informed manipulation of species composition by shifting the opportunities for seedling establishment in one direction or another.

As research continues on the regenerative mechanisms of plants throughout the world, the data base will become sufficiently comprehensive to allow comparisons between national and local floras with respect to the relative importance of seed bank types (cf. the life form spectra of Raunkiaer, 1934). This will allow rigorous tests for the existence of geographical patterns in seed bank types related to latitude, variation in habitat productivity, and frequency of vegetation disturbance (Livingston and Allessio, 1968; K. Thompson, 1978). As an illustration of this approach, Table 2 classifies 171 common British plants into the seed bank Types I–IV of Thompson and Grime (1979). The species in each category are further classified according to whether they are currently increasing or decreasing in abundance in Britain. The data reveal that the capacity for expansion is strongly consistent with the hypothesis (Grime, 1981) that species of seed bank Type III combine the potential for rapid expansion of populations into disturbed habitats with the capacity to persist as dormant seeds when circumstances are not immediately favorable to colonization. Species of seed bank Type III are clearly encouraged by the increasingly disruptive patterns of land use now applied to the British landscape. Table 2 also indicates declining abundance in species with Type II seed banks, which characteristically involve synchronous germination of relatively large seeds in the early spring. This reflects the continuing loss of native broad-leaved deciduous woodland in which seed bank Type II predominates in the tree, shrub, and herbaceous components.

In an even wider context, the increasing confidence with which plant ecologists can now identify and elucidate the various types of seed banks provides a botanical contribution to the current effort to devise a universal functional classification of organisms. Already there are pointers (e.g., Noble and Slatyer, 1980; Grime, 1987) to the way in which seed banks can be incorporated into general models of vegetation succession and cyclical response to habitat perturbation. Later, as ecological theory solidifies into hard predictive science, seed banks will surely find their place within the assembly rules of ecosystems.

PART 1

Introduction

CHAPTER 1

Seed Banks: General Concepts and Methodological Issues

Robert L. Simpson

School of Science
William Paterson College
Wayne, New Jersey

Mary Allessio Leck

Biology Department
Rider College
Lawrenceville, New Jersey

V. Thomas Parker

Department of Biology
San Francisco State University
San Francisco, California

I. Overview

II. Methodological Considerations

III. Questions Related to Seed Banks

I. Overview

All viable seeds present on or in the soil or associated litter constitute the soil seed bank. Each has spatial and temporal dimensions. Seeds display both horizontal and vertical dispersion, reflecting initial dispersal onto the soil and subsequent movement. Seed banks may be either transient, with seeds that germinate within a year of initial dispersal, or persistent, with seeds that remain in the soil more than 1 yr. This persistent component of the seed bank embodies a reserve of genetic potential accumulated over time that simultaneously represents proximate genetic diversity for the population and ultimate genetic expression on which natural selection can act.

Seed bank input is determined by the seed rain (Fig. 1). Within a community, local dispersal predominates, but inputs from distant seed sources occur and may make a major contribution to the vegetation. Local dispersal may be passive, by mechanical ejection of the seed from the fruit, or by fire, wind, water, and animals; the latter three agents are also important for long-range dispersal. Seed bank losses result from: genetically controlled physiological responses to environmental cues, including light, temperature, water, oxygen tension, and chemical stimulants, leading to germination; from processes leading to deep burial or redispersal; from interaction with animals and pathogens leading to death; and from natural senescence leading to physiological death (Fig. 1). These inputs and outputs directly control seed density, species composition, and genetic reserve; other life history processes indirectly influence these parameters. Shifts in the relative importance of these processes over time govern seed bank dynamics.

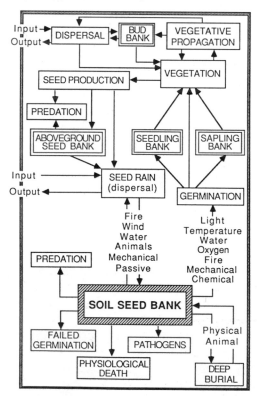

Figure 1.   General model of seed bank and vegetation dynamics.

Ultimately the vegetation is perpetuated in two ways, through production of seeds or vegetative propagation (Fig. 1). Indeed, for some species perpetuation is exclusively by vegetative means. Vegetative propagation can give rise to a bud bank (*sensu* Harper, 1977) with an accumulation of bulbs, bulbils, and buds on rhizomes, corms, and tubers. Unlike the seed bank, the bud bank conserves already established genotypes. As with seeds, however, vegetative propagules may be dispersed through time and space and may require environmental cues such as temperature or moisture to break dormancy or initiate growth.

While seeds are largely produced in an aerial environment, they are generally found in the soil, where water and minerals are most readily available. In some instances, however, it is advantageous to store the genetic reserve elsewhere. Thus, ecologically equivalent aboveground seed banks occur in fire-prone areas, sapling banks occur in forests, and seedling banks occur in a variety of communities (Fig. 1). Responses are environmentally cued, fire results in seed dispersal from aboveground banks, increased light created by disturbance stimulates sapling growth, and favorable temperature and moisture initiate continuation of seedling growth.

II. Methodological Considerations

The data on seed banks have been collected and presented in a myriad of forms, often making comparisons across studies and communities extremely difficult or impossible. Among the commonly encountered problems are heterogeneity of sampling methods, insufficient number of samples, failure to sample throughout the year and for more than 1 yr, inappropriate sample design, failure to consider dormancy breaking and germination requirements, lack of controls, inadequate provision of summary data, and failure to analyze the data statistically.

While it is unreasonable to expect standardization of sampling techniques across ecosystems, several general considerations emerge. A good basic knowledge of the community to be sampled, including the life history characteristics of the species, especially the dominants, is essential. What are the major controlling environmental variables for the system? How heterogeneous is the environment? The principal attributes of seed banks must be a central focal point for base line information. These attributes include how many species are found in the seed bank and in what densities. What are the spatial characteristics, both horizontally in the community and vertically in the soil column? What proportion of the seed bank of any species is transient or persistent and what are the important germination cues? What are the important temporal changes, especially timing of dispersal with respect to germination and changes in density due to other processes such as predation?

Perhaps the most critical question to be addressed when initiating seed bank studies is project design. Ideally, the first year should establish procedures based upon clearly focused goals (Hutchings, 1986). Initial information required for a successful study includes sample collecting and handling techniques, germination requirements, species profiles, variability, sample size and number, and unique aspects of the system. The actual study should then be performed in subsequent years.

While project design will vary with the requirements of the study, its essential features must be described in sufficient detail to permit comparability of data from one study to another. The size of the sample area and the location where samples were collected within the sample area should be presented. It is also important to describe neighboring habitats, as they may have an important influence on the seed bank of the site under study. Distances from these habitats to the sample site should be mentioned.

A description of how sample sites were selected (e.g., random, systematic, stratified random, and clustered) is essential. Usually several factors, including the purpose of the study, the expected physical heterogeneity of the site, the precision desired, and the time and labor involved, will dictate the procedure used. In addition, the practical problem of available space (e.g., greenhouse or growth chamber) to accommodate samples must be considered. If heterogeneity is expected, stratified random sampling offers advantages because separate values of the mean are possible for different parts of the sample and a more even distribution of sampling units results (Sampford, 1962). Several recent works (e.g., Roberts, 1981; Benoit, 1986; Bigwood and Inouye, 1988) consider these issues in detail.

The size of the sampling unit (area and depth), the number of samples collected, and the distance between sample sites when samples are collected along transects must be presented. The volume of soil collected should be large relative to the size of seeds in the soil (Sampford, 1962) and should include the litter layer, which may contain many seeds. Optimal size of sampling units can be determined by species area curves (e.g., Liew, 1973; Kellman, 1978; Hutchings, 1986). A large number of small sampling units appears more appropriate than a few large ones for the same total volume of soil sampled (Roberts, 1958, 1970, 1981; Kropáč, 1966; Dospekhov and Chekryzhov, 1972; Benoit, 1986; Bigwood and Inouye, 1988). The number of samples collected is a function of the expected seed density and the degree of precision desired, rather than the size of the area surveyed (Goyeau and Fablet, 1982; Forcella, 1984; Pratt *et al.*, 1984). The number of samples needed increases when rare as well as common species are of interest or if seeds are aggregated (e.g., Goyeau and Fablet, 1982). Preferably, the appropriate number of samples is established statistically from an initial pilot study (Hutchings, 1986). This information along with estimates of sample variation should be presented routinely for seed bank studies.

Two techniques are available to establish seed bank densities once samples are collected, direct counting and seedling emergence. Direct counting, often utilizing flotation, sieving, or other separation methods, determines the total number of seeds in the soil, but gives no information about viability (Kropáč, 1966; Roberts, 1981), which must be subsequently established using tetrazolium or germination tests. In contrast, seedling emergence techniques provide an estimate of viable seeds in the soil based on germination of seeds maintained under conditions favorable to germination. These requirements are seldom completely met, as germination patterns are very sensitive to light (Baskin and Baskin, Chapter 4, this volume), fluctuating temperatures (Grime et al., 1981), oxygen availability (Leck and Simpson, 1987a), soil texture (Harper et al., 1965; Keddy and Ellis, 1985), and other factors. Thus seedling emergence techniques may greatly underestimate viable buried seed abundances. Despite this limitation, for community level studies, especially where the potential number of species is high, seedling emergence is desirable because direct counting is extremely tedious and also requires that viability be tested. A combination of seedling emergence and direct counting provides a more precise estimate of seed bank size than either technique alone (Conn et al., 1984).

III. Questions Related to Seed Banks

While there are several recognizable reproductive banks in the community (Fig. 1), by virtue of its size and importance in life history, the soil seed bank assumes overwhelming importance. The existence and importance of the soil seed bank has been recognized at least since Darwin's time (Darwin, 1859), but seed banks have received intensive study only within the last two decades. The subsequent chapters in this volume describe seed bank dynamics, focusing on the following set of questions:

1. Are there differences in seed bank characteristics among congeneric species, among genera within a family, etc.? Are there underlying ecological explanations for differences?

2. Are there differences in seed bank dynamics that characterize particular selective pressures? Are there interactions among

stress and disturbance gradients and seed physiology that result in the observed dynamics?

3. Are there adaptive syndromes that arise repeatedly, among certain life forms, among species characterizing particular ecological positions, or among different communities?

4. How and why are seed banks of major vegetation types similar or different?

The differentiation of seed bank dynamics among species in a community is central to short- or long-term changes in composition. At the population level the timing of emergence from the seed bank may determine survival rates (Cook, 1979). Differential survival and response to a variety of disturbances can be critical to maintaining species diversity (Grubb, 1977a; Pickett and White, 1985a). Exploration of the role seed banks play in population or community processes is just beginning.

Chapters 2–5 consider general processes that organize and influence seed banks in every community: life history and reproductive strategies (Baker, Chapter 2), predation (Louda, Chapter 3), dormancy and germination (Baskin and Baskin, Chapter 4), and evolution of seed banks using models (Venable, Chapter 5). Seed dispersal is discussed in the context of individual vegetation associations; recent reviews (e.g., Howe and Smallwood, 1982; Murray, 1987a) consider this topic in depth. Chapters 6–13 examine major environmental constraints, size and composition of the seed bank, and adaptive strategies displayed by seed bank species in major ecosystems: arctic and alpine (McGraw and Vavrek, Chapter 6), coniferous forests (Archibold, Chapter 7), temperate deciduous forests (Pickett and McDonnell, Chapter 8), tropical forests (Garwood, Chapter 9), grasslands (Rice, Chapter 10), chaparral and other Mediterranean environments (Parker and Kelly, Chapter 11), deserts (Kemp, Chapter 12), and wetlands (Leck, Chapter 13). Chapters 14–16 address issues related to management of soil seed banks: arable lands (Cavers and Benoit, Chapter 14), vegetation management (van der Valk and Pederson, Chapter 15), and species conservation (Keddy *et al.*, Chapter 16). The final chapter considers directions for future seed bank research (Parker *et al.*, Chapter 17). Collectively these chapters provide an extensive, though not exhaustive, overview of the role seed banks play, and raise important questions about the nature of seed banks that require further elucidation.

CHAPTER 2

Some Aspects of the Natural History of Seed Banks

Herbert G. Baker

Department of Botany, University of California,
Berkeley, California

I. Introduction

As a generalist student of plant reproductive biology, I have some concern with seeds and seed banks, but I will not attempt a comprehensive and intensive review of published work; the specialists in this book are

substantial claims of exceptionally prolonged viability are some very fascinating stories, but an element of doubt exists about the accuracy of dating the seed in most cases (Godwin and Willis, 1964; Godwin, 1968).

There is a paradox in reports of extended longevity of seeds. Herbarium material can be dated accurately, but the seeds have existed for many years in the unnatural environment of the herbarium. On the other hand, seeds that are taken from seed banks in the soil are only indirectly datable. The buried seed experiments are intermediate in their naturalness.

C. Herbarium Studies

The herbarium experiments provide clues for retention of viability in nature. Consistent laboratory studies of seed viability seem to have been pioneered by Ewart (1908), who carried out an enormous investigation of herbarium specimens largely of Australian species.

The most famous herbarium studies are Becquerel's (1907, 1934) attempts at the Museum d'Histoire Naturelle in Paris to germinate seeds taken off herbarium sheets. They were mostly collected in the first half of the nineteenth century, but one specimen (*Cassia multijuga*, Leguminosae) dated from 1776. Although this specimen was 158 yr old at the time of testing, it showed 100% germination. Another *Cassia* species (*Cassia bicapsularis*) showed 40% germination after 115 yr. Most species that showed longevity ranging from 55 to 158 yr were members of the Leguminosae.

During World War II, the herbarium of the British Museum (Natural History) in London was hit by a fire bomb during an air raid in September, 1940. Hoses were used to extinguish the fire; as a result, some seeds of *Albizia julibrissin* (Leguminosae) were moistened and germinated (Ramsbottom, 1942a,b). Plants were grown from these seedlings and appeared quite normal. The seeds had been collected in China in 1713.

D. Collections from Nature

The claimant to the longest seed life of all is *Lupinus arcticus*, also in the Leguminosae. Seeds of this species were collected in a rodent burrow beneath permafrost in the Yukon Territory, Canada (Porsild *et al.*, 1967). There was no opportunity to carbon date the seeds. The authors based the age of the seeds (Pleistocene, circa 10,000 yr ago) on the age of animal remains alongside them. These seeds produced healthy progeny.

Another famous taxon that has figured in several claimed longevity records is the genus *Nelumbo* (or *Nelumbium*) (Nymphaeaceae), represented by the viable nutlets of Indian or Sacred lotus (*Nelumbo nucifera*) from a peat layer under loess in what was an old lake bottom in southern Manchuria (Ohga, 1923, 1926a). Estimates of seed ages varying from 150 to several thousand years were deduced from geological considerations (see Godwin, 1968). In addition, viable *Nelumbo nucifera* seeds were taken from an archaeological item, a boat estimated to be 3000 yr old, found in a lake near Tokyo, Japan (Ohga, 1926b). Robert Brown, between 1843 and 1853, germinated seeds of *Nelumbo nucifera* from the British Museum herbarium that were more than 150 yr old. Seeds from the same collection were still germinable after 237 yr (Ramsbottom, 1942a,b).

It is not generally publicized that Ohga (1926c) visited the British Museum and experimented with *Nelumbo* nutlets from the same source that Brown had used (from Sir George Stanton's mission to China in 1793), but could not make them germinate. They had apparently suffered fungal contamination. Yet Ramsbottom (1942b) reported successful germination of other seed from this collection.

Ødum (1965) reported that seeds, mostly of weedy species, remained viable under a church and in other archaeological sites in Denmark and Sweden for periods ranging from 600 to more than 1700 yr. However, it is likely that these were modern seeds that may have entered the soil during the excavations (Godwin, 1968).

Other evidence from Argentina may be more convincing. A seed of *Canna compacta* (Cannaceae) was used as a pellet in a rattle made from *Juglans australis* (Juglandaceae) fruit shell that was estimated to be 550 yr old (Sivori *et al.*, 1968). This archaeological evidence was supported by ^{14}C dating of the walnut shell, suggesting a 600-yr age for the seeds (Lerman and Cigliano, 1971).

Spira and Wagner (1983) tested the longevity of seeds of 40 species that had been taken from unfired bricks in California and northern Mexico collected by Hendry in the 1920s (Hendry, 1931) (also see Burcham, 1957). These bricks were from buildings approximately 200 yr old. Spira and Wagner tested the germinability of these seeds; only one seed of *Medicago polymorpha* germinated, but the vitality of seeds of 7 species (tested by tetrazolium staining) was demonstrated. This storage might not have been as stressful as it appears, because the adobe clay bricks provided insulation from summer heat. Even so, the habitat is not a typical one for the species concerned.

Youngman (1951) makes the valid point that, ecologically, the proportion of seeds of a species that remain viable for an adequate time, rather than a record-breaking performance by a single seed, is what is important.

E. Experiments

Early interest in the longevity of seeds actually buried in soil was generated by Beal in 1879 (Beal, 1905) and by Duvel (1904). The work of these investigators included long-term experiments with buried seeds (in sand in upturned, unstoppered bottles and in soil in earthenware pots, respectively). They were sampled at intervals. The seeds buried by Beal (23 species) were tested for germinative power by Darlington (1922, 1931), Darlington and Steinbaur (1961), and Kivilaan and Bandurski (1973, 1981). Those buried by Duvel (107 species) were tested by Toole and Brown (1946). There were large differences between the performances of species.

In the Beal experiment, it is notable that *Trifolium repens* (Leguminosae) seed lasted for only 5 yr and seed of *Euphorbia maculata* (Euphorbiaceae) and *Agrostemma githago* (Caryophyllaceae) lasted for less than that. In contrast, *Verbascum blattaria* showed 42% germination after 100 yr of burial; *Oenothera biennis* and *Rumex crispus* both showed low germination at 80 yr (Kivilaan and Bandurski, 1981).

Results of the Duvel experiment (Toole and Brown, 1946) showed that some seeds of most species (27 out of 48) were still capable of germination at the end of the 39-yr study. However, several species, e.g., *Agrostis vulgaris* (Gramineae), *Saponaria vaccaria* (Caryophyllaceae), *Neslia paniculata* (Cruciferae), *Sisymbrium altissimum* (Cruciferae), *Lycopersicon esculentum* (Solanaceae), and *Grindelia squarrosa* (Compositae), retained capability of germination for less than 10 yr.

Turner (1933) reported experiments with stored seeds collected in the Isle of Wight (England) and kept in glass jars (not hermetically sealed). These showed a low proportion of germination in 7 species (all Leguminosae) at ages 81 to 90 yr. When they were 100 yr old (Youngman, 1951), 900 seeds of 3 of the species were sown. Less than 1% of *Trifolium pratense* and *Lotus uliginosus* germinated but *Sarothamnus* (*Cytisus*) *scoparius* did not germinate.

F. Seed Size and Longevity

A common assumption is that large seeds (with presumably more food in storage) will be more likely to show extended longevity. The long-lived large seeds of Leguminosae and nutlets of *Nelumbo* and Labiatae, compared with the short-lived tiny seeds of *Salix* (Brinkman, 1974), would seem to support this assumption. However, some of the long-lived seeds are small (e.g., *Verbascum blattaria* and many of the Caryophyllaceae), while the huge seeds of *Aesculus* (Rudolf, 1974) and acorns of *Quercus* (Olson, 1974) are relatively short-lived (see also Cook, 1980). The death of seeds is not usually from exhaustion of food reserves, but from a failure of the supply of enzymes that mobilize food reserves. DNA repair systems may also decline (Bewley and Black, 1985).

III. Occurrences of Seed Banks

A. Weed Seed Banks

Weed seed banks have been studied longer and more intensively than other kinds because of their agricultural significance. Studies of weeds in pastures were made by Brenchley (1918), Brenchley and Warington (1930, 1933, 1936), and Chippendale and Milton (1934), and in arable lands by Lewis (1973), Roberts and Feast (1973a), and many other authors. Seeds of pioneer or early successional stages (including weeds) have seed banks which may persist to climax vegetation in temperate deciduous forests (e.g., Oosting and Humphreys, 1940; Livingston and Allessio, 1968; Nakagoshi, 1984a) and in tropical rain forests (e.g., Enright, 1985, for New Guinea; Putz and Appanah, 1987, for Malaysia).

Seeds of weed species in cropland have adapted to reach reproductive maturity at the most opportune time in relation to harvest. In relatively primitive agricultural systems, weed seed is harvested accidently with crop seed. The classic case is *Camelina sativa* (Cruciferae) in flax fields of Eastern Europe (Baker, 1974). Its adaptations include phenological and reproductive attributes that make the separation of weed seeds from crop seeds almost impossible without machinery. Until the advent of modern seed-cleaning machinery, weed seeds would have been sown with the crop seeds at the beginning of the next season. Now, the selective pressure on the weeds favors genotypes that cause flowering and seed setting to occur just before the crop is harvested, producing at least a temporary seed bank in the soil of the crop field (Baker, 1974; also see Salisbury, 1961).

B. Nonweedy Seed Banks

Seed banks may be found in a variety of other habitats, such as annual grasslands (e.g., in California; Major and Pyott, 1966), pastures (containing perennial grasses and forbs), hay meadows, cultivated land, waste places, and forests. They can also be found in wetlands. Intuitively one might expect wetland seed banks to be less frequent or less species rich in freshwater and saltwater marshes where scouring occurs. However, where scouring is not too severe, substantial seed banks may occur (Leck and Simpson, 1987a, and references therein). Grime (1979) points to the immense seed banks maintained under water by *Juncus* species in wetland habitats. Salt marsh seed banks were investigated by Milton

(1939). In warm, subtropical waters, sea grass beds, containing *Halodule wrightii* and *Syringodium filiforme* (Potamogetonaceae), have seed banks with seeds that may remain viable for at least 3 yr (McMillan, 1981, 1983). Lakes and ponds that dry up unpredictably tend to possess a seed bank of opportunistic rapidly growing and fruiting species (Salisbury, 1942).

The sterility of very acid sphagnum bogs may account for the good survival of seeds shed into this kind of habitat. Some seeds (of moist habitats; e.g., wetlands and wet tropics) survive better if they are kept moist. Seeds of most other terrestrial species usually show increased retention of viability if they are kept at a certain low level of moisture (see Bewley and Black, 1985). Went (1969) carried out an experiment with seeds of species from dry places in California in which the seeds were kept in a vacuum for 5 yr. He found a greater retention of their germinative power than in those that were under comparable conditions but lacking the vacuum. The ecological significance of this is not obvious.

C. Geocarpic Species

Some species have underground ovaries derived from partially or totally underground flowers or resulting from the burial of fruits by bending and extension of the floral axis, e.g., *Oenothera* (*Camissonia*) *ovata* (Onagraceae), *Trifolium subterraneum*, and *Arachis hypogaea* (Leguminosae), as well as species of *Stylochiton* and *Biarum* (Araceae) and *Ficus* (Moraceae). These create at least temporary seed banks in the soil, but I know of no evidence of how long they can persist in a dormant condition. The Israeli flora has been studied with geocarpy in mind (Zohary, 1937); several species of desert plants (mostly Leguminosae) show both aerial and subterranean seed production. A succinct consideration of geocarpy is given by van der Pijl (1969).

IV. Physiology of Seed Banks

A. Dormancy and Its Breakdown

The dormancy of the buried seeds may originally be primary (inherent) in the newly deposited seed, but may become secondary (imposed) with storage. The physiology of dormancy and germinability is well treated in Cook (1980), Bewley and Black (1985), Baskin and Baskin (Chapter 4, this volume), and other reference works on seeds.

Dormancy imposed by fruit coat or seed coat should be considered separately from embryo dormancy, the latter being chemically imposed.

Because of the complex nature of dormancy and its elimination by after-ripening, all seeds of one species may not behave alike. The collective response of different species results in even greater variability at the community level.

Dormancy of seeds in soil may be broken by several factors, often acting in concert, that change with depth in the soil. These include an appropriate chilling followed by favorable temperatures, availability of oxygen, release from chemical inhibitors (including ethylene and carbon dioxide), and light regimes, e.g., photoperiod, spectral quality, and intensity. In addition, the water supply must be adequate and pH and salinity must be within certain limits.

Usually germination is enhanced at the soil surface. However, there are species in which seeds exposed to light at the surface are inhibited. This is the case for *Citrullus colocynthis* and *Calligonum comosum* (Cucurbitaceae) in sandy soils in the Negev desert (Meeuse, 1974), where germination on the soil surface would be a risky business. This may be a generalization for arid habitat plants (Mayer and Poljakoff-Mayber, 1975).

Ploughing, which brings seeds to the surface, is a routine agricultural practice for weed control. However, there are other kinds of disturbances in which dormant seeds can be brought to a level at which they can germinate. The famous poppy (*Papaver* spp.) displays in the Belgian battle grounds of World War I followed the disturbance of the ground by exploding shells and the digging of trenches and graves (Hill, 1917). *Matricaria chamomilla* (Compositae) and *Sinapis arvensis* (Cruciferae) were less abundant. Earthworms are more modestly effective in bringing buried seeds to the soil surface, but they can also bury seeds (Darwin, 1897).

A different form of release from dormancy is shown by the hard seeds of leguminous shrubs and trees of genera such as *Cercidium*, *Dalea*, and *Olneya* that grow in or by desert washes (Went, 1957). The seed bank that exists under the tree is broken up when a flash flood sweeps down the wash, carrying seeds and the sand that abrades hard seed coats. When the flood ceases, the seeds are in germinable condition in a moist environment. Rapid growth of a tap root keeps pace with the drying out of the surface layers of the soil. However, Jaeger (1940) regarded the flooding of desert washes by storms as primarily destructive to seeds of *Parosela* (*Dalea*) *spinosa*.

B. Chemical Interactions between Plants and Seeds

There may be chemical control of the germination of seeds in a seed bank. Seeds of the parasitic species of *Striga* (Scrophulariaceae), which are tropical and subtropical invaders of cereal crop fields, are apparently stimulated to germinate by chemicals exuded by the roots of the potential host (Vallance, 1950; Brown and Edwards, 1944, 1948). For *Striga lutea*, Brown *et al.* (1949) identified the substance as a *d*-xyloketose. Worsham *et al.* (1962), working with *Striga asiatica*, claimed that coumarin derivatives were active, but Cook *et al.* (1972) identified a stimulant as a lactone. Indeed, there are probably many stimulants from different host plants. Sunderland (1960) concluded that a variety of chemicals could produce the effect, sometimes with synergistic actions, in both *Orobanche minor* and *Striga*.

Orobanche minor has an apparently almost unlimited range of host plants. However, the closely related *Orobanche hederae* is restricted to *Hedera helix*. The role of exudates may be further complicated by others being required for production of haustoria by the parasites. In *Striga* and *Orobanche* infestation, the agricultural tactic is to plant a crop which will stimulate the germination of the *Striga* but will not be susceptible to parasitization.

Another chemical influence of an opposite kind occurs when exudates from the aerial portions of a plant wash into the soil, inhibiting the potentially germinable seeds of that species. This has been postulated for *Hypericum perforatum* (Hypericaceae). Apparently oil from the capsules (and to a lesser extent from the leaves) inhibits germination, but does not kill the seeds (Clark, 1953). Competition between the perennial parent and its offspring is reduced, but the seed is available and freed from inhibition if the parent plant dies or is removed. A similar autostatic action by *Typha latifolia* is discussed by Grace (1983) with, however, the conclusion that the claimed effect is not on the seed but on the seedling. This would not have a conservational influence and may only be a mechanism of avoiding intraspecific competition by thinning the population.

V. Demography of Seed Banks

Not all deposits in the seed bank are withdrawn by germination. Some seeds are destroyed by predators or decomposers, others lose viability through natural causes (i.e., an age-related loss of vitality). Susceptibility to destruction (or, alternatively, resistance) seems to be genetically controlled, but is markedly influenced by environmental conditions (Bewley and Black, 1985; Priestley, 1986).

California grasslands were modified by grazing pressure in the nineteenth century. The annual Mediterranean grasses and forbs that replaced much of the bunch grass cover (Burcham, 1957) developed large seed banks (Major and Pyott, 1966). This seems to have followed the usual pattern that pioneer species, be they weeds or native species, tend to be well represented in seed banks, for which their abundant seed production is initially responsible. The large seeds of climax (or, at least, late successional) taxa (e.g., *Quercus* or *Aesculus*) are rather moist, and this may be associated with their short survival if ungerminated. But this correlation is not absolute, because, as already mentioned, tiny seeds of Salicaceae are notoriously short-lived (Brinkman, 1974).

Differential seed rains and differential survival of seeds in the soil must result in a difference between the species of the seed bank and those of the standing vegetation. This is recognized for temperate forests (e.g., a recent paper by Nakagoshi, 1984a) and other habitats such as subarctic forests (Johnson, 1975).

VI. Seed Chemistry

The chemistry of the seeds clearly must play an important role in their differential behavior. Is the longevity of leguminous seeds partly due to resistance to decay or predation by use of toxic substances, as well as hard coats? Alkaloidal and cyanogenic glycosidal contents are common in the Leguminosae. Some alkaloids are extremely fungistatic (McKey, 1979). Herbaceous plants that lack condensed tannins in leaves may still possess them in seed coats (Bate-Smith and Ribereau-Gayon, 1958), where "their fungistatic and bacteriostatic properties may continue to confer a net selective advantage" (McKey, 1979).

The cyanogenic glycosides have mostly been selected out of seeds of cultivated species of beans (*Phaseolus* spp.), although they were significantly present in their ancestors (Conn, 1979). Nonprotein amino acids are also found in leguminous seeds. Probably the high concentration of the nonprotein amino acid *l*-dopa in the seeds of tropical vines of the genus *Mucuna* (Leguminosae) prevents seed predation by animals whose nervous systems are upset by this powerful drug (Rosenthal and Bell, 1979).

Lectins (phytohemagglutinins), also found in leguminous seeds, may have a protective function. The long list of other substances that may contribute to the resistance to predation and decay in the soil includes toxic seed lipids, glucosinolates, sesquiterpene lactones, saponins, proteinase inhibitors, flavonoid pigments, and even analogs of insect hormones and antihormones. Determination of whether seed bank species have seeds that are protected by these or other chemicals is a fertile field for further research.

VII. Genetical Considerations

A seed bank conserves genetic variability. According to Templeton and Levin (1979), seeds in a seed bank provide a population with a "memory" of the selective conditions that prevailed in the past as well as more recent conditions.

There is a difference between annuals and perennials in the effects of a seed bank on the evolution of adaptations to changeable environments. The soil seed bank of annuals will be disproportionally representative of genotypes that were successful in good years when large quantities of seeds were formed. In contrast, the seed bank of perennials will be derived from plants that have persisted through (and were presumably adapted to) both good times and unfavorable times.

Evidence of an ecologically significant seed bank in the soil is provided dramatically by the vernal pools of California, in which many seeds remain ungerminated if there is insufficient rainfall in the autumn or long drought periods (Jain, 1976). For example, *Limnanthes douglasii* vars. *rosea* and *nivea* (Limnanthaceae) may be entirely missing from the floral displays of areas in the drier parts of the species' range for several years. Many of these populations are gynodioecious (having both hermaphrodite and male-sterile plants in the population). It appears that the male-sterile genotype in *Limnanthes* is preserved in the population, despite lesser contribution to the seed bank, by the superior longevity of the seeds that contain it in dry soil conditions (H.G. Baker, unpublished).

The case of *Delphinium gypsophilum* and its hybrids with *Delphinium recurvatum* and *Delphinium hesperium* var. *pallescens* (thought to be parents of *Delphinium gypsophilum*) in California is particularly interesting. Lewis and Epling (1959) followed a number of populations. They found that the aboveground populations (usually flowering) varied enormously in numbers from year to year. However, there was always a reserve of rootstocks in the soil which did not put up aerial shoots. The genotypes of the plants that flowered in different years appeared to be genetically different samples arising from the underground stocks. In this manner, genetic diversity in the population as a whole would be maintained, although at any one time only plants with particular genotypes would be actively flowering. Perhaps the same can be substantiated for seed banks.

Modern techniques such as the use of electrophoresis in enzyme analyses of seeds may give us a picture of seed bank genetical variation from season to season, as they already have for populations of seedlings and mature plants (Hamrick *et al.*, 1979).

VIII. Conclusions

For too long ecologists and other field workers have tended to give greater emphasis to aboveground phenomena and pay less attention to the hidden life below the soil surface. But the corpus of information now being assembled promises to provide a basis for understanding the interactions of seeds in the banks as well as the influence of the roots (and, to a lesser extent, the shoot systems) of plants growing in the soil.

This emphasizes that the laboratory and herbarium experiments, which were useful in identifying taxa whose seeds are capable of retaining germinability long enough to be constituents of seed banks, are now less appropriate than more natural experiments and observations. Even the long-term experiments by Beal, Duvel, and others, who buried containers of seeds, need to be followed up by even more natural experiments with the influence of soil factors taken into account.

An appreciation of the stability or instability of seed banks through time may be as influential ecologically as are studies of the breeding systems, pollination biology, analysis of seed dispersal, and associated physiology and biochemistry in natural and artificial ecosystems.

Some means will surely be found to discover the genetical patterns of seed bank populations. These may be indicative of any changes that have taken place in the populations of seed-producing plants, i.e., the so-called "memory" that they provide. The furnishing and replenishing of a seed bank is a dynamic process, and, if technological details can be solved, should provide evolutionary information of great value as well as generating data of use to conservationists.

PART 2

Seed Bank
Processes

CHAPTER 3

Predation in the Dynamics of Seed Regeneration

Svata M. Louda

School of Biological Sciences
University of Nebraska
Lincoln, Nebraska

I. Introduction

A. Overview

Evaluation of predation in seed bank dynamics is either trivial or surprisingly difficult. Seed consumption is a major gustatory strategy that causes significant seed losses (e.g., Janzen, 1971a; Harper, 1977; Crawley, 1983; Hendrix, 1988). So, in the trival sense, it is obvious that seed predation, like any mortality factor causing a consistent loss of young (Lubchenco, 1979), will influence population ecology and evolution. Consistent losses have a potential impact on plant abundance, distribution, competitive status, life cycle traits, or other adaptations (e.g., Harper, 1977; Silvertown, 1982; Fenner, 1985; Sih *et al.*, 1985). Differences in damage among individuals or between species can be significant even where the magnitude of the loss is small (Fox and Morrow, 1986). Clearly, seed predation can matter.

However, clear-cut generalizations for when and where seed predation actually does have a major influence have been difficult to make (Harper, 1977). The descriptive data required to compare seed predation intensities among plant communities are far from complete. The comparative, experimental data that are necessary to determine the general contribution of seed predation to the dynamics and evolution of plant regeneration from seed are uncommon (Harper, 1977; Crawley, 1983, 1988a; Fenner, 1985). More information is required on the character, magnitude, and ecological variation of these interactions.

There are several direct tests of the effect of seed predation on the abundance and distribution of seeds in the soil or on the output of seedlings from the seed bank. These studies suggest that the relative contribution of predation to seed occurrence and plant dynamics varies over several orders of magnitude among plant species and among communities. However, we cannot yet predict the circumstances under which the high and variable predation losses become significant (i.e., for what species, with what frequencies, or under what environmental conditions). By significant, I mean specifically situations in which predation changes distribution, abundance, or characteristics of seed or plant populations from patterns that occur in its absence.

Here I examine evidence for seed predation and suggest five generalizations: (1) seed predators help create the difference between the transient and persistent seed pools; (2) the environment influences and helps determine the net effect of the seed losses; (3) the major impact of predators on plant and community dynamics often occurs through the differential losses among species; (4) fugitive species, ones that depend on seed for recolonization and persistence in moderately disturbed environments, are particularly vulnerable to the demographic and distribution effects of seed predators; and (5) all consumers, including those feeding before and after the seed phase of the life cycle, can influence the release and survival of seeds, and thus the input–output balance of the seed bank. The scarcity of experimental data with which to evaluate and compare these tentative generalizations for most plant communities suggests that a substantial gap exists in our knowledge of predation in the dynamics of plant regeneration from seed.

B. Basic Patterns

Seed is important both for the maintenance and growth of existing populations and for the initiation of new populations. However, the relative importance of seed recruitment, especially from the seed bank, differs among plant species and among communities (Harper, 1977; Hickman, 1979; Silvertown, 1982; Fenner, 1985). The relationship between inputs and outputs of seed to the soil determines the numbers, longevity, and characteristics of the seeds that are present (Harper, 1977; Cavers, 1983). Input reflects the seed rain, which itself is a function of plant density and per capita seed release. Output includes germination and losses to parasitism, predation, and other hazards. Predators at various stages in the life cycle can influence the input or output to the seed bank (Fig. 1).

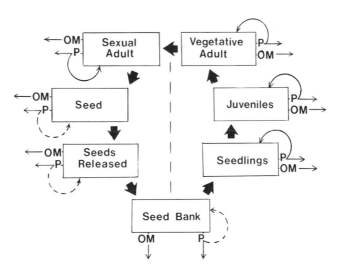

Figure 1.   The dynamics of seeds in the soil in relation to the key phases of plant growth and regeneration. P, Predation mortality; OM, other mortality; back arrows indicate a feedback effect on the condition of the state variable, solid arrows are well documented and generally observed, whereas dashed ones are possible but less frequently documented.

Thus we need to examine (1) the traits of seeds entering versus those persisting in the soil, (2) the factors determining the rates of seed input and output, and (3) the potential role of predation in the observed ecological and evolutionary characteristics of seeds and seed banks.

Comparison of traits for transient versus persistent seeds provides a clue to the cumulative effects of factors, including predators, determining output. Key traits include morphology, species composition, species richness and evenness, dormancy, mean longevity, density, and spatial dispersion of seeds (e.g., Harper, 1977; Cook, 1979, 1980; Naylor, 1984).

The persistent seed bank is characterized by a preponderance of the very small, light-weight seeds, especially those of weedy and ephemeral species (Harper, 1977; Thompson and Grime, 1979; Fenner, 1985; Nakagoshi, 1985a). The persistent seeds usually do not have specialized dispersal mechanisms (e.g., Harper et al., 1970) or chemical defenses (Janzen, 1969, 1977). However, dormancy mechanisms are common (Jain, 1982; Baskin and Baskin, Chapter 4, this volume) and four strategies have been identified (Thompson and Grime, 1979; Grime, Foreward, this volume). Individual seeds of some species have extreme potential longevity, but the contribution of these seeds to population dynamics appears negligible (Cook, 1980). However, dormancy gives the seeds in the soil both age and genetic stucture. Such structure tends to stabilize population dynamics by spreading risk and diminishing erratic fluctuations in response to short-term environmental perturbations (Harper, 1977; Templeton and Levin, 1979; Cavers, 1983; Venable and Brown, 1988).

Persistent seeds represent only a subset of the plant species in the community over time. The long-term seed bank is generally composed of (1) the characteristic ruderal and ephemeral species (Harper, 1977) or (2) the persistent monospecific dominant plant (Silvertown, 1982). The ruderal species are typically short-lived, especially annual weeds that occur very early in succession. The ephemerals are predominantly annuals of harsh or periodic environments (Grime, 1979; Begon et al., 1986).

At the other extreme, species with large seeds generally do not have dormant seeds resident in the soil (Janzen, 1969, 1971a, 1982a; Fenner, 1985; Foster, 1986). Large seeds tend to attract high levels of predation pressure (Janzen, 1971a; Foster, 1986). As communities mature, and long-lived species that tend to recruit from quiescent juveniles increase (Silvertown, 1982), the composition and relative abundances of species in the seed bank versus in the aboveground vegetation become increasingly different (Thompson and Grime, 1979; Naylor, 1984; Fenner, 1985; Nakagoshi, 1985a). Such divergence has been documented for dry grasslands (Major and Pyott, 1966), marshes (van der Valk and Davis, 1976), subalpine meadows (Whipple, 1978), tropical forests (Hall and Swaine,

1980), prairies (Rabinowitz and Rapp, 1980; Rabinowitz, 1981), and pine forests (Pratt *et al.*, 1984).

In sum, recruitment from a persistent seed bank tends to be by small-seeded species whose populations experience frequent or predictable local extinctions, such as cosmopolitan weeds, early successionals, and ephemerals of harsh environments (Silvertown, 1982; Fenner, 1985; Nakagoshi, 1985a). The question then is how might seed predation be involved as a component process of these patterns?

II. Effects of Seed Predation

A. Demographic and Evolutionary Change

Conceptually, plant populations could exhibit two major types of response to significant seed losses: short-term ecological changes and long-term evolutionary ones. Some evidence exists for both.

Over the short term, seed and plant consumption becomes important in plant population dynamics or dispersion when it (1) depresses plant density generally (Harper, 1977; Louda, 1982a; Fenner, 1985), (2) compresses or shifts the observed ecological niche by differentially decreasing plant survival or fecundity in part of the potential distribution (Janzen, 1970; Louda, 1982b; Brown *et al.*, 1975; Louda *et al.*, 1989a), or (3) causes differential losses among competing plants (e.g., Tevis, 1958b; Harper, 1969; Chew, 1974; Whittaker, 1979; Dirzo, 1984, 1986; Risch and Carroll, 1986; Louda, 1988). The data available suggest that each of these effects occurs.

Observations of heavy and variable seed losses are well documented (e.g., Cameron, 1935; Salisbury, 1942; Matthews, 1963; Janzen, 1969, 1970, 1971a,b,c, 1972a,b, 1974a, 1975a,b,c, 1976, 1977, 1978a,b, 1980, 1982a, 1983; Breedlove and Ehrlich, 1968, 1972; Gashwiler, 1970; Bohart and Koerber, 1972; Harris, 1973; Platt *et al.*, 1974; Harper, 1977; Louda, 1978, 1982a,b,c, 1983; Higgens, 1979; Hubbell, 1980; Crawley, 1983; Dirzo, 1984, 1986; Mittelbach and Gross, 1984; Fenner, 1985; Jolivet, 1985, 1986; Andrew, 1986; Hendrix, 1988; T.J. Smith, 1987a,b, 1988). Granivory is common and often intense. Also, seed consumers are selective (Fowler, 1979). For example, predation pressure on large seeds, especially those with thin seed coats, is generally severe and much higher than that on small or tough seeds (Janzen, 1969, 1971a, 1982a; Boucher and Sork, 1979; Silvertown, 1982; Denslow and Moermond, 1982; Nelson and Johnson, 1983; Sork, 1983; O'Dowd and Gill, 1984; Becker and Wong, 1985; Fenner, 1985). Seed predation is sometimes density or frequency dependent, with the densest crop or most common species being used disproportionately (e.g., Tevis, 1958b; Miller *et al.*, 1984;

Greenwood, 1985). Other examples of differential losses abound (e.g., Rosenzweig and Sterner, 1970; Dolinger *et al.*, 1973; Janzen, 1971a,b, 1975a,c, 1977, 1978a,b, 1983a; Pulliam and Brand, 1975; Reichman, 1979; Inouye *et al.*, 1980; Heithaus *et al.*, 1980; Heithaus, 1981; Louda, 1982b, 1983; Davidson *et al.*, 1985; Marshall *et al.*, 1985; Auld, 1986a; Borowicz and Juliano, 1986; Foster, 1986; T.J. Smith, 1987a,b; Crawley, 1988a,b). Thus losses to seed consumers are generally unequal among individuals, species, and years. In fact, Crawley (1988b) suggests that some species can be considered "ice cream plants" because they are disproportionately attractive and damaged by consumers.

At least three responses appear to be selectively reinforced by seed predation. Masting should be selected and reinforced by high intensities of seed predation on large seeds by persistent seed feeders (e.g., Griffin, 1971; Janzen, 1971a, 1974a, 1976; Silvertown, 1981, 1982; Sork, 1983; O'Dowd and Gill, 1984; Ballardie and Whelan, 1986; Foster, 1986). Predator satiation strategies, including masting or episodic flushes of annuals, require synchrony. Augspurger (1981) has shown that natural enemies can select strongly against temporally asynchronous individuals.

Alternately, selective consumption should increase the frequency of traits that facilitate escape, such as dispersal by vertebrates prior to predation by specialized insects or destructive vertebrates (O'Dowd and Hay, 1980; Howe and Smallwood, 1982; Howe and Westley, 1986; Howe, 1986). In one case the frequency of seeds with defenses appropriate to particular predators changed inversely with the relative abundances of predators (Pulliam and Brand, 1975). Another response is increased reliance on alternative methods of recruitment, e.g., from long-lived quiescent seedlings (Silvertown, 1982).

Additionally, in some cases a spatial pattern in seed loss has been identified. If such predation pressure is consistent, it will influence the relative favorability of different portions of the environment for seed survival, seedling recruitment, and adult plant occurrence (Wilson and Janzen, 1972; Janzen, 1975a; Louda, 1978, 1982b, 1983; Louda *et al.*, 1987a,b,c,d). Potentially significant variation in seed loss to predators has been observed along several major environmental gradients, including variation with latitude, elevation, soil topography, soil moisture, habitat, patch size, and position (e.g., Janzen, 1969, 1970, 1971a,b,c, 1972a,b, 1975a,b,c, 1978b; Wilson and Janzen, 1972; J.H. Brown *et al.*, 1975; Louda, 1978, 1982b, 1983; J.N. Thompson, 1978, 1985; Reichman, 1979; Whittaker, 1979; V.K. Brown, 1984; Fenner, 1985; Goldberg, 1985; Louda *et al.*, 1987a,b,c,d). These data suggest that predators often contribute to the extensive spatial heterogeneity observed in seed rain and seed banks.

Natural perturbations and controlled experiments also show spatial patterns of predator effect. Several experiments demonstrate decreased densities of seeds or seedlings caused by seed feeders (e.g., Louda, 1978, 1982a,b, 1983; Reichman, 1979; J.H. Brown *et al.*, 1975, 1979a; Inouye *et al.*, 1980; J.H. Brown and Munger, 1985; Kinsman and Platt, 1984; Andersen, 1988). Other studies substantiate differential consumption among plants, especially on large-seeded and preferred species (e.g., Reichman, 1979, 1984; J.H. Brown *et al.*, 1975; Inouye *et al.*, 1980; Abramsky, 1983; Davidson *et al.*, 1985; Louda and Zedler, 1985). The few experiments that have analyzed seed predation along an environmental gradient show a differential effect of predators on number of seeds or seedlings recruited in a specific portion of the gradient (Janzen, 1971a, 1975a; J.H. Brown *et al.*, 1975; Louda, 1982b, 1983, 1988). In one case the evidence is strong enough to suggest that seed predators compressed and shifted the observed distribution of abundance for the whole population, significantly modifying the ecological niche relative to the potential one (Louda, 1982b). Additionally, predation is often more intense in one of two adjacent occupied habitats (Janzen, 1971b, 1975b; Louda and Rodman, 1983a,b; Louda, 1988), or in some environmental patches but not others (Rissing, 1986; Andersen, 1988).

Over longer time frames, increased adaptation and coevolution should occur in response to consistent predation pressure. Differential survival among genetic variants underlies increased adaptation, compensatory responses and coevolutionary potential (e.g., Ehrlich and Raven, 1964; Gilbert and Raven, 1975; Breedlove and Ehrlich, 1968, 1972; Thompson, 1982; Futuyma and Slatkin, 1983). Thus, when predation pressure varies among genetically differentiated plants, the losses will be selective (e.g., Cook *et al.*, 1971; Berenbaum *et al.*, 1986; Simms and Rausher, 1987). The more intense and consistent the pressure, the more likely the accumulation of adaptations to avoid, tolerate, or minimize seed predation (Ehrlich and Raven, 1964; Janzen, 1969; Harper, 1977; Thompson, 1982; Fenner, 1985). Circumstantial evidence for predator-driven adaptation is provided by variation in seed or fruit traits correlated with differences in consumer pressure, in mutualist defense, or in dispersal agents (e.g., Janzen, 1969, 1971a; Harper, *et al.*, 1970; Carroll and Janzen, 1973; C.C. Smith, 1970, 1975; Orians and Janzen, 1974; Pulliam and Brand, 1975; Bradford and Smith, 1977; Levin and Turner, 1977; Mitchel, 1977; Herrera, 1984a; Beattie, 1985; Foster, 1986). The limited evidence available from natural perturbations and field experiments also suggests a selective effect by predators on seed traits (e.g., Cook *et al.*, 1971; Janzen, 1971a, 1975b; Pulliam and Brand, 1975; O'Dowd and Hay, 1980; Turnbull and Culver, 1983).

Overall, the effect of consistent predation on seed banks should be twofold and cumulative. The first effect should be demographic, with decreased abundance and modified distribution of seeds in the soil, especially for plants with large, nutritious seeds. Losses to herbivores are usually differential, with consumption potentially altering population structure (Waloff and Richards, 1977; Louda, 1978, 1982a,b,c, 1988), succession (Cates and Orian, 1975; V. K. Brown, 1984, 1985; Carroll and Risch, 1984), or competitive interactions and plant community composition (e.g., Harper, 1969; Janzen, 1970; Connell, 1971; Whittaker, 1979; Bentley et al., 1980; Dirzo, 1984, 1986; V.K. Brown et al., 1987; Joern, 1987; Louda et al., 1989a). The second effect should be selective (Janzen, 1971a; Foster, 1986), resulting in traits adjusted to optimize seed survival in the soil despite predation.

Both demographic and selective effects should continue concurrently for at least two reasons. First, the interaction occurs within the constraints established by multiple, concurrent selective pressures (e.g., Inouye, 1982; Louda, 1982c; Nelson and Johnson, 1983; Benkman et al., 1984; Horvitz and Schemske, 1984; Wright, 1986). Noise should persist in the adjustment to any one pressure (e.g., Lewontin, 1974; Louda, 1982c; Horvitz and Schemske, 1984; Rathcke and Lacey, 1985). Second, the predator–seed interaction exists within heterogeneous and changing environments, modifying the conditions of the interaction and forcing continued readjustment. Thus, the contemporaneous demographic effects are observed in population and community dynamics, and the evolutionary effects of predation are reflected in plant and seed traits.

B. Selective Seed Losses and Relocation

Postdispersal use of seeds and fruits causes the most obvious and direct consumer impact on the survival and dispersion of seeds in the soil. Animals handling and consuming released seeds range from rodents, bats, and birds to ants, ground beetles, and earthworms (see Crawley, 1983; D. M. Murray, 1987a). The literature on postdispersal use of seeds is voluminous (e.g., see Janzen, 1971a; Willson, 1973, 1983; McKey, 1975; Harper, 1977; Petal, 1978; J.H. Brown et al., 1979b; Fleming and Heithaus, 1981; Howe and Smallwood, 1982; Jordano, 1983; Naylor, 1984, 1986; Fenner, 1985; Hobbs, 1985; Howe, 1986; Howe and Westley, 1986; Price and Jenkins, 1986; Hendrix, 1988). These studies show that consumption and relocation of seeds are common and often appear important in plant community structure.

For these potentially antagonistic interactions to persist, the costs must be outweighed by the benefits to the plant. Identified costs are (1) high probability of immediate mortality, (2) reduced long-term seed viability, (3) decreased relative fitness of more susceptible individuals, and (4) enhanced evolutionary vulnerability to extinction when dispersers are more specialized.

Benefits associated with vertebrate dispersers for some plants include (1) scarification of surviving seeds (e.g., Janzen, 1971a, 1980, 1983a; Harper, 1977; Howe and Smallwood, 1982; Crawley, 1983; Fenner, 1985); (2) "planting" of seeds, hiding larger ones, and relocating cached seeds to better germination sites (e.g., Reichman, 1979; Howe and Smallwood, 1982; Turnbull et al., 1983; Beattie, 1985; Howe, 1986; Howe and Westley, 1986; D.M. Murray, 1987a; Price and Jenkins, 1986); and (3) rapid removal from the vicinity of the parent and thus from specialized or density-dependent seed predators (e.g., Janzen, 1970, 1983a; Connell, 1971; Boucher and Sork, 1979; O'Dowd and Hay, 1980; Heithaus, 1981; Heithaus et al., 1982; Crawley, 1983).

The costs of postdispersal predation losses are generally high. Consumption by specialists, especially insects, is usually totally destructive (e.g., Janzen, 1969, 1971a, 1972b; Howe and Smallwood, 1982; Crawley, 1983). Both relative specialists and generalists frequently cause large reductions in seed numbers, frequently 90–100% of a seed crop (see Janzen, 1969, 1971a; Crawley, 1983). In some cases, postdispersal reduction in seed by predators can be severe enough to negate compensatory, density-dependent responses in subsequent life history stages (Harper, 1977; Hickman, 1979; Fenner, 1985).

Alternately, the benefits of dispersal by generalized vertebrates can be high for the seeds that survive the handling (e.g., Howe, 1986). Invertebrates also sometimes influence dispersal and seed survival (e.g., Beattie, 1985; Kjellsson, 1985). For example, ants harvest seeds (e.g., Davidson, 1980; Davidson and Morton, 1981; Davidson et al., 1985), consuming some seeds but changing the spatial distribution especially of myrmecochorous seeds (Buckley, 1982; Beattie, 1985; Keeler, 1988). The impact of ants varies with species and geographic area (e.g., Tevis, 1958b; Abbott and Van Heurck, 1985; Morton, 1985). Experimental evidence demonstrates that this movement can improve the relative survival of adapted or unconsumed seeds (e.g., Turnbull and Culver, 1983; Turnbull et al., 1983; Beattie, 1985). Handel et al. (1981) noted that a large proportion of the herbaceous flora of a northeastern deciduous forest is myrmecochorous.

Extensive work has been done on the interaction of seeds with granivorous small mammals and ants in deserts, grasslands, and pastures. Small mammals are selective and are often responsible for very heavy seed losses (e.g., Rosenzweig and Sterner, 1970; Janzen, 1971a; J.H. Brown et al., 1979b; Price and Heinz, 1984). Similarly, ants are selective and maintain strong yet variable pressure on the seed supply in the soil in grasslands, deserts, and even eucalyptus forests (e.g., Tevis, 1958b; Pulliam and Brand, 1975; Mares and Rosenzweig, 1978; Whitford, 1978; Ashton, 1979; Davidson and Morton, 1981; Buckley, 1982; Collins and

Uno, 1985; Morton, 1985; Andersen, 1988). To what extent and under what conditions such losses actually change seed or adult plant density remains an open question.

Experimental studies of postdispersal seed predation in seed and plant dynamics demonstrate several key points. One of these points is that postdispersal seed predation can determine density and relative species abundance in grasslands. For example, in annual grasslands of California, rodents depleted the seed bank of the dominant annual grasses (Borchert and Jain, 1978), changing the relative abundance of the four main seed species present. Seed predation is heavy in prairie grasslands of North America (e.g., Platt, 1976; Louda et al., 1989b), reducing seed regeneration and apparently reinforcing dependence on vegetative reproduction for persistence by the dominants. These studies are suggestive of the potential impact of postdispersal seed predation on the dynamics of grasslands. More experimental studies of the consequences of seed predation are needed.

In the deserts of North America, granivorous species, such as heteromyid rodents and harvestor ants, (1) reduce seed densities 30–80% (Nelson and Chew, 1977; Brown et al., 1979a,b; Reichman, 1979, 1984; Inouye et al., 1980; Davidson et al., 1985; Tevis, 1958b; Price and Jenkins, 1986); (2) reduce overall plant density (Inouye et al., 1980); (3) select among seeds and deplete seed pools of the larger seeded shrubs and herbs (Rosenzweig and Sterner, 1970; J.H. Brown et al., 1979a,b; J.H. Brown and Munger, 1985); (4) contribute to the spatial heterogeneity of seed distribution (Reichman, 1979, 1984); and (5) differentially decrease seed and plant abundance of some ruderals, such as Erodium cicutarium, but not of most of the characteristic ephemerals (Inouye et al., 1980). In fact, Inouye et al. (1980) found that ants depended on numerically dominant plants whereas rodents used seeds of biomass dominants.

Seed losses of forbs can be as high as those recorded for grasses. In British pastures, rodents removed 50% of the seeds of Ranunculus repens within 6 months and 35–50% of those of the related Ranunculus acris and Ranunculus bulbosa over 14 months (Sarukhan, 1974). Differential choice, coupled with the high rate of removal, affected relative abundances of the species in the seed pool, and potentially in the plant community.

Depletion of seed numbers in the soil by consumers also occurs in forests and chaparral. In temperate forests, trees that have relatively large seeds (e.g., oaks and hickories) or seeds concentrated in cones (e.g., conifers) lose large portions of their seed crops to small mammals and a variety of seed-feeding insects (e.g., Gashwiler, 1970; Griffin, 1971; Marquis et al., 1976; Sork and Boucher, 1977; Sork, 1983; Boucher and Sork, 1979). Although such trees are long-lived and iteroparous, seed losses sometimes influence their relative regeneration success (e.g., Mellanby, 1968; Sork, 1983; T. J. Smith, 1987a,b), especially in a series of dry or nonmast years. Seed predation can also be severe in tropical forests and can influence tree recruitment (De Steven and Putz, 1984; Becker and Wong, 1985; Marquis and Braker, 1988). On tropical Pacific islands (e.g., Enewetak Atoll), intense predation on tree seeds by crabs and insects resulted in rapid destruction of large seeds and loss of most larger seeds from the seed bank within a year (Louda and Zedler, 1985).

High seed death rates also occur in the chaparral of Mediterranean scrub ecosystems (e.g., Keeley and Hays, 1976; Louda, 1983, 1987; Mills, 1983). For obligate seeders, species that depend upon seed recruitment after fire, this loss can be locally critical. For two fugitive *Haplopappus* shrubs occurring along an environmental gradient, seedling establishment is limited by seed predators (Louda, 1982a,b, 1983).

Thus, predators reduce and redistribute the number seeds in soil of many communities. The demographic and distributional consequences of such losses will vary with (1) degree of dependence on seed regeneration; (2) variation in the probability of seed escape through dispersion, dispersal, and predator satiation; and (3) oppportunity for subsequent compensation for the loss. Experimental tests of these predicted conditions, however, are still required in most systems and for species of varied life histories.

C. Decreased Input to the Seed Bank

1. Literature

Traditionally, the analysis of predation in seed banks has been restricted to assessing the fate of seeds already incorporated into the soil. However, the consumption of flowers and developing ovules can also influence seed input and change seed shadows (e.g., Janzen, 1971a,b,c, 1975a,b,c; Rockwood, 1973; Green and Palmblad, 1975; Harper, 1977; Hickman, 1979; Louda, 1978, 1982a,b, 1983, 1988; Edwards and Wratten, 1980; Hodkinson and Hughes, 1982; Silvertown, 1982; Fenner, 1985; Hendrix, 1988). Therefore, predispersal seed predation also needs to be considered as a process in seed bank dynamics.

Feeding on flowers, ovules, seeds, and developing fruits usually causes several changes in seed production. These include: (1) delay in seed release; (2) change in the number of seeds matured and released; (3) increase in the heterogeneity of seed size, seed release, and seed shadows; (4) selection among differentially susceptible genotypes; and perhaps (5) compensatory production of seeds that differ in key traits (Janzen, 1971a; Harper, 1977; Hendrix, 1979, 1984, 1988; Louda, 1982a,b, 1983; Crawley, 1983; Fenner, 1985; Keeley *et al.*, 1986).

Predispersal consumption generally reduces the number of viable seeds (Janzen, 1971a). The numerous studies since Janzen's review substantiate the generalization that significantly fewer viable seeds are released to the seed bank when predispersal seed predation occurs (Janzen, 1971b,c, 1972a,b, 1974a, 1975a,b,c, 1977, 1978b, 1980; Bohart and Koerber, 1972; Slater, 1972; Ueckert, 1973; Platt *et al.*, 1974; Vandermeer, 1974; Green and Palmblad, 1975; Simpson *et al.*, 1976; Baskin and Baskin, 1977; Risch, 1977; Waloff and Richards, 1977; Hawthorne and Hayne, 1978; Moore, 1978a,b; Louda, 1978, 1982a,b,c, 1983, 1988; J.N. Thompson, 1978, 1983, 1985; Inouye and Taylor, 1979; Kamm, 1979; Lamp and McCarty, 1979, 1981, 1982; Hare, 1980; Harris, 1980; Schlising, 1980; Zimmerman, 1980a; De Steven, 1981, 1983; Johnson, 1981; Haddock and Chaplin, 1982; Inouye, 1982; McCarty and Lamp, 1982; Ohmart, 1982; Scott, 1982; Auld, 1983, 1986a; Fowler and Whitford, 1983; New, 1983; Bond, 1984; Hainsworth *et al.*, 1984; Keeley *et al.*, 1984, 1986; Zedler *et al.*, 1983b; Zammit and Hood, 1986; Zammit and Westoby, 1988a; Louda *et al.*, 1987b,c; Windus, 1987). Seed losses may actually be much greater than usually estimated because usual methods underestimate the total damage (Andersen, 1988). Seed damage and losses, however, may not have a major influence on plant recruitment and dynamics relative to other limiting factors (Beattie *et al.*, 1973; Wapshere, 1974; Louda, 1983; Duggan, 1985; Louda *et al.*, 1987c). Timing of the loss, resources available for compensatory recovery, and frequency of "safe sites" contribute to the impact of predispersal seed predation (Harper, 1977; Cox and McEvoy, 1983; Crawley, 1983; Rathcke and Lacey, 1985; Williams, 1985; Collinge and Louda, 1988b).

Compensation for early loss is common (see Hendrix, 1979, 1988), but not universal (e.g., Louda, 1982b,c, 1983; Louda *et al.*, 1987b). When nutrient resources are not limiting, early removal of immature fruits may lead to increased size of the remaining fruit and seeds, or it may stimulate production of additional flowers and fruits (e.g., Janzen, 1971a; Maun and Cavers, 1971a,b; McNaughton, 1983; Crawley, 1983). Under agricultural conditions, compensation for early defloration often occurs in *Apium graviolens* (Scott, 1970), *Phaseolus vulgaris* (Binnie and Clifford, 1980), *Zea mays* (DuRant, 1982), and *Glycine max* (R.H. Smith and Bass, 1972; Hallman *et al.*, 1984). Under natural conditions compensation has been observed for *Ipomopsis aggregata* (Paige and Whitham, 1987), but not for *Haplopappus squarrosus* (Louda, 1982a,b) or *Haplopappus venetus* (Louda, 1983). More work in natural systems would be profitable.

Pollination and seed predation often interact to determine the number of viable seeds (Platt *et al.*, 1974; Zimmerman, 1980a,b; Augspurger 1981; Davi, 1981; Heithaus *et al.*, 1982; Arnold, 1982; Louda, 1982b,c, 1983; Zimmerman and Gross, 1982; Evans and Smith, 1988). If pollination determines the size of the seed crop and resources are limited, then the abortion or loss of fruits should improve resource use and success of seed release (Stephenson, 1980, 1981). The net effect of predispersal seed predation on input to the seed bank will be determined by time, resources available, pollination, and other constraints.

2. Case Histories

I will use examples from my own work to illustrate patterns and their implications for seed bank and plant population dynamics. These serve as a basis for five generalizations, presented as ideas for further testing in the next section.

(a) **Annual.** *Cleome serrulata* is a fugitive species sporatically common in the relatively dry, short grass prairies of western Nebraska and eastern Colorado. We observed plants along a local topographic gradient of soil moisture (Louda *et al.*, 1987b). Plant abundance, methylglucosinolate (a mustard oil precursor), and insect feeding damage varied along the gradient (Table 1). Plants were generally larger and initiated more flowers and seeds in the lower, moister part of the gradient (Table 1; Farris, 1985). However, seed predation was also greater at the moist end and inversely proportional to the concentration of methylglucosinolate. Thus, seed release and potential recruitment by this annual were com-

Table 1
Variation in plant size, methyl glucosinolate concentration, and seed production for *Cleome serrulata* after predispersal seed predation[a]

Parameter measured[b]	$\bar{x} \pm$ SE
Plant size by gradient position **($r = -0.42$ to -0.71, $p < 0.05$ for all dates)**	
Wet half of gradient ($N = 7$)	72 ± 4.8
Dry half of gradient ($N = 6$)	33 ± 4.0
Methyl glucosinolate concentration by gradient **position ($r = 0.56$ to 0.77, $p < 0.05$ for all dates)**	
Wet half of gradient ($N = 11$)	12.0 ± 3.83
Dry half of gradient ($N = 10$)	41.7 ± 10.27
Viable seeds per plant (square-root-transformed **data, $p < 0.001$)**	
Wet half of gradient ($N = 8$ of 49)	9.0 ± 2.80
Dry half of gradient ($N = 8$ of 37)	22.5 ± 3.33

[a]Methylglucosinolate concentration in mg g^{-1} dry weight; dispersion is along a 30-m soil moisture gradient in short-grass prairie.

[b]Data from Louda *et al.*, 1987b, site 2.

pressed and concentrated into the drier end of the local gradient by differentially intense foliage, flower, and seed predation at the wet end.

(b) Monocarpic Perennial.   *Cirsium canescens* is a native plant species that is fugitive in disturbances (Lamp and McCarty, 1979, 1981, 1982) of the sandhills prairie in western Nebraska (Keeler *et al.*, 1980; Barnes *et al.*, 1984). The shifting nature of open areas means that this short-lived perennial is dependent on seed regeneration for population maintenance like other fugitive species (e.g., Platt *et al.*, 1974; Platt, 1976; Werner, 1977; Kinsman and Platt, 1984; Platt and Weis, 1985).

We tested the impact of seed predation on plant recruitment and population dynamics by excluding predispersal predators with insecticide and postdispersal predators with 1-cm mesh hardware cloth cages (Louda *et al.*, 1989b). The decrease in seed input into the seed bank caused by predispersal seed predators reduced recruitment probabilities and lowered average individual fitness, even when postdispersal hazards were not altered (Table 2). In addition, postdispersal seed predators decreased seedling recruitment. Some dormancy allowed low germination in the following season, but only if postdispersal seed feeders were excluded. The controls show that the losses due to seed predation reduced seedling density below the level set by safe sites. Consequently, seed predation reduced plant density and individual fitnesses significantly.

(c) Fugitive Shrubs.   *Haplopappus squarrosus* and *H. venetus* are small shrubs that reach their highest abundances 3–5 yr after disturbance (usually fire) in the chaparral of southern California (Munz and Keck,

Table 2
Fate of *Cirsium canescens* treated with insecticide (Isotox) or water[a]

Parameter per plant	Insecticide exclusion ($\bar{x} \pm SE$)		Water control[b] ($\bar{x} \pm SE$)	
Seeds initiated	716	$\pm$ 77.9	577	$+$ 60.9
Viable seeds released	105	$\pm$ 11.2	41	$\pm$ 5.6***
Seedlings	3.0	$\pm$ 0.48	0.5	$\pm$ 0.19***
Monocarpic adults established	0.37	$\pm$ 0.11	0.07	$\pm$ 0.03**

[a]Plants were treated with Isotox in water to exclude insects from developing capitula; control plants were treated with water. The experiment was conducted at Arapaho Prairie in western Nebraska beginning in May, 1984 and 1985. Data are averaged for 1984 and 1985 trials, and adults established and flowering are as of May 1988 (Louda *et al.*, 1989b).

[b]**, $p < .01$; ***, $p < 0.001$.

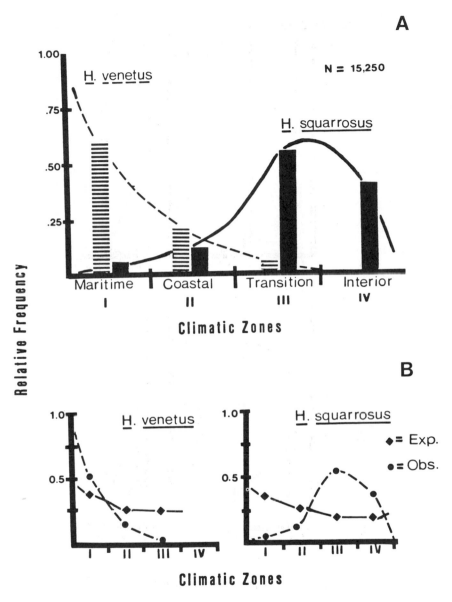

Figure 2. (A) Observed frequency of occurrence and replacement of *Haplopappus venetus* by *Haplopappus squarrosus* over the complex environmental gradient from ocean to mountains, San Diego County, California, (B) observed distribution of species, compared to their expected distribution based on growth and flower production of established individuals (see Louda, 1978, 1982b, 1983).

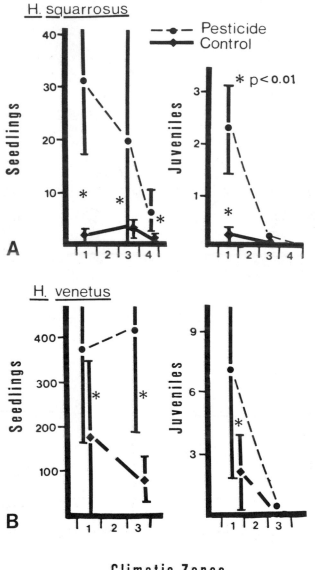

Figure 3. Seedling establishment and survival to juvenile stage ($\overline{X} \pm$ SE) for *Haplopappus squarrosus* and *Haplopappus venetus* in the predispersal seed predator exclusion experiments, at four sites along the complex environment gradient from ocean to mountains (Fig. 2). Zone 1, maritime; zone 4, interior (data from Louda, 1982b, 1983).

1970). In San Diego County, they replace each other along an 80- to 100-km gradient from coast to inland mountains; *Haplopappus venetus* is coastal whereas *Haplopappus squarrosus* predominates inland (Fig. 2A). Both species have relatively large seeds and no permanent seed banks (Louda, 1978). Plant size and growth for both species are greatest toward the coast (Louda, 1982b, 1983). Adults persist in closed chaparral but no recruitment occurs.

For the inland species, *Haplopappus squarrosus*, predispersal seed

Experimental exclusion of seed feeders at sites along the gradient from coast to mountains demonstrated two relevant points. First, predispersal seed predation limited seed input to the soil, and subsequently determined local seedling recruitment for both species. Second, the net effect of these losses on adult plant distribution along the gradient changed as physical conditions and impact of other consumers varied.

For the inland species, *Haplopappus squarrosus*, predispersal seed predation was the most important factor explaining both local recruitment and adult distribution over the gradient (Louda, 1982a,b). Seedling recruitment was proportional to uneaten seeds; predation lowered seed release differentially and was most severe at the coast (Louda, 1982a,b). For control plants (with seed predators), the distributions of both seedlings and juveniles (Fig. 3A) corresponded with the observed adult dis-

Table 3

Fate of *Haplopappus venetus* in two tests
during the year after seed predator exclusion[a]

Test	Number	Desiccated in place (%)	Missing or removed (%)
Spray exclusion of seed predators			
Maritime (coastal) zone (1)			
Open plots	677	77.9	22.1
Transition (inland) zone (3)			
Open plots	782	26.4	73.7
Cage exclusion of seedling predators			
Maritime zone (1)			
Full cages	8	87.5	12.5
Partial cages	20	65.0	35.0
Open plots	16	75.0	25.0
Transition zone (3)			
Full cages	15	93.3	6.7
Partial cages	23	4.3	95.7
Open plots	21	23.8	76.2

[a]From Louda, 1978, and unpublished data; ANOVA on ranks shows significant differences among treatments, $p < 0.05$.

tribution (Fig. 2A). But, for plants with seed predators excluded, the distributions of both seedlings and juveniles (Fig. 3B) corresponded to the expected adult plant distribution along the gradient (Fig. 2B). So for *Haplopappus squarrosus*, predispersal seed predators both limited local recruitment and confined plant abundance to the inland portion of its potential niche (Louda, 1982b).

For the coastal *Haplopappus venetus*, seed predators also restricted seedling establishment (Fig. 3). In addition, seedling mortality, caused primarily by herbivores (Louda, 1978), was disproportionately severe inland (Table 3). Together, higher seed losses and higher seedling mortality in the inland area restrict the observed distribution of *Haplopappus venetus* to the coastal portion of its potential range (Louda, 1983).

Species replacement along the gradient was, therefore, caused by differential consumption. These results provide an alternative explanation and challenge the usual competitive hypothesis for the elevational replacement of species (e.g., MacArthur, 1972; Cody, 1978).

(d) Dominant Understory Perennial Herb.   *Aster divaricatus* is a major understory species of North American eastern deciduous forests. Ex-

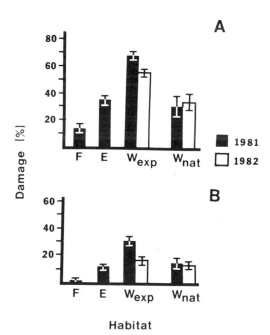

Figure 4.   Insect damage ($\overline{X}\%$ ± SE) to *Aster divaricatus* naturally in the woodland habitat (W_{nat}), as well as for plants experimentally transplanted into the woodland (W_{exp}), into the adjacent field (F), and into the intermediate ecotone (E) between woods and field, in both 1981 and 1982: (A) capitula damaged, (B) total seeds damaged (from S. M. Louda, S. N. Handel, and J. Mischken, unpublished data).

amining both naturally occurring plants and transplants within and between habitats, Louda *et al.* (1987c), found that insect damage to flowers and seeds was higher in the usual woodland habitat, where density was greatest, than it was in either the adjacent field or the ecotone between (Fig. 4). Consequently, it is unlikely that seed predation is significant in determining local distribution. Wood and Andersen (1988), working on vegetation recovery at Mt. St. Helens, also concluded that predispersal seed predation did not depress seed release by *Aster ledophyllus*, whereas elk browsing did.

Two key points emerge from these studies. First, seed predation can be an extremely important influence on plant density and distribution. Second, predation appears more important in the population dynamics of fugitive species than in those of more dominant species.

III. Emerging Generalizations

A. Predation Influences Seed Bank Strategy

The evidence suggests that seed predators select seeds differentially, determining the average value of key characteristics of seeds that remain in the soil. By finding and using clumped and larger seeds, predators will reinforce other pressures that select for seed traits characteristic of persistent seed banks, including small seed size and hard seed coats. Additional experiments are needed to evaluate the trade offs among processes [e.g., germination success versus vulnerability to predation (Stebbins, 1971; Silvertown, 1981, 1982)] that affect seed size and shape under the range of environmental conditions encountered by a species.

Additionally, fruit and seed consumers change seed distribution by eating, moving, or caching propagules. They sometimes increase germinability and recruitment of the seeds that escape destruction, counteracting some of the selective effect of predation for small size. The net effect must be determined by the balance between costs and benefits of the plant–animal interaction in relation to other concurrent processes. The next step is to test this proposed relationship in suites of species chosen to compare different life history strategies, phenologies, and dispersion patterns, as well as plant communities with a range of disturbance regimes and varying nutrient availabilities.

B. Environment Determines Variation in the Impact of Predation

The case histories suggest that the impact of predation on individual plants and on populations of a species depends on their location within the potential environmental range. Predation typically influences distribution by varying in space and being more severe and significant in one part of a gradient. Other examples of variation in seed predation, among habitats and along environmental gradients, include differences among patches of differing densities, among plants in different habitats, and along a montane elevational gradient (e.g., Janzen, 1975a,b,c; McKey, 1978; Reichman, 1979, 1984; Louda *et al.*, 1987c; S. M. Louda, J. Mulroy, and R. D. Wulff, unpublished data). Interestingly, Krusi and Debussche (1988) show no differences in seed predation between habitats; predation on *Cornus sanguinea* in southern France was similar in an olive grove, adjacent deciduous forest, and in the ecotone.

More work on spatial variation in consumption has been done with insect damage to foliage. These studies suggest that a critical aspect is the plant's ability to compensate for the loss, which is a function of the timing of the loss, the resources available for recovery, and the plant's physiological condition. Resources and plant condition are often related to the plant's position within its potential environmental range (Cox and McEvoy, 1983; Williams, 1985; Louda *et al.*, 1989a). For example, Cox and McEvoy (1983) showed experimentally that recovery by *Senecio jacobaea* from defoliation increased with soil moisture.

Thus, I expect that the same amount of damage or seed predation will have different consequences in different parts of the plant's range. Few tests of this hypothesis exist (e.g., Parker and Root, 1981; Louda, 1982b, 1983; Louda *et al.*, 1989a). Corollaries that also require testing include the predictions that losses will (1) tend to vary along relevant gradients, (2) be important in a particular portion of the range of interaction with the predator, and (3) have strong effects on fugitive species. The generality of these predictions needs to be evaluated. I suggest that we have enough evidence to refine our focal question. It is no longer: "do consumers have an impact?"; rather it has become: "under what circumstances would consumers be predicted to have a significant effect?"

C. Major Impact Often Occurs through Differential Losses

The evidence demonstrates that predation losses among coexisting plants vary greatly. Within a species, the effect of variation in loss among individuals will be correlated with the heritability of differences that determine resistance to predators. Among co-occurring species, differential resistance will influence relative abundance and relative competitive ability in resource-limited communities (Harper, 1977; Whittaker, 1979; Crawley, 1983; Dirzo, 1984, 1986; Louda *et al.*, 1989a). Thus, if resources are limited or if compensatory responses are constrained by plant response to the environment, selective differential predation could be critical in the plant's response and its relative position in the community.

Relative growth rates and differential losses do influence densities of coexisting plants in some communities (Tevis, 1958b; Borchert and Jain, 1978; Whittaker, 1979; Bentley *et al.*, 1980; Inouye *et al.*, 1980; V.K. Brown, 1984; Dirzo, 1984, 1986; Cottam, 1986; Cottam *et al.*, 1986; V.K. Brown *et al.*, 1987), but most of the experimental work has been with foliage consumers rather than with seed predators.

At least one striking implication reoccurs in experiments excluding seed predators from annual grasses (Borchert and Jain, 1978), from desert annual communities (Inouye *et al.*, 1980; Davidson *et al.*, 1985), and from fugitive chaparral shrubs (Louda, 1978, 1982a,b,c, 1983, 1987), that is, that relative, as well as absolute, seed loss may be important in plant dynamics. More experimental studies that simultaneously evaluate both seed predation and plant competition in the dynamics of co-occurring plants are needed.

D. Fugitive Species Have High Vulnerability to Predator Effect

Prediction of relative plant susceptibility to predator impact is difficult. Because the role of consumers in plant population size, density, and genetic structure has not been analyzed experimentally for most species, the controversy over the role of consumers in plant dynamics continues (e.g., Owen and Wiegert, 1981; Belsky, 1986). We need testable predictions for (1) the types of plants that are likely to have their dynamics influenced by predators, and (2) the conditions under which significant predation would be expected.

One such prediction is suggested by examining the underlying assumptions of the widely held ecological truism that, on the average, a plant only needs one successful seed to replace itself. Key assumptions required for this to hold include (1) stable, constant population size, (2) homogeneous environment, (3) absence of significant environmental change, and (4) no differential reproductive success by co-occurring genotypes or species (Harper, 1977; Silvertown, 1982). The dictum holds true only under static conditions; so, predation should be expected to have an influence primarily on the dynamics of expanding populations (Harper, 1977).

Two groups of species with predictable periods of rapid population expansion and significant predator impact were evident in the literature and case histories reviewed. First, seed predation changed density and relative abundance of dominant species that had annual life histories (e.g., grasses of annual grasslands in California; Borchert and Jain, 1978), or that had high dependencies on seed recruitment for population maintenance and recovery from disturbance (e.g., mangroves; T. J. Smith, 1987a,b, 1988). Second, seed predation influenced recruitment, occurrence, and distribution of moderately large-seeded plants with fugitive life histories (Fig. 5; e.g., Platt *et al.*, 1974; Platt, 1976; Louda, 1982a,b,c; Kinsman and Platt, 1984; Louda *et al.*, 1987b, 1989b). Dependence of fugitives on seed for population maintenance and the influence of predators has been modeled recently (Solbreck and Sillen-Tullberg, 1986a,b). The model also suggests that a fugitive strategy, especially in maturing, saturating communities, tends to increase the significance of differences in predator-caused seed loss.

The risk of predator impact for these species increases as the canopy matures. Growth of the dominant species, with accompanying closure of the canopy and decreased nutrient availability, guarantees environmental change. Predictable environmental change increases the frequency with which populations of fugitives will be in the colonizing and expanding phases of growth. The trade offs required for plant persistence as colonizers of disturbances create a group of species that are generally intermediate in their traits between the larger, more competitive species and small, short-lived ruderal or ephemeral species. Seed size, time to maturation, germination date, generation time, etc., repre-

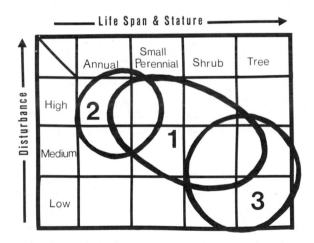

Figure 5. Predicted relative vulnerabilities to seed predators by plants with different life histories. Group 1 is composed of fugitive and other species whose life history traits can be considered intermediate between group 2, the ruderals and ephemermals specialized for frequent disturbances or harsh environments, and group 3, long-lived perennials adapted for competitive, stable, low-nutrient environments. Increasing populations of species in group 1 appear most vulnerable to contemporaneous demographic and distributional effects of seed predation. (I call these "resource speculators.")

sent coevolved compromises (Silvertown, 1981) between those selected for highly disturbed or highly competitive environments. These species are usually relatively short-lived, early-intermediate colonizers within their association. Many are herbaceous, but not necessarily, because small fugitive trees or shrubs, such as *Prunus pensylvanica* (Marks, 1974) or possibly *Vaccinium angustifolium* (Wesley et al., 1986), seem to fit this category.

The fugitive group often includes small monocarpic or short-lived iteroparous perennials that have refuges in time or in space. Although fugitive species are directly dependent on seed regeneration for persistence, the temporal and spatial refuges buffer extinction probabilities associated with severe seed predation and allow the interactions to continue.

Life history and seed traits of such species are intermediate between "resource trackers," the r-selected or ruderal species (Grime, 1979), and "resource cultivators," the K-selected or competitor species (Grime, 1979). Instead, these species could be called "resource speculators," the species that invest in a strategy with high potential for rapid resource gains in changing environments but also with high risks. One of the key risks appears to be differentially increased vulnerability to the effects of high and variable seed predation in part of their environmental range (e.g., Louda, 1982b, 1983).

E. Consumers throughout the Life Cycle Affect Seed Dynamics

Finally, the data suggest that the consequences of seed loss are determined by the influence of the biotic and physical environment on plant performance and compensatory ability. Consumers affect seed dynamics and plant performance both (1) directly, by removal of reproductive investment in seeds or seedlings, or (2) indirectly, by modification of net primary production and subsequent flowering and seed maturation. The impact of such consumption varies in relation to plant capacity to compensate for the loss, immediately or subsequently. This suggests that the study of seed banks needs to be placed more firmly within the framework of whole life cycles, as suggested by Harper and White (1974) for plant dynamics in general. It is clear that the significance of a particular mortality factor can only be evaluated in relation to the performance and mortality of subsequent stages (Fig. 1).

Grazers, browsers, and insect foliage feeders contribute to significant variation in plant growth, size, and reproductive output. Consumption of roots, stems, and leaves usually decreases growth rate, plant size, maturation rate, and size or number of seeds (Harper, 1977; Morrow and LaMarche, 1977; Louda, 1984; Stamp, 1984; Crawley, 1985;

Begon *et al.*, 1986; Belsky, 1986; Louda *et al.*, 1989a), as well as dispersal distance of propagules (Harper, 1977; Louda, 1982b, 1983). Moreover, defloration or defoliation can increase average seed weight and seed coat thickness (Maun and Cavers, 1971a,b) and result in compensatory shifts in seed numbers and weight (Hendrix, 1979, 1984; Wulff, 1986a). Although increased seed size generally improves seedling survival (Harper, 1977; Weis, 1982; Nelson and Johnson, 1983; Gross, 1984; Stanton, 1984, 1985; Crawley and Nachapong, 1985; Foster and Janson, 1985; Wulff, 1986b), it also tends to increase seed vulnerability to predators (see Foster, 1986). Therefore, these interactions change patterns of seed production and longevity and, subsequently, the input–output balance for the seed bank.

Multiple sequential effects are documented between competitors, herbivores, pollinators, seed predators, seed dispersers, and seedling predators (Platt *et al.*, 1974; Platt, 1976; Simpson *et al.*, 1976; Sork and Boucher, 1977; McDade and Kinsman, 1980; O'Dowd and Hay, 1980; Stephenson, 1980; Davis, 1981; Scott and Black, 1981; Arnold, 1982; Heithaus *et al.*, 1982; Louda, 1982c, 1983; Herrera, 1984b,c; Horvitz and Schemske, 1984; Manzur and Courtney, 1984; Reader and Buck, 1986). Our studies of foliage consumption of *Cardamine cordifolia* in the Rocky Mountains illustrate this. Herbivory is generally greater in sunny microhabitats than in adjacent willow-shaded ones (Louda and Rodman, 1983a,b; Louda, 1988; Collinge and Louda, 1987, 1988a). Foliage loss to herbivores increases on moderately stressed plants (Louda, 1986, 1988), and decreases seed production significantly (Louda, 1984, 1988). The effects of consumers are consequently additive, influencing the levels of seeds in soil and transition probabilities into the soil bank.

IV. Recommendation: Path Analysis Plus Field Experiments

Some models are available for the evaluation and prediction of the influence of seed predation, especially in the postdispersal phase of plant life histories (e.g., Vandermeer, 1975; Caswell, 1978; Crawley, 1983; Greenwood, 1985; Solbreck and Sillen-Tullberg, 1986b; J.S. Brown and Venable, 1986, 1987). However, developing a general analytical model for the demographic and distributional effect of seed predation for stage-structured, probabilistic sequences in life histories is difficult (Hubbell and Werner, 1979). The present state of knowledge suggests that path analysis provides a potentially attractive, alternative approach (see Price and Jenkins, 1986).

Fate diagrams in this type of analysis give a visual display of the various possible paths from seed to seed generations. Such analyses then direct attention to the links for which data on transition probabilities are still required. Once those data are accumulated experimentally, path analysis allows the calculation of the relative contribution of each factor, such as seed consumption, to net recruitment and population dynamics (e.g., Hubbell and Werner, 1979; Price and Jenkins, 1986). Application of this approach, including the development of a process model using path analysis and the experimental determination of transition probabilities in the field, is needed for plants (1) with contrasting life history strategies, (2) under the full range of relevant environmental conditions for an interaction, and (3) among ecosystems of differing structure. This should increase our understanding and improve our capacity to answer questions regarding the role of various processes, including predation, on seed bank and plant dynamics.

V. Summary and Conclusions

Predation reinforces selection for specific traits characteristic of seeds that form the permanent, compared to the transient, seed bank. Clearly, the most vulnerable seeds are large, conspicuous, high-energy packages, the opposite of seeds that persist in the seed bank (see Foster, 1986; Hendrix, 1988). Many seed characteristics and key differences between transient and permanent seed banks can be understood, at least in part, as adaptive responses to the quantitative and selective effects of consumers. In addition, seed predation has been shown to influence the input, survival, dispersion, composition, and longevity of seeds in the soil. The circumstantial evidence is voluminous. Direct experimental evidence is less common, but it also suggests that seed predation has significant effects on both the seed bank and the plant dynamics for species that depend on seed regeneration for persistence in the vegetation, including both renewing dominants such as annual grasses (e.g., Borchert and Jain, 1978; T. J. Smith, 1987a,b, 1988) and colonizing fugitive species (e.g., Platt et al., 1974; Platt, 1976; Lamp and McCarty, 1979, 1981, 1982; Louda, 1982a,b; Kinsman and Platt, 1984; Louda et al., 1989b).

A second generalization is that environmental context is essential to the prediction of predator effect. The importance of consumption for plant dynamics depends both on the frequency and intensity of the interaction and on the availability of time and resources with which to compensate and recover from losses (Janzen, 1974a; Whittaker, 1979; Dirzo, 1984; Coley et al., 1985; Hendrix, 1988; Louda et al., 1989a). Specifically, the effect of consumer-imposed seed loss was related to (1) the availability of resources and the contraints on resource allocation, (2) the extent to which compensatory opportunities occur, and (3) the relative intensity of preceding, concurrent, and subsequent mortality factors influencing regeneration probabilities. These were correlated with the plant's position within its potential environmental range.

Third, seed predation studies need to focus on differential losses. The consistent inequities in seed mortality that are observed among co-occurring plants can potentially predict species composition, as well as densities, dispersions, turnover rates, and traits of seeds remaining in the soil. Few studies have focused directly on such comparative analysis. Heterogeneity is a conspicuous and characteristic aspect of seed bank data (Harper, 1977; Grime, 1979; Silvertown, 1982). We know that heterogeneity can influence plant population and community dynamics (e.g., Harper, 1977; Louda, 1982b, 1983; Begon et al., 1986), and that predation contributes to this heterogeneity (e.g., Janzen, 1970; Connell, 1971; Reichman, 1979, 1984; Hendrix, 1988). However, we still need to characterize the consequent dynamics, including their phenotypic, genotypic, numerical, and distributional ramifications.

Fourth, my results suggests that fugitive species, which I call "resource-speculators," have a high vulnerability to significant seed predator impact in the determination of their abundance and distribution. These species are characterized by a relatively high dependence on seed regeneration, but also by intermediate responses to the other simultaneous selective pressures determining persistence.

Finally, consumers at each stage of plant development can influence seed production and input to the seed bank. Consequently, the effects of consumption at other stages of the life cycle deserve more consideration in seed bank studies. The traditional assessment of seed predation concentrates on the direct interactions of seed consumers with postrelease seeds. Underlying this emphasis is the implicit assumption that seed production is an independent variable in plant dynamics. Major progress in understanding predation in the determination of both the regeneration niche (Grubb, 1977b) and in the adult or persistence niche (Grime, 1979) will be made by evaluating the influence of consumption within the context of the whole life cycle, stage by successive stage. Path analysis, coupled with field experiments, provides a powerful approach toward this end.

Acknowledgments

I am grateful to all of those who have helped in the research cited and to those who have discussed aspects of these ideas, or inspired them with their work, especially G.A. Baker, B.D. Collier, J.H. Connell, M.J. Crawley, L.R. Fox, R.D.

Goeden, J.L. Harper, G.B. Harvey, D.H. Janzen, A. Joern, K.H. Keeler, R.F. Luck, P.A. Morrow, W.W. Murdoch, R.W. Otley, R.B. Root, and P.H. Zedler. My research program has been generously supported, financially and logistically, by San Diego State University grants and fellowships, Sigma Xi Grant-in-Aid, The National Science Foundation (Dissertation Improvement Grant, DEB80-11106, DEB82-07955, BSR84-05625, BSR85-16515), the University of Nebraska—Lincoln (Research Council, Dean's Special Fund, and Cedar Point Biological Station), and the Rocky Mountain Biological Laboratory. This paper is dedicated to the memory of Deborah Rabinowitz, whose work and vibrance will continue to inspire, in spite of her untimely death.

CHAPTER 4

Physiology of Dormancy and Germination in Relation to Seed Bank Ecology

Jerry M. Baskin
Carol C. Baskin

School of Biological Sciences
University of Kentucky
Lexington, Kentucky

I. Introduction

Reserves of buried, viable seeds have been found in many habitats, and they are of considerable ecological importance. Thompson and Grime (1979) distinguished four types of seed banks: Type I, transient seed bank present during summer; Type II, transient seed bank present during winter; and Types III and IV, persistent seed banks. In Type III, a large proportion of the seeds germinate following dispersal, and only a small fraction becomes buried in the soil. In Type IV, few seeds germinate before they are incorporated into the soil. Thus, in a species with a transient seed bank, none of the seeds persists for more than 1 yr, while in a species with a persistent seed bank, at least some seeds live longer than 1 yr. This chapter focuses on the physiology of dormancy and germination of seeds in persistent seed banks. Further, because a high proportion of seeds in persistent seed banks are buried (Thompson and Grime, 1979), we will concentrate on the physiology of seeds buried in soil.

To understand why seeds do or do not germinate while they are buried or after they are brought to the soil surface, we need to know the types of dormancy in seeds, how and when each type of dormancy is broken, and if, when, and how dormancy is reimposed. Also, information is required on how seeds in various states of dormancy respond to temperature, light (or darkness), oxygen (particularly in flooded and waterlogged soils), and moisture conditions in the burial and nonburial environments.

II. Types and States of Dormancy in Seeds

We define a dormant seed as one that will not germinate under any set of normal environmental conditions. There are five general types of dormancy exhibited by seeds at maturity (Table 1). These are distinguished on the basis of (1) permeability or impermeabilty of the seed coat to water, (2) whether the embryo is fully developed or underdeveloped (i.e., incomplete development of the embryo at seed maturity, sensu Grushvitzky, 1967), and (3) whether the embryo is physiologically dormant or nondormant. Potentially, seeds with all types of dormancy enter soil seed banks, but most of those found in seed banks in temperate regions have physiological dormancy, with physical dormancy being second in importance (Baskin and Baskin, 1985a; C.C. Baskin and Baskin, 1988).

As seeds with physiological dormancy afterripen, they pass through a series of states known as conditional dormancy (Vegis, 1964; Karssen,

Table 1
Types, causes, and characteristics of seed dormancy

Type	Cause(s) of dormancy	Characteristics of embryo
Physiological	Physiological inhibiting mechanism of germination in the embryo	Fully developed; dormant
Physical	Seed coat impermeable to water	Fully developed; nondormant
Combinational	Impermeable seed coat; physiological inhibiting mechanism of germination in the embryo	Fully developed; dormant
Morphological	Underdeveloped embryo	Underdeveloped; nondormant
Morphophysiological	Underdeveloped embryo; physiological inhibiting mechanism of germination in the embryo	Underdeveloped; dormant

1980–1981) before finally becoming nondormant. In the transition from dormancy to nondormancy, seeds first gain the ability to germinate over a very narrow range of environmental conditions. As afterripening continues, seeds become nondormant and thus can germinate over the widest range of environmental conditions possible for the species (Fig. 1). If, however, unfavorable environmental conditions (e.g., darkness) prevent germination of nondormant seeds, subsequent changes in environmental conditions (e.g., low or high temperatures) cause them to enter secondary dormancy. As seeds enter secondary dormancy, the range of conditions over which they can germinate decreases until finally they cannot germinate under any set of normal environmental conditions (Fig. 1). Thus, seeds exhibit a continuum of changes as they pass from dormancy to nondormancy and from nondormancy to dormancy (Baskin and Baskin, 1985a, and references cited therein).

In seeds with physical dormancy, germination is prevented by lack of imbibition of water. After the seed coat becomes permeable and the seed imbibes water, the seed is capable of germinating over a wide range of temperatures in light and darkness (Baskin and Baskin, 1984b). Unlike seeds with physiological dormancy, lack of suitable environmental conditions, primarily temperature, apparently does not cause the embryo to enter secondary dormancy. Seed coat permeability, however, has been reported to be reversible in some Leguminosae (Hagon and Ballard, 1970).

It is not known if seeds with combinational and morphophysiologi-

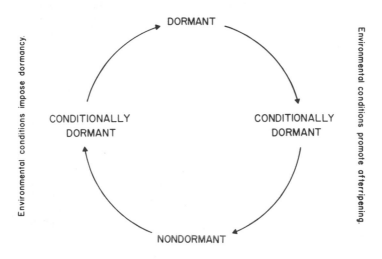

Figure 1.   Changes in the dormancy states of seeds with physiological dormancy; the seeds are dormant at maturity and go through all the possible stages of the dormancy cycle.

cal dormancy go through dormancy cyles, or if the embryo in seeds with morphological dormancy can be induced into physiological dormancy.

III. Germination Ecology of Buried Seeds

A. Seeds with Physiological Dormancy

1. Changes in Dormancy State

Many seeds in buried seed banks, such as those of obligate winter annuals and spring-germinating summer annuals, exhibit annual dormancy/nondormancy cycles. In obligate winter annuals (Fig. 2a), such as *Phacelia dubia* (Baskin and Baskin, 1973), *Arabidopsis thaliana* (Baskin and Baskin, 1983a), and *Lamium purpureum* (Baskin and Baskin, 1984a), seeds become nondormant during summer and germinate in autumn if light and soil moisture are nonlimiting. Seeds that fail to germinate in autumn reenter dormancy during late autumn and winter; they become nondormant the following summer. Dormant seeds of winter annuals must be exposed to high summer temperatures (20–30°C) to afterripen sufficiently to germinate at autumn temperatures (20/10 to 15/6°C, maximum/minimum) in autumn (Baskin and Baskin, 1986a). If nondormant seeds of winter annuals are subjected to low winter temperatures (1–5°C), they are induced into dormancy. Simulated autumn and spring temperatures (15/6 and 20/10°C), however, do not induce dormancy (Baskin and Baskin, 1984a; J.M. Baskin and C.C. Baskin, unpublished data).

In summer annuals (Fig. 2b) such as *Ambrosia artemisiifolia* (Baskin and Baskin, 1980) and *Polygonum aviculare* (Courtney, 1968), which germinate only in spring, seeds become nondormant during winter and germinate in spring if light and soil moisture are nonlimiting. Seeds that fail to germinate in spring reenter dormancy in late spring to early summer; they become nondormant the following winter. Dormant seeds of spring-germinating summer annuals must be exposed to low temperatures (5 to 15/6°C) to afterripen sufficiently to germinate at early spring temperatures (15/6°C) in spring (Baskin and Baskin, 1987). In *Ambrosia artemisiifolia*, it has been shown that seeds enter dormancy as tempera-

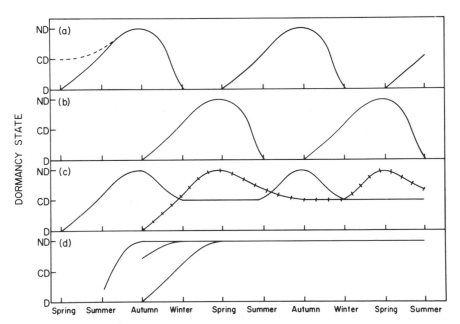

Figure 2. Patterns of changes in seeds with physiological dormancy. (a) Obligate winter annual with annual dormancy/nondormancy cycle. Freshly matured seeds in some species are dormant (————) and in others conditionally dormant (--------). (b) Spring-germinating summer annual with annual dormancy/nondormancy cycle. (c) Facultative winter annual (————) and spring- and summer-germinating summer annual (+++) with annual conditional dormancy/nondormancy cycles. (d) Perennials with no changes in dormancy state after seeds come out of dormancy.

tures increase in spring and early summer (Baskin and Baskin, 1980).

Other seeds in buried seed banks are conditionally dormant or dormant initially and become nondormant and thereafter exhibit annual conditional dormancy/nondormancy cycles (Fig. 2c). Seeds with this pattern of change in dormancy state include (1) facultative winter annuals, such as *Lamium amplexicaule* (Baskin and Baskin, 1981a) and *Aphanes arvensis* (Roberts and Neilson, 1982a), which become conditionally dormant during winter and nondormant during summer; (2) some spring- and summer-germinating summer annuals, such as *Solanum sarrachoides* (Roberts and Boddrell, 1983), which become conditionally dormant during late summer to autumn and nondormant during winter; and (3) the monocarpic perennials *Verbascum thapsus* and *Verbascum blattaria*, which become conditionally dormant during summer and nondormant during winter (Baskin and Baskin, 1981b).

Still other seeds in buried seed banks are conditionally dormant or dormant when they enter the seed bank, and they become nondormant and remain nondormant (Fig. 2d). Examples of species exhibiting this pattern of changes are *Rumex crispus* (Baskin and Baskin, 1985b), *Portulaca smallii* (Baskin et al., 1987), *Oenothera biennis*, *Potentilla recta*, and *Penthorum sedoides* (J.M. Baskin and C.C. Baskin, unpublished data).

2. Prevention of Germination during Burial

A primary reason nondormant seeds do not germinate while buried is that many of them have a light requirement for germination (e.g., Baskin and Baskin, 1983a, 1984a, 1985b). However, seeds of some species undergo annual changes in their light requirement. For example, seeds of *Ambrosia artemisiifolia* (Baskin and Baskin, 1980), *Lamium amplexicaule* (Baskin and Baskin, 1981a), and *Verbascum blattaria* (Baskin and Baskin, 1981b) gain the ability to germinate in darkness during winter, but lose it during spring/summer. In winter, low temperatures prevent germination. Sometimes there is an overlap in early spring between temperatures of the soil and those required for germination in darkness, and a small proportion of the seeds in the buried seed bank germinate. This leads to a gradual depletion of the seed bank.

An inhibitor produced by seeds of *Spergula arvensis* buried under sterile conditions prevented germination until a light requirement was induced (Wesson and Wareing, 1969b). Holm (1972) found that germination of buried seeds of *Ipomoea purpurea*, *Abutilon theophrasti*, and *Brassica kaber* was inhibited by production of volatile metabolites and by low oxygen levels.

3. Germination of Exhumed Seeds

When seeds are brought to the soil surface and are exposed to unfiltered sunlight and adequate moisture, changes in dormancy states and germination requirements of nondormant seeds interact to control the timing of germination. Dormant seeds that reach the soil surface must afterripen before they can germinate. If seeds are nondormant or conditionally dormant, habitat temperatures must be within the range of those required for germination; otherwise, germination will be delayed.

From information on changes in dormancy states, it is possible to predict when exhumed seeds of a species will germinate. An obligate winter annual with its annual dormancy/nondormancy cycle will germinate only in autumn, while a facultative winter annual with its conditional dormancy/nondormancy cycle will germinate in both autumn and spring. A spring-germinating summer annual with its annual dormancy/nondormancy cycle can germinate only in spring, while a spring- and summer-germinating summer annual with its conditional dormancy/nondormancy cycle potentially can germinate throughout most of the growing season. Also, seeds that become nondormant after burial and remain nondormant can germinate at any time throughout the growing season.

Timing of germination of seeds brought to the soil surface may not necessarily depend on the physiological state of the seeds if they are under a canopy of green leaves. Sunlight filtered by leaves has a lower red/far-red photon flux ratio than sunlight (Smith, 1982), which results in a low proportion of Pfr, the active form of phytochrome (Franklin, 1980). Numerous studies have shown that leaf-filtered sunlight inhibits germination of nondormant seeds (e.g., Górski *et al.*, 1977; Fenner, 1980a; Silvertown, 1980a; Pons, 1983; Washitani and Saeki, 1984). The inhibiting effect of light quality would be alleviated with leaf abscission. However, by the time the light environment becomes nonlimiting for germination, the seeds may have reentered dormancy.

B. Seeds with Physical Dormancy

1. Mechanism of Hard Seed Dormancy

Impermeable or hard seed coats occur in seeds of the Anacardiaceae, Cannaceae, Convolvulaceae, Cucurbitaceae, Geraniaceae, Leguminosae, Malvaceae, Nympheaceae, Rhamnaceae, Sapindaceae, Sterculiaceae, Tiliaceae (Shaw, 1929; Barton, 1934; Floyd, 1976; Rolston, 1978; Mann *et al.*, 1981), and perhaps other plant families. The impermeability of the seed coat is not due to a waxy cuticle, but to a palisade layer of macrosclerid or malpighian cells impregnated with water-repellent substances, including suberin, cutin, and lignin (Rolston, 1978). Seeds do not become permeable until an opening or passage through the palisade layer is made somehow.

The older ideas of how hard seeds become permeable in nature include ingestion and thus acid scarification in the digestive tract of various animals, attack by bacteria and/or fungi, and scarification by being rubbed against rocks or other hard objects. Available data showing that seeds are softened by these agents in nature are limited and contradictory. For example, fungi are effective natural scarifiers of hard seeds of *Albizia julibrissin* (Gogue and Emino, 1979), but the impermeable seeds of *Abutilon theophrasti* possess antimicrobial compounds that inhibit the growth of soil bacteria and fungi (Kremer, 1986). Thus, these mechanisms for breakdown of hard seed coats have been seriously questioned (Ballard, 1973).

Anatomical studies of several species have shown that a specific region of the seed coat, rather than the whole seed coat, becomes permeable. In the Malvaceae, water enters through an opening in the palisade layer of the chalazal region (Christiansen and Moore, 1959). In *Gossypium*, the dense parenchymatous plug (Winter, 1960) moves and water enters after the pallisade cells surrounding the plug break down. In *Sida spinosa*, the seed coat becomes permeable when palisade cells pull away from subpalisade cells in the chalazal region (Egley *et al.*, 1986). The strophiole is the site of water entry into seeds of Leguminosae. In the subfamily Papilionoideae, this is a zone of weakness in the seed coat near the hilum where macrosclerids pull apart (Hamley, 1932), while in the subfamily Mimosoideae [e.g., *Albizia lophantha* (Dell, 1980) and *Acacia kempeana* (Hanna, 1984)] high temperatures disrupt the strophiolar plug. In the Convolvulaceae, water enters the seed when a small pluglike structure near the micropyle is dislodged or removed (Koller and Cohen, 1959).

2. Softening of Hard Seeds

Temperature is probably the most important environmental factor regulating timing of breakdown of the special anatomical region of the seed coat and the entry of water into the seed, but this phenomenon is not fully understood. For example, although many hard-seeded species in temperate regions germinate in spring, little is known about how low winter temperatures cause softening of the seed coat. *Pediomelum subacaule* seeds disperse in May, and germination occurs the following March and April. Preliminary studies indicate that low temperature (5°C) increases the percentage of permeable seeds, but more seeds germinate if high summer temperatures precede low winter temperatures (J.M. Baskin and C.C. Baskin, unpublished data). In *Lespedeza cyrtobotrya*, 40% of the seeds that overwintered on the ground under snow became permeable, compared to only 12% of those on dead, upright plants (Iwata, 1966). However, alternate freezing and thawing had little or no effect on softening seeds of *Medicago sativa*, *Cuscuta* sp. (Midgley, 1926), and *Sesbania* spp. (Graaff and van Staden, 1983).

It is well known that high temperatures cause hard seed coats to become permeable. Drastic treatments such as immersion into boiling water for short periods of time (Iwata, 1966; Gratkowski, 1973; Clemens et al., 1977) or exposure to dry heat at 60 to >100°C (Rincker, 1954; Williams and Elliott, 1960; Narang and Bhardwaj, 1974; Datta and Sen, 1982) result in nearly 100% permeability. In nature, fire is the most extreme example of high-temperature stimulation of germination of hard-seeded species (e.g., Stone and Juhren, 1951; Floyd, 1976; Auld, 1986c). *Acacia suaveolens* seeds become permeable when exposed to temperatures between 60 and 80°C; below 60°C they remain impermeable, and above 80°C seed death is likely (Auld, 1986c). The maximum soil depth at which buried seeds become permeable depends on the intensity of the fire (Auld, 1986c). This also is true for seeds of other species (e.g., Pieterse and Cairns, 1986).

Although fire is important, more hard-seeded species are rendered permeable by high summer temperatures. High fluctuating temperatures resulting from daily heating and cooling of the soil during summer, rather than high temperatures per se, are responsible for the breakdown of the special anatomical region of the seed coat. Daily alternating temperatures (maximum 30–60°C, minimum 15–20°C) soften seeds of *Lupinus digitatus, Lupinus lutens, Medicago tribuloides, Trifolium subterraneum* (Quinlivan, 1961), *Neptunia oleracea* (Sharma et al., 1984), and *Stylosanthes* spp. (McKeon and Mott, 1982, 1984). Drying associated with summer habitat conditions also is a factor in the softening of seeds of *Abutilon theophrasti* (LaCroix and Staniforth, 1964), *Geranium carolinianum* (Baskin and Baskin 1974), and *Stylosanthes* spp. (McKeon and Mott, 1984). Alternate wetting and drying at high temperatures likewise can be beneficial (Baskin and Baskin, 1984b).

3. Germination of Hard Seeds in Seed Banks

Because daily temperature fluctuations, heat from fires, and wetting and drying are more intense at than below the soil surface, seeds at or near the surface have a greater probability of germinating than do those buried several centimeters. Most hard-seeded species germinate in light and darkness over a wide range of temperatures (e.g., Baskin and Baskin, 1974, 1984b), and seedlings can emerge from depths of 10–15 cm (e.g., Mann et al., 1981). Thus, seeds do not have to be on the soil surface to germinate, but only close enough to it for environmental conditions to render them permeable.

Habitat modifications can result in changes in temperature and other factors that cause buried hard seeds to soften and germinate. For example, because *Heliocarpus donnell-smithii* (Tiliaceae), a component of secondary tropical forests, has impermeable seed coats, few seeds of this species germinate under the closed canopy where there is little difference between day and night temperatures. If a gap is created in the forest, day temperatures are 10–15°C above night temperatures, and seeds germinate immediately (Vázquez-Yanes, 1981; Vázquez-Yanes and Orozco Segovia, 1982a). Hard seeds of *Ulex europaeus* (Ivens, 1978) and *Rhus javanica* (Washitani and Takenaka, 1986) also germinate in response to increased daily temperature fluctuations when canopy vegetation is removed. In *Trifolium subterraneum* (Taylor, 1981) and *Sida spinosa* (Baskin and Baskin, 1984b), preconditioning at lower temperatures causes a more rapid increase in permeability after seeds are transferred to higher temperatures. In nature, this could happen if seeds are brought to the soil surface or if the plant canopy is removed from a site.

Evolution of a specific anatomical region of the seed coat that becomes permeable in response to certain temperature regimes argues against the general importance of soil microbes or of scarification as natural mechanisms of seed coat breakdown. Release of seed dormancy by microbial activity or scarification is not an effective survival mechanism for seeds that need to germinate at a certain time of the year (e.g., autumn or spring) or in response to habitat disturbance (e.g., formation of gaps in vegetation) for the plant to become established and complete its life cycle. On the other hand, response to low, high, or fluctuating temperatures is an adaptive mechanism for restricting the germination of a hard-seeded species to a certain season of the year or to the changed conditions following habitat disturbance.

C. Seeds with Combinational Dormancy

Some seeds have a combination of physiological and physical dormancy, and thus they have impermeable seed coats and dormant embryos. In these seeds, high fluctuating temperatures usually are required to render the seed coat permeable, and either low winter temperatures or high summer temperatures are required to overcome dormancy of the embryo. Examples of combinational dormancy that require a cold treatment to overcome embryo dormancy are *Tilia americana* (Barton, 1934), *Cercis canadensis* (Afanasiev, 1944), *Ceanothus* spp. (Quick, 1935; Radwan and Crouch, 1977), and *Parkia pendula* (Rizzini, 1977). Embryos of some hard-seeded winter annuals, such as *Trifolium subterraneum* (Quinlivan and Nicol, 1971) and *Geranium carolinianum* (Baskin and Baskin, 1974), are dormant at maturity, but they quickly afterripen at high summer temperatures, before the seed coat becomes permeable.

In a study of the seed banks in three mature coniferous forests in Idaho, Kramer and Johnson (1987) found more seeds of *Ceanothus velutinus*, a species with combinational dormancy (Quick, 1935), than any other species. Seeds of this species are believed to remain viable in forest seed banks for more than 500 years (Zavitkovski and Newton, 1968).

D. Seeds with Morphological Dormancy

In seeds with morphological dormancy, the embryo is underdeveloped but nondormant. Completion of embryo growth occurs after the seeds are dispersed from the parent plant. When the embryo is fully developed, seeds are ready to germinate. Thus, there is no afterripening requirement.

This type of dormancy occurs in species of tropical (e.g., Magnoliaceae, Degeneriaceae, Winteraceae, Lactoridaceae, Canellaceae, and Annonaceae; Grushvitzky, 1967) and temperate (e.g., Umbelliferae and Ranunculaceae; Baskin and Baskin, 1984c; J. M. Baskin and C. C. Baskin, unpublished data) plant families. Apparently, in tropical species embryos grow slowly following seed dispersal (Grushvitzky, 1967). In *Isopyrum biternatum*, a temperate mesic woodland herb, seeds are dispersed in mid to late May, but embryo growth and germination are delayed by high temperatures until September and October (Baskin and Baskin, 1986b). In *Conium maculatum*, a temperate monocarpic perennial, seeds mature in early summer, and embryos in imbibed seeds can grow over a wide range of temperatures in both light and darkness (J.M. Baskin and C.C. Baskin, unpublished data). In contrast, *Apium graveolens* seeds require light for embryo growth (Jacobsen and Pressman, 1979). The long-term persistence of seeds with morphological dormancy in buried seed banks depends on a light requirement for embryo growth; otherwise, seeds would germinate within a year or less. It is likely that morphologically dormant seeds with a light requirement eventually will be found in persistent seed banks.

E. Seeds with Morphophysiological Dormancy

Seeds with morphophysiological dormancy have underdeveloped embryos that must complete growth before germination can occur. Like seeds with morphological dormancy, completion of embryo growth occurs after dispersal. However, unlike seeds with morphological dormancy, there is an afterripening requirement for embryo growth and/or germination. At least eight types of morphophysiological dormancy have been identified (Nikolaeva, 1977; Baskin and Baskin, 1983b, 1984c, 1985c, unpublished data). In all types, seeds must be exposed to high (summer) and/or low (winter) temperatures before they can germinate (Nikolaeva, 1977; Baskin and Baskin, 1983b, 1984c, 1985c; J.M. Baskin and C.C. Baskin, unpublished data).

Morphophysiological dormancy occurs in a number of plant families

in the temperate zone, including the Araceae, Araliaceae, Aristolochia-
ceae, Berberidaceae, Fumariaceae, Liliaceae, Papaveraceae, Ranunculaceae, and Umbelliferae. Many herbs of mesic (especially deciduous)
temperate forests have morphophysiological dormancy (C.C. Baskin and
Baskin, 1988). Thus, mesic temperate forests seem to be the best habitat
in which to determine whether seeds with this type of dormancy form
seed banks. Seed bank studies by Roberts *et al.* (1984), Morash and
Freedman (1983), Naka and Yoda (1984), and Nakagoshi (1984a,b) found
seeds of genera that are known to exhibit morphophysiological dormancy, whereas those by Oosting and Humphreys (1940), Olmsted and
Curtis (1947), Livingston and Allessio (1968), Kellman (1974a), and
Hill and Stevens (1981) did not. Buried viable seeds of *Osmorhiza chilensis*
(Strickler and Edgerton, 1976), and *Osmorhiza depauperata* (Whipple,
1978) were found in coniferous forests. It is reasonable to think that
seeds of these species have morphophysiological dormancy, because
seeds of *Osmorhiza longistylis* (Baskin and Baskin, 1984c) and *Osmorhiza
claytonii* (J.M. Baskin and C.C. Baskin, unpublished data) do have this
type of dormancy. Recently, Kramer and Johnson (1987) reported the
presence of seeds of *Hydrophyllum capitatum* and *Trillium ovatum* in seed
banks of mature coniferous forests. The seeds of other species of *Trillium*
(Barton, 1944) and *Hydrophyllum* (Baskin and Baskin, 1983b, 1985c),
whose germination requirements have been studied, have morphophysiological dormancy, suggesting these also do.

Our studies on germination phenology indicate that seeds of some
species with morphophysiological dormancy [e.g., *Hydrophyllum appendiculatum* (Baskin and Baskin, 1985c), *Trillium flexipes*, *Trillium sessile*,
Asarum canadense, *Osmorhiza longistylis*, *Osmorhiza claytonii*, and *San-
guinaria canadensis* (C.C. Baskin and Baskin, 1988)] can remain dormant
and viable for ≥3 yr on soil under leaves in a nonheated greenhouse
before they germinate. Thus, it seems that seeds of these species have
the potential to form seed banks. It is not known if long-persisting seeds
with morphophysiological dormancy undergo annual changes in dormancy states.

IV. General Patterns of Germination
Responses in Seed Banks

Review of the literature indicates that most species in the seed bank
have physiological dormancy; it appears that physical, combinational,
morphophysiological, and morphological dormancy follow in descending order of importance. In any habitat the seed bank is composed of
species with a variety of germination strategies. These seeds may have

been produced at the site and/or brought there from other habitats by various dispersal agents.

If we focus on seeds produced in the habitat, a knowledge of habitat factors may provide some clues as to what germination responses may be represented among the seeds in the seed bank. This is particularly true for seeds with physiological dormancy. If the habitat has predictable periods of stress and the seedlings are not drought tolerant, many species will exhibit annual dormancy/nondormancy cycles, with the seeds being dormant during the period of stress. Obligate winter annuals native to summer-dry, winter-wet habitats illustrate this response. If the habitat is unpredictable, many species will be conditionally dormant or dormant when they enter the seed bank; they become (and remain) nondormant or exhibit annual conditional dormancy/nondormancy cycles. Examples of these include mesic summer annuals that germinate on river banks in summer following soil disturbance associated with flooding, drought-tolerant summer annuals that germinate on rock outcrops after a summer rain, and annuals and perennials that germinate on mud flats in summer after the water recedes. Seeds with physical dormancy occur in the seed banks of predictable (Narang and Bhardwaj, 1974) and unpredictable (Auld, 1986c) habitats. In habitats that are not subjected to periodic disturbances (e.g., temperate mesic forests or tropical rainforests), it remains to be seen whether or not the "climax" species form much of a seed pool. In both temperate mesic deciduous forests (Olmsted and Curtis, 1947) and tropical rainforests (Hopkins and Graham, 1984), most of the species found in the seed banks thusfar are secondary species or species that grow elsewhere; their seeds have been dispersed into the forests.

V. Summary and Conclusions

Seeds can exhibit physiological, physical, combinational, morphological, or morphophysiological dormancy, and all types potentially can form seed banks. A high percentage of temperate zone seed bank species have some degree of physiological dormancy; many display annual cycles in germination requirements in response to seasonal temperature changes. However, seeds of some species are dormant or conditionally dormant when they enter the seed bank and do not exhibit further changes in the dormancy state after they become nondormant.

A primary reason seeds do not germinate in the soil while nondormant is that most have a light requirement for germination. Soil temperature, moisture content, volatile metabolites, and decreased oxygen levels also may contribute to lack of germination of nondormant seeds in the soil. Seeds on the soil surface will germinate only if they are conditionally dormant or nondormant, and temperature, soil moisture, and light quality (red/far-red ratio) are nonlimiting. If seeds are dormant when they are brought to the soil surface, they must afterripen before they can germinate.

Seeds with physical dormancy have specific regions of the seed coat that become permeable to water; the entire seed coat does not become permeable. Temperature is the most important environmental factor causing hard seeds to become permeable. High temperatures and high daily fluctuations of temperature cause softening of the hard seed coats of species that germinate in late spring, summer, or autumn, whereas low temperatures during winter appear to be responsible for softening of hard seeds that germinate in early spring. Changes in amplitude of the daily fluctuations of temperature that result when canopy vegetation is removed or when seeds are brought to the soil surface are an effective trigger for seed softening and germination of many hard-seeded species. This response of hard seeds to elevated temperatures is a mechanism for detection of environments favorable for growth. Fire is an important factor in the breakdown of hard seed coats in fire-prone habitats. There is little evidence that coats of hard seeds are broken down by the action of soil microbes or by natural scarification.

Seeds of several genera of woody plants have a combination of physical and physiological dormancy, and thus require high summer temperatures to soften the seed coat followed by low winter temperatures to overcome embryo dormancy. In hard-seeded winter annuals, the embryo may be physiologically dormant at maturity. However, unlike embryos of woody plants, these become nondormant at high summer temperatures, and this occurs before the seed coat becomes permeable.

It is doubtful that seeds with morphological dormancy persist in the soil for more than a few months unless they have a light requirement for embryo growth. Seeds with morphophysiological dormancy may remain viable under leaf litter for several years, but their importance in seed banks is poorly understood.

The level of habitat stress and its predictability provide some clues to the types of dormancy mechanisms and/or germination responses represented among the species in the seed bank.

CHAPTER 5

Modeling the Evolutionary Ecology of Seed Banks

D. Lawrence Venable

Department of Ecology and Evolutionary Biology,
University of Arizona, Tucson, Arizona

I. Why Model the Population Dynamic Consequences of Seed Banks?

At the symposium motivating this volume, the term "seed bank science" was used by Jim McGraw with humorous intent. The point was that, rather than being a coordinated research program with well-defined goals, approaches, and funding, the study of seed banks has been a collection of isolated works using varied approaches, often undertaken as side projects in investigations with other primary foci. This state of affairs justifies the following remarks on the relevance of modeling to the development of our empirical understanding of seed banks. In a more mature science these comments would hopefully be superfluous.

One role of modeling is to provide interpretation. For example, we know that there is a predominance of annual and early successional plant species represented in the seed banks of even late successional communities in temperate deciduous forests (Pickett and McDonnell, Chapter 8, this volume). Does this pattern have something to do with the relative importance of dispersal in space versus time for plants of different life history strategies? Does it have something to do with the relative importance of seed regeneration versus perennation? Mathematical models can formalize the different hypotheses and double check their logical coherence. In the process, additional mechanisms and selective forces often become apparent.

A second role of models is that they can suggest appropriate data to collect. We might, for example, be interested in the relative importance of dispersal and dormancy in the persistence of early successional species in a particular habitat. From a careful consideration of models of these phenomena, it becomes clear that we need to understand patterns of spatial and temporal variation in the expected fecundity of germinating seeds; correlations in temporal, spatial, and temporal–spatial variation; the survival of dormant seeds; and the proportion of a plant's seed output that can be expected to move to microsites of different quality. Attempts to measure some of these parameters can be incorporated in subsequent research designs.

Models can often be tailored to capture the essential elements of a specific experimental system. Then the model can be manipulated to generate predictions that can be tested with the experimental system. For example, I am interested in the evolution of seed proportions in the seed heteromorphic composite, *Heterosperma pinnatum*, in central Mexico (Venable *et al.*, 1987). Each plant produces seed morphs with differences in within-year timing of germination, correlated differences in dispersal, no between-year seed bank, and proportions that vary among populations. Elements of some of the more general models discussed below combine easily to mimic this system and predict how environmental changes favor different seed proportions. Predicted differences in seed proportions can then be checked against those observed in different environments in which this species is found.

A final role of models, often forgotten in our search for testable predictions, is that they permit exploration of questions that cannot be easily answered experimentally at present. We may have only 4–5 yr of data on germination patterns, the demographic success of germinating seeds, and the survival of dormant seeds. We might want to know what the population dynamic consequences of such variation would be after 100 or 1000 yr, or how certain environmental changes might favor different germination strategies. While we may be able to collect data on the results of environmental differences along some environmental gradient, the actual selective mechanisms and the operation of population dynamic forces cannot be reasonably measured over long periods of time. The value of models for quickly answering, at least in a hypothetical way, many of our "What if . . . ?" questions should not be underestimated, particularly in the developmental stages of a young science.

II. Issues Relevant to Understanding Seed Banks

First, I shall briefly outline three basic issues of seed biology relevant to understanding seed banks: the within-year timing of germination, the production of a between-year seed bank, and predictive or plastic germination. I shall restrict my discussion to density-independent models of annual plants because such models are fairly easy to understand and construct. Modeling annuals avoids the difficulties of perennation, vegetative propagation, and other complications.

A. The Within-Year Timing of Germination

The within-year timing of germination affects the study of seed banks because it determines how many seeds will be in the soil at different times of the year. What factors determine how natural selection shapes the within-year germination schedule; that is, at what season(s) should seeds germinate? The simplest case is if there is no seed bank carry-over between years. In this case the question becomes, given that it is this year or never, when should seeds germinate to yield highest fitness? In general seeds will germinate according to some distribution within one or more germination seasons of a single year (Fig. 1). The question is, what proportion of seeds should germinate in each of the possible windows of opportunity (Fig. 1)? The simplest case is with a choice of two windows of opportunity. For example, some temperate zone annuals germinate in the fall, some germinate in the spring, and some germinate in both periods (e.g., *Papaver dubium*, Arthur *et al.*, 1973; *Bromus tectorum*, Mack and Pyke, 1983). Similarly, desert winter annuals around Tucson, Arizona, often germinate predominantly in October–November, December–January, or in both periods.

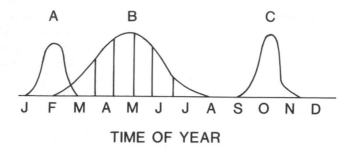

TIME OF YEAR

Figure 1.   Possible within-year germination schedules. What proportion of seeds should germinate during each census interval to yield highest fitness? Some annuals germinate predominantly in the fall (C), some in the winter (A), and some during both (A and C). Some species have fairly synchronous germination (e.g., A or C), while others spread germination over a longer season (e.g., B).

Consider a population described by the following simple growth equation:

$$\lambda = S^E q + (1 - q)\, S^L$$

where λ is the finite rate of increase, q is the proportion of seeds germinating early, and S^E is sum of $l_x b_x$ for early germinating seeds (i.e., the average seed yield per early germinating seed). In a constant environment, either S^E would be higher, in which case all seeds should germinate early, or S^L (where L stands for late) would be higher, in which case all seeds should germinate late (if $S^E = S^L$, it does not matter when seeds germinate). However, year-to-year variation in the success of early and late cohorts can favor the simultaneous production of both. In some years early germination might yield higher fitness, while in others, late germination might be better. We can incorporate this into the model by allowing S^E and S^L to vary from year to year (define S_i^E to equal $\Sigma\, l_x b_x$ in year type i). Now, rather than consider a single λ, we are interested in the geometric average of λ after a series of years (we use the geometric mean because population growth is multiplicative through time; see Leon, 1985):

$$\text{GEO}(\lambda) = \Pi (q S_i^E + (1 - q) S_i^L)^{p_i} \qquad (1)$$

where p_i is the probability of year-type i occurring.

   We can now ask what germination strategy, q, yields the highest geometric mean fitness (Fig. 2). If the highest geometric mean fitness is at $q = 0$, selection favors all seeds germinating late; if the maximum is at $q = 1$, all seeds should germinate early; if the maximum is between 0 and 1, multiple germination cohorts are favored. Algebraic manipulation of

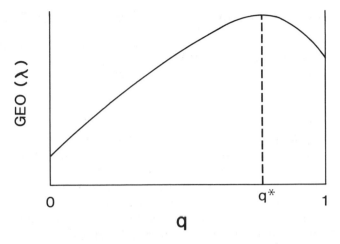

Figure 2. To determine what proportion of seeds should germinate in each of two windows of opportunity, we determine what proportion, q, of early germination and what proportion, $1 - q$, of late germination yield highest geometric mean fitness, GEO(λ), where geometric mean fitness is described by Eq. (1) in the text. If q^*, the proportion of early germination that yields highest fitness, is 0, selection favors only late germination. If q^* is 1, selection favors only early germination. Multiple cohorts are favored when $0 < q^* < 1$.

the model shows that multiple cohorts within years (or more generally the spread of germination within years) are favored by a high year-to-year variation in the seed yield per germinating seed, and by a low or negative correlation between the seed yield per seed of early and late cohorts (i.e., if years that the early cohort does well tend to be years that the late cohort does poorly and vice versa). For multiple cohorts to be favored, the expected seed yield per seed must be greater for each cohort in some year types (obviously if early germination always resulted in lower seed set, it would not be favored). Finally, more germination in one or another cohort is favored by a greater probability of favorable conditions for that cohort. Equation (1) was analyzed in Venable (1985) where the two sources of reproduction referred to two seed morphs; the same algebra and conclusions apply for early and late germination. A simple elaboration of this model would be to consider multiple windows of opportunity and the proportion of seeds that should germinate in each to maximize plant fitness. Leon (1985) has outlined a mathematically more sophisticated approach to the within-year timing of germination, considering the optimal continuous germination schedule.

B. The Production of a Between-Year Seed Bank

The second issue relevant to understanding seed banks is the production of a between-year seed bank. This issue can be addressed in a similar manner. Consider the following equation:

$$\lambda = GS + R(1 - G)$$

where G is the proportion of seeds germinating, $(1 - G)$ is the proportion remaining dormant between years, and R is the survival of dormant seeds. In what follows I will often use the term "dormancy" in the ecological sense of not germinating for whatever reason. This usage should not be confused with the more physiological usage meaning unable to germinate for some physiological or morphological reason (see Baskin and Baskin, Chapter 4, this volume). If we let the seed yield per seed vary, the geometric mean growth rate becomes

$$\text{GEO}(\lambda) = \Pi(GS_i + R(1 - G)^{p_i} \tag{2}$$

If the germination fraction resulting in highest geometric mean fitness is 1, selection favors no between-year seed bank; if it is less than 1, the production of a seed bank is favored (Fig. 3). Analysis of this model

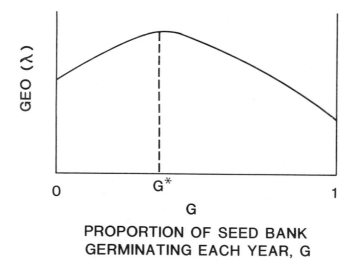

PROPORTION OF SEED BANK
GERMINATING EACH YEAR, G

Figure 3.   To determine what proportion of seeds should enter a between-year seed bank we determine what proportion, G, of the seed bank should germinate each year to yield highest geometric mean fitness, $\text{GEO}(\lambda)$, where geometric mean fitness is described by Eq. (2) in the text. If G^*, the germination fraction that maximizes fitness, is 1, selection favors complete germination (i.e., no seed bank). If G^* is less than 1, the production of a seed bank is favored.

shows that a between-year seed bank is favored by (1) a low probability of high seed yield, (2) high year-to-year variance in seed yield, and (3) a high survival rate of seeds in the soil. This model was first analyzed by Cohen (1966) and has been recently summarized in Leon (1985) and Brown and Venable (1986). For a particular application more complicated possibilities can be imagined. For example, one could simultaneously explore within- and among-year timing of germination by putting together Eqs. (1) and (2) and finding the optimal combination of G and q.

C. Predictive Germination

Missing from the above discussions is dormancy plasticity, or the modification of the timing or fraction of seeds germinating in response to environmental cues. This may be an important source of variability in seed bank dynamics and it must be dealt with in any complete discussion of the ecology of dormancy. The approach in Sections II,A and II,B was to determine the one "correct" germination strategy and see how different conditions select for a different fixed germination strategy. Plants of many habitats produce seeds whose germination behavior varies considerably, depending on environmental cues. For example, the germination fraction of desert annuals depends critically on temperature

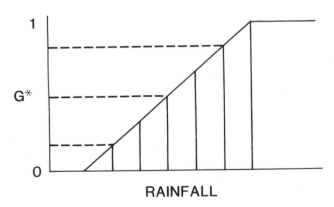

RAINFALL

Figure 4.   Predictive dormancy. How should the germination fraction that maximizes fitness, G^*, vary with different cues that provide information as to the probability of favorable conditions? For example, for desert annuals, the amount of rainfall (and temperature) at the time of germination is a predictor of the expected fecundity of germinating seeds. The fitness-maximizing germination fraction is determined, where fitness is described by Eq. (2), but the probabilities of different year types (environmental conditions) are replaced by the conditional probabilities of different year types given the cue. Thus there may be a different G^* for each value of the cue.

and the amount of rainfall (Capon and van Asdall, 1966; Freas and Kemp, 1983). This was first modeled by Cohen (1967), who showed that the problem is neatly solved by performing the same analyses as described above, but making the probabilities (the p_i values) conditional on the cues (summarized in Leon, 1985). The question changes from knowing the probability of a high seed-yield year to knowing the probability of a high seed-yield year *given* that, at the time of germination, there are specific environmental cues, such as a temperature of 24°C and 20 mm of precipitation. Thus rather than one optimal germination fraction, there are many, one for each cue (Fig. 4). One prediction of this kind of model is that for an ordered set of cues (that predict increasingly favorable conditions, as in Fig. 4), reliable cues favor an abrupt shift in germination strategy while unreliable cues favor a more gradual shift in germination strategy. In the extreme, the best germination strategy is insensitivity to cues that give no information about the probability of favorable and unfavorable conditions. This is because the conditional probability is the same as the probability in the absence of the cue. Thus, in the example above, different amounts of rainfall would result in the same germination if rainfall were an uninformative cue, because the conditional probabilities would all be the same.

D. Research Questions and Field Tests

The models presented above are simple and straightforward, yet they generate numerous questions in need of empirical examination. What are the expected seed yields of plants from early and late cohorts in different years? How constant is the partition into early and late germination? What cues regulate this partition and how do they correlate with the probability of favorable and unfavorable conditions (i.e., how reliable are the cues)?

To illustrate how such models can guide the collection of field data and aid in its interpretation, I present 4 yr of demographic data on two common desert winter annuals, *Plantago patagonica* and *Schismus barbatus*, growing near Tucson, Arizona (Table 1). This project was motivated by a desire to understand patterns of demographic variability thought to shape the evolution of seed dormancy. Plants emerging following each rainfall event during these years were mapped and their survivals and fecundities were recorded to determine the average seed yield per germinating seed, a critical variable in the models above. In most years there were major germination cohorts in October or November and in December or January; seed was set in March or April. Seed bank samples were collected in February at the end of the germination season to estimate the density of ungerminated viable seeds with the potential for surviving to future seasons. Fitness topographies were gen-

Table 1
Demographic data for two species of desert winter annuals at the Desert Laboratory, Tucson, Arizona, 1982–1986

Species	Year	Cohort	Initial germination	Survival to reproduction (%)	Mean $b_x{}^a$	$l_x b_x$
Schismus *barbatus*	82/83	E	71	70.4	196.20	138.20
		L	25	76	221.10	168.00
	83/84	E	1045	35.7	10.49	3.74
		L	83	42.2	10.37	4.37
	84/85	E	6	0	—	0
		L	61	47.5	70.90	33.70
	85/86	E	135	32.6	95.10	30.99
		L	27	18.5	21.00	3.89
Plantago *patagonica*	82/83	E	580	66.4	25.30	17.82
		L	123	57.7	15.50	9.70
	83/84	E	2646	54.7	5.59	3.05
		L	92	51.1	4.40	2.25
	84/85	E	76	25.0	26.60	6.65
		L	365	42.2	24.70	10.26
	85/86	E	205	15.1	65.05	9.83
		L	260	48.5	24.04	11.65

$^a b_x$, the average fecundity of plants surviving to reproduce; $l_x b_x$, the expected fecundity of an emerging seedling (i.e., the product of survival and fecundity).

erated from these data using a simple model that allows plants to germinate early (October–November) or late (December–January) and to have a proportion of seeds remaining dormant between years (i.e. a combination of the within- and between-year dormancy models presented above). Seed yields were calibrated by the proportion of seeds produced in one yr that could be accounted for in the next year (germinating seed densities plus seed bank densities) to correct seed yield per germinating seed for seed mortality between the April seed production and the following germination season. The proportion of early germination and the proportion of seeds remaining dormant between years were arbitrarily varied through the full range of possible values, and the geometric mean fitness for each species was calculated (Fig. 5).

This analysis provides several insights not readily apparent in the original data. First, it appears that fitness is much more sensitive to changes in the within- and between-year timing of germination in *Schismus* than it is in *Plantago* because the fitness topography for *Plantago* is flatter. This difference in the shape of the fitness topography results from the greater seasonal and between-year variation in reproductive success of *Schismus* cohorts. *Schismus* cohorts ranged from an

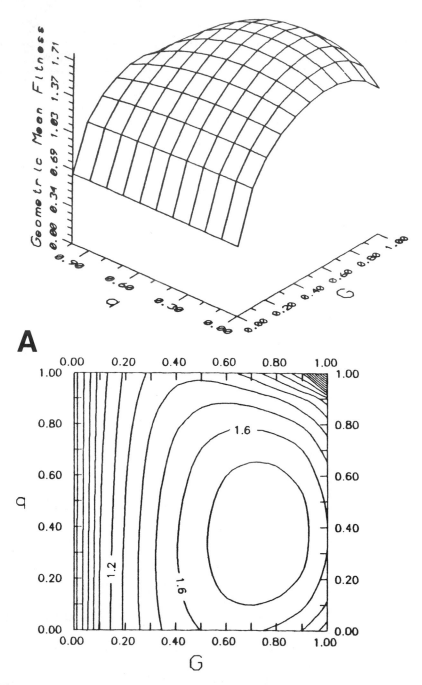

Figure 5. Fitness topographies and corresponding fitness contour maps generated from 4 yr of field data on the expected fecundity of germinating seedlings for early and late cohorts of (A) *Schismus barbatus* and (B) *Plantago patagonica* and a simple model combining within- and

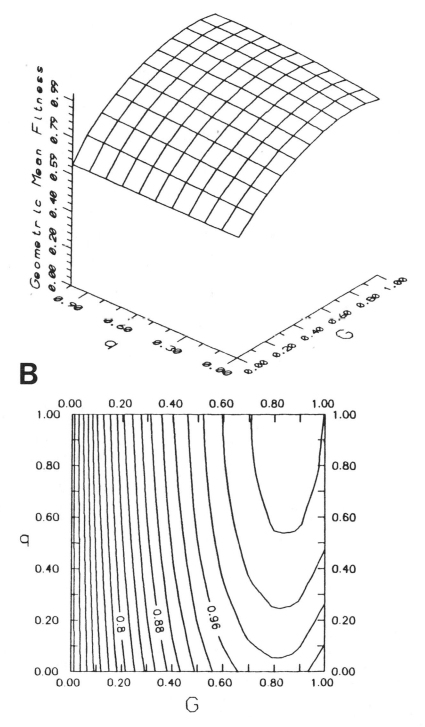

B

between-year dormancy. The survival between years of nongerminating seeds was assumed to be 65% for these calculations (see Section II,D for details).

average of 0–221 seeds per adult, while *Plantago* cohorts ranged from 4.4–65 seeds per adult. It is tempting to attribute this difference between species to the variance-reducing properties of larger seeds (Venable and Brown, 1988) (0.08 mg for *Schismus* versus 0.8 mg for *Plantago*). Another prediction generated by the positions of the fitness optima is that selection should favor a percentage of early (i.e., risky) germination greater for *Plantago* than for *Schismus*. The observed germination schedule is in general agreement with this prediction; *Plantago* had more early germination than did *Schismus* in 3 of 4 yr. The positions of the optima also suggest that *Schismus* should have greater between-year dormancy than should *Plantago*. The closest estimate of between-year dormancy available for this study is the proportion of viable seeds in the seed bank that do not germinate by the end of the germination season. The grand mean of this is 15% for *Plantago* and 37% for *Schismus*, indicating general agreement with the prediction.

These simulations searched for the best constant partition into early and late germination, yet the actual partition varies considerably between years (Table 1). Whether the observed plasticity in germination time is capable of increasing fitness can be ascertained by calculating the geometric mean population growth rate using the actual variable partition into early and late germination and comparing it to that for the best constant partition. When this is done, the mean growth rate for *Schismus* with variable germination time is 2.18 compared to 1.78 calculated for the "optimal" constant partition into early and late germination (for *Plantago* the respective growth rates are 1.12 versus 1.03). These differences suggest that the observed variation in the proportion of early germination is adaptive in that it tends to correlate germination with the best time for successful establishment and reproduction. This "predictive germination" is, however, less than perfect; if all germination is assigned to the cohort with highest expected fecundity, growth rates are higher yet (2.43 and 1.18 for *Schismus* and *Plantago*, respectively). Apparently these species can use cues that permit them to adaptively modify germination time from year to year. The information, however, is not perfect; it is not beneficial to germinate only at the predicted best time. The imperfect nature of predictive germination can be seen in the *Schismus* data, where the majority of the seeds actually germinated in the less successful cohort in 2 yr.

Thus, the simple models discussed in this chapter can aid in the interpretation of empirical data on seed behavior and generate insights into the nature of seed banks. They also suggest further data requirements. For *Schismus* and *Plantago* there is clearly a need for data on the proportion of seeds remaining dormant between years and their viabilities.

Venable (1985) and Silvertown (1988), reanalyzing published experimental and field data, have shown a reasonable fit of empirical data to model predictions (very reasonable considering the simplicity of the models and the shortcomings of the data sets for this purpose). For two seed heteromorphic annuals (*Heterotheca latifolia* and *Gymnarrhena micrantha*) Venable (1985) found that dormancy, seed bank survival, and variation in reproductive success of germinating seeds favored seed heteromorphism and the production of a between-year seed bank when fitted to a model similar to those in Eqs. (1) and (2). Silvertown (1988) analyzed two species for within-year timing of dormancy and found that *Bromus tectorum* had population dynamics at one site that should favor multiple cohorts, while *Avena sterilis* dynamics favored only fall germination. These predictions mimic actual germination pattern for these two species. He applied a model of between-year dormancy to demographic data for 10 annual species (*Androsace septentrionalis*, *Avena barbata*, *Bromus tectorum*, *Carrichtera annua*, *Emex australis*, *Erophila verna*, *Sorghum intrans*, *Avena sterilis*, *Medicago polymorpha*, and *Spergula vernalis*); the model correctly predicted the dormancy characteristics of the first 7 species. A study by Freas and Kemp (1983) demonstrated that two Chihuahuan Desert winter annuals that receive low amounts of unreliable rain (i.e., unfavorable conditions) had innate dormancy ($G < 1$), while a summer annual that receives more abundant and predictable rainfall (i.e., favorable conditions) did not ($G = 1$). All three species exhibited plasticity in germination fraction, with germination increasing with the amount of rain above a threshold of ~15 mm. The basic models presented here could be combined or made more elaborate for comparison with data sets from organisms with more complicated life histories.

III. How Seed Banks Affect the Way Selection Operates on Other Traits

The basic models presented in the previous section can be used to explore how seed banks affect and are affected by other traits. For example, we can demonstrate that natural selection in high seed-yield years may have a disproportionate impact on the direction of evolutionary change (Templeton and Levin, 1979; Brown and Venable, 1986). This occurs because the large influx of seeds in such years constitutes a population memory of selective events. Up to now the seed yield per germinating seed, S_i, has varied depending on the environment (i.e., it takes different values in different year types with different environmental conditions), but it has not been able to evolve. However, the values of this demographic parameter were important for determining how the within- and among-year timing of germination evolved.

Now let the seed yield per germinating seed be a function of some trait with heritable variation that is selected for under favorable environmental conditions but selected against under unfavorable conditions (e.g., low root/shoot ratio). For visualization, consider only two year types, favorable and unfavorable. We can plot the seed yield per germinating seed in favorable years on the x axis and the yield in unfavorable years on the y axis of a two-dimensional graph (Fig. 6). Seed yield, which varies under different conditions, would be represented as a point on this graph (for more year types it would be a point in some higher dimensional space). For illustration, let root/shoot ratio be a heritable trait that affects the seed yield per seed in the manner depicted in the curve drawn in Fig. 6. According to this curve (which we will call a "constraint set"), a higher root/shoot ratio gives a higher seed yield per germinating seed in unfavorable (e.g., dry) conditions. This would occur because greater allocation to roots makes it less likely for plants to die from desiccation. However a high root/shoot ratio would lower seed yield per germinating seed in favorable (wet) conditions because nutrients and energy that could have been creating more leaf area (resulting in more photosynthesis, growth, and ultimately seed set) were unnecessarily allocated to roots. We can now let root/shoot ratio evolve (resulting in changes in seed yield in favorable and unfavorable years) subject to the constraint set in Fig. 6.

It is possible to describe the slope of the constraint curve illustrated in Fig. 6 for the root/shoot ratio (and thus seed yields) of highest fitness

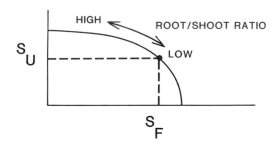

Figure 6.   The constraint set for different feasible root/shoot ratios is plotted on axes of seed yield per germinating seed in favorable (S_F) and unfavorable (S_U) conditions. A higher root/shoot ratio results in higher seed yield per seed in unfavorable (dry) years because greater allocation to roots makes it less likely for plants to die from desiccation. It results in lower seed yield per seed in favorable (wet) years because nutrients and energy that could have been allocated to photosynthetically active tissue are unnecessarily allocated below ground. A particular root/shoot ratio is represented by a point on the curve. Evolution of root/shoot ratio is represented by movement along the curve of feasible root/shoot ratios.

for a given set of parameters (germination fraction, probability of favorable and unfavorable years, etc.). Then we can determine how this slope changes with a change in germination fraction (see Brown and Venable, 1986, for the mathematical details). As dormancy is increased (more and more seeds diverted into the seed bank), the slope becomes more negative, resulting in a lower root/shoot ratio and an increase in adaptation to favorable years.

But notice that the seed yields per germinating seed in the different environmental conditions experienced in different years also determine whether selection favors between-year dormancy. Thus the production of seed banks affects selection on traits like root/shoot ratios that in turn determine how selection operates on dormancy. This coevolutionary dynamic has been explored in detail by Brown and Venable (1986).

I will use this principle of selective interactions between seed banks and other traits to explore seed size, seed banks, and dispersal as an adaptive syndrome. To consider dispersal we must add complexity to the model. Consider a species or population distributed over a number of patches that have varying environmental conditions in space and time. In any single year, plants in some patches are experiencing favorable conditions while plants in others are experiencing unfavorable conditions (e.g., due to local disturbance or local predation). Also, conditions in each patch vary over time. There could actually be many different environmental conditions but for simplicity I will discuss only "favorable" and "unfavorable" conditions that result in high versus low expected seed yield per germinating seed.

Larger seed size is favored in conditions of shading or drought, which are unfavorable for the establishment and growth of plants (e.g., Baker, 1972; Gross, 1984; Wulff, 1986b). Small seeds are superior in moist open habitats where seed size is less critical for establishment, and where more seeds per unit energy or nutrients can be produced. Thus we can draw a constraint set for the evolution of seed size similar to that previously used for root/shoot ratio (Fig. 7). The basic expression describing population growth in a patch is almost the same as was used in Section II, B to model the evolution of between-year seed bank dormancy:

$$[GS_{ij} + R(1 - G)]$$

The extra subscript on S_i indicates that seed yield per germinating seed experiencing environmental conditions i is occurring in patch j. Summing over all patches ($j = 1, \ldots, n$) we have

$$\lambda = \sum_{j=1}^{n} \{\rho_j[GS_{ij} + R(1 - G)]\} \tag{3}$$

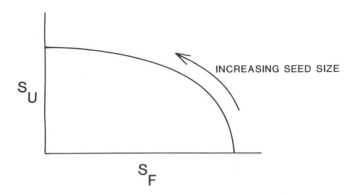

Figure 7.   The constraint set for different feasible seed sizes plotted on axes of seed yield per germinating seed in favorable (S_F) and unfavorable (S_U) conditions. Larger seeds are assumed to result in higher seed yield per seed in unfavorable conditions while small seeds are assumed to result in higher seed yield per seed in favorable (open, moist) conditions.

where λ equals the finite rate of increase during a growing season for which we have specified the environmental conditions in each patch and the proportion of the seed bank in each patch. The proportion of the seed bank in each patch is designated by ρ_j. Finally, consider that there might be some dispersal-related mortality. If a is the survival of dispersing seeds and D is the proportion of seeds dispersing out of the parental patch, then the average number of seeds produced in all patches must be discounted by $1 - D(1 - a)$:

$$\lambda = [1 - D(1 - a)] \sum_j \{\rho_j[GS_{ij} + R(1 - G)]\} \qquad (4)$$

Because this equation is the population growth rate for only 1 yr with a particular set of conditions, we must take the geometric average over all possible λ, each weighted by its probability of occurrence p_i.

Because of the difficulty of calculating geometric mean λ, a simpler procedure is to establish a number of patches in the computer and simulate population growth for any particular set of model parameters. Seed size can be varied by varying values of S_F and S_U, which are assumed to be determined by seed size, along a constraint set like that in Fig. 7. Dispersibility is determined by the proportion of seeds leaving the parental patch. Dispersing seeds could either be uniformly spread among all patches after suffering some dispersal related mortality, or they could be restricted to a few neighboring patches. To find the seed syndrome that yields the highest fitness in a particular environment, one searches for the three-dimensional adaptive peak by simultaneously varying dispersibility, germination fraction, and seed size and calculating the geometric mean fitness after 1000 yr. Then different environmental parameters (such as the probability of favorable conditions or the pattern of spatial and temporal autocorrelation) can be varied one at a time to see in what direction the optimal three-dimensional seed syndrome shifts (see Venable and Brown, 1988, for more details on the simulations).

This procedure provides several important insights as to how selection simultaneously operates on seed size, dormancy, and dispersal. First, the risk-reducing properties of dispersal, dormancy, and seed size in variable environments are partly substitutable. That is, for a given set of environmental conditions, if one of the seed traits is arbitrarily shifted from the optimum, correlated selective pressures are generated on the other two traits to compensate. For example, if seed size is arbitrarily increased, a shift toward less between-year dormancy and less dispersibility is favored. The implication of this is that the optimal germination fraction (and thus the production of seed banks) depends on the dispersibility and the size of the seeds.

While the three seed traits are partly substitutable adaptations to variable environments, they are also complimentary in that they reduce risk in slightly different ways. Different environmental changes often preferentially favor one trait over another. Selection favoring one trait may elicit a correlated response in the other two traits, in addition to any other direct selective effects on them. For example, if the number of patches is increased, there is less overall risk [variability in Eq. (4)] from which to escape. With only one patch, the environment is either favorable or unfavorable in a given year; but with many patches, it is more likely to be favorable in some places and simultaneously unfavorable in others. Less overall risk favors less dispersal, less dormancy, and smaller seed size. Yet the presence of more patches with different environmental conditions creates a further opportunity for dispersal to lower risk by evening out the distribution of seeds among the varying patches. The net effect is selection for more dispersal. Because this further reduces variability in λ and risk, there is correlated selection for even further reduction in dormancy and seed size.

We also can explore how selection impinges on the seed syndrome when the probability of favorable conditions is increased, when the distance traveled (in terms of patches) by dispersing seeds is varied, and when the spatial and temporal autocorrelations of environmental variability are varied (Table 2; see Venable and Brown, 1988, for more details).

One important implication of selective trade-offs among seed traits is that it may be necessary to study traits in an integrated fashion. For example, in surveys of seed size, seed banks, and dispersal in which

Table 2
Risk-reducing responses of seed size, dormancy, and dispersal to different environmental changes[a]

Environmental change	Selective effect on seed syndrome[b]		
Increasing the number of patches	Dispersal ↑	Dormancy ↓	Seed size ↓
Increasing the probability of favorable conditions	Dispersal*	Dormancy ↓	Seed size ↓
Decreasing the effective dispersal radius	Dispersal ↓	Dormancy ↑	Seed size ↑
Increasing positive spatial autocorrelation	Dispersal ↓	Dormancy ↑	Seed size ↑
Increasing positive temporal autocorrelation	Dispersal ↓	Dormancy ↓	Seed size ↑

[a]See Venable and Brown, 1988.

[b] ↑, Increases; ↓, decreases; *, dispersal increases with variance in environmental conditions.

attempts are made to explain patterns in terms of habitat factors or growth habits of the plants, there is a considerable amount of unexplained variance. Perhaps some of the variance can be accounted for by variation in correlated traits that share the job of risk reduction.

It is important to realize that the selective correlations discussed above are the result of fitness interactions due to the fact that they all reduce risk. Usually correlations between seed traits such as dispersal, dormancy, and seed size are discussed in terms of biophysical constraints. For example, it has been suggested that large seeds are less dispersible because of their greater weight or inertia (Salisbury, 1975; Fenner, 1985). Likewise, a correlation between seed bank production and seed size has been proposed based on the following scenario involving biophysical constraints. Small seeds are unable to emerge from great burial depth and as a result they often have evolved a light requirement for germination that leads to the production of a between-year seed bank (Thompson and Grime, 1979; Cook, 1980). The models presented here demonstrate that, regardless of any weight constraints, large seeds should be less dispersive because seed size and dispersal are complementary ways of reducing risk. Small seeds should have greater dormancy independently of any problems involved with burial. The question now concerns the importance of biophysical constraints in determining trait correlations as opposed to selection for trait correlations resulting from interacting population dynamic effects, such as traits sharing in risk reduction.

IV. Density Dependence and Kin Selection

By restricting my attention to density-independent models, I have focused on seed banks, dispersal, and seed size as avenues of adaptation to risk. These seed traits can also function as adaptations to escape the effects of local crowding or to escape the effects of competition with siblings (via kin selection). Crowding and sibling competition (which must be modeled using game theoretical approaches or explicitly genetic models because of their density/frequency-dependent nature) can also give rise to fitness interactions between seed traits. It can be shown that the density-escaping properties of seed banks, seed size, and dispersal can evolve even in the absence of uncertainty [in the sense of year-to-year variation in λ averaged over all patches; Eq. (4)] as long as there is local spatial and temporal variation in environmental conditions, and thus opportunities to escape the effects of crowding. The selective interactions are similar but not identical to the ones generated for the risk-reducing function of seed size, dispersal, and dormancy (J.S. Brown and D.L. Venable, unpublished data). These three seed traits can also evolve to escape the effects of sibling competition even in the absence of environmental variation, as long as there is some spatial structure to the environment (J.S. Brown and D.L. Venable, unpublished data). This occurs because, under competitive conditions, to germinate here and now rather than elsewhere or later is more likely to remove some fitness from a relative even if a germinating seed has the same expected fitness near the parent as elsewhere (or now, as at some future date). These matters are considered in Venable and Brown (1988). Ellner (1985a,b) analyzes the case for seed banks evolving to escape the effects of density and global uncertainty in a single patch environment. Levin *et al.* (1984) explore the interaction of dispersal and seed banks evolving to escape the effects of density in a globally risk-free environment. Hamilton and May (1977), Comins *et al.* (1980), and Schoen and Lloyd (1983) all examine how dispersal can evolve via kin selection even if the expected fecundity is the same in all patches at all times. Ellner (1987a) does the same for the evolution of a seed bank.

V. Community Seed Bank Patterns

By restricting consideration to single species models, I have ignored how seed banks may contribute to the coexistence of species and conversely how coexisting species may contribute to the evolution of seed banks. Chesson (1986) has explored how factors such as seed banks can create a "storage effect" that permits coexistence in temporally varying environments of species that would otherwise not coexist. The basic idea is that if interspecific competition mostly affects the seed yield component of the life cycle (but not the survival of dormant seeds), an annual can persist through many competitively (or abiotically) unfavorable years via a seed bank that is replenished in occasional years, when competition is escaped and when underlying abiotic variability is favorable. Thus a number of competing species can coexist if each can replenish its seed bank at some time when the other species are at low density.

Ellner (1987b), Cohen (1987), and Schmida (personal communication) have considered how species interactions affect the evolution of dormancy strategies in what may be called "the community evolutionary stable strategy." The basic idea can be encapsulated in the following scenario. If a dominant species achieves its optimal dormancy strategy in a variable environment, it will tend to survive and reproduce more in some years and less in others. If a second species is competitively inferior, otherwise favorable years will be unfavorable for it because of the high density of the dominant species. Thus the optimal dormancy strategy of the second species will be shifted to utilize the temporal variation differently and have peak years that do not coincide with those of the dominant. If two species are coevolving in this way, one may become a low-risk species with a large seed bank, producing a few seeds even in the worst years and never very many in good years, because most of the seed bank remains dormant. The other species may then become a high-risk species that is only successful in a subset of very favorable years; the high germination fraction results in boom and bust years. This divergence in the use of temporal variation creates more times when each species escapes the competitive effects of the other. Models of this nature suggest that if competition is strong enough, the seed bank biology of similar species in similar environments, which might otherwise be expected to be convergent, may diverge. These ideas could be generalized to the community evolutionary stable strategy of seed size, dispersal, and dormancy syndromes for a set of species.

VI. Summary and Conclusions

Many questions regarding the ecology and evolution of seed banks are relatively easy to model and require fairly simple mathematics and programming skills. I have focused on the questions of the timing of germination within years, of when it is adaptive to produce a between-year seed bank and what proportion of seeds should remain dormant between years, of how the timing of germination within and among years may be expected to vary with different environmental cues, of how fitness interactions are likely to provide links between seed bank ecology and other life history traits (e.g., root/shoot ratio, dispersibility, and seed size), of some of the implications of density dependence and kin selection for the evolution of seed banks, and of how interactions among species may affect community patterns of seed bank production.

Many interesting questions remain to be modeled, the most obvious being combinations of the factors discussed above. How does the presence of a between-year seed bank alter selection for within-year timing of dormancy? How do interspecific interactions favor different dormancy/dispersal strategies? How do various aspects of the perennial life cycle (e.g., vegetative reproduction) affect the evolution of dormancy/dispersal strategies?

Models can be made less heuristic and to mimic a particular system more closely. This can often be done without increasing conceptual complication, although a shift from analysis to simulation or to a bigger computer may be required. Some simple applications to particular annual plant species were analyzed and provided insights into the plants' seed bank biology and suggested new questions in need of empirical exploration.

A close interaction between theoretical and empirical studies will sharpen our insights of seed bank functions. The models presented here suggest numerous patterns and mechanisms to be tested empirically. For example, can the explicit consideration of dispersal and seed size significantly improve our ability to explain the ecological patterns of seed bank formation? Do seed bank species have traits that tend to specialize them more (compared to nonseed bank relatives) for favorable subsets of conditions? Are there density-dependent or individual–fecundity-dependent shifts in dispersal, dormancy, or seed size? To what extent can population dynamic data, fit to simple models, accurately predict actual patterns of within- and between-year timing of dormancy and predictive dormancy? Indeed, many of the interesting ecological patterns in seed bank production documented in the different chapters of this book beg for models of the possible mechanisms explaining them.

Acknowledgments

I thank J.S. Brown for his collaboration in developing many of the models presented here, and A.C. Caprio and M. Pantastico for their help in collecting and analyzing the data presented in Table 1. This work was supported by National Science Foundation Grant BRS 8516971.

PART 3

Seed Banks and Vegetation Type

CHAPTER 6

The Role of Buried Viable Seeds in Arctic and Alpine Plant Communities

James B. McGraw

Milan C. Vavrek

Department of Biology, West Virginia University,
Morgantown, West Virginia

I. Introduction

The presence of buried viable seeds in cold environments has been a topic of debate ever since Porsild *et al.*'s (1967) claim of the discovery of >10,000-yr-old viable seeds of *Lupinus arcticus* [nomenclature here and throughout follows Hultén (1968)] in the central Yukon. Despite the interest sparked by this paper, few attempts have been made in the past 20 yr to duplicate such a discovery or to study seed banks in arctic or alpine areas. In part this neglect is explained by the skepticism with which Porsild *et al.*'s paper was received (Godwin, 1968; Villiers, 1973; Bewley and Black, 1982; Priestley, 1986). Nevertheless, the phenomenon of long-term seed viability in cold environments deserves more attention than it has received. For example, improved preservation of valuable crop seeds may be possible by studying the mechanisms already possessed by plants from cold environments. Attempts to revegetate disturbed areas in fragile tundra ecosystems will benefit from knowledge of natural seed banks.

Arctic and alpine vegetation is defined here as that treeless vegetation above (in alpine areas) or north of (in the arctic) the tree line having plant species characteristic of cold climates. In this chapter, the following questions are addressed: How large and how variable are arctic and alpine seed banks? Are certain growth forms or taxa better represented than others in the dormant seed pool? What is the relationship between the buried seed community and the aboveground community? What selective forces favor delayed germination as a life history strategy? What is known about seed bank dynamics and seed longevity in tundra seed banks? Finally, what are the needs for future research on buried seeds in tundra environments?

II. Buried Seed Densities in Arctic and Alpine Soils

Johnson (1975) noted a tendency toward a poleward decrease in buried seed densities, but no quantitative community level studies of seed banks were available from arctic or alpine tundra to refute or support this generalization. Subsequently, seven studies covering 18 sites have characterized seed banks of arctic and alpine areas. All reported finding seed banks, many comparable in size to those of temperate forests. Thus, the poleward decrease in seed bank size does not extend to tundra communities. Perhaps the postulated decrease applies only to forested communties or to seed banks within certain forest community types.

Too few tundra sites have been examined to make generalizations about seed banks of specific, recognized tundra community types. However, the 18 sites examined thus far are evenly divided between hydric, mesic, and xeric sites. Data were tabulated separately for the three categories because such sites have contrasting soil properties that might influence seed bank properties. All studies employed a germination assay under greenhouse or growth chamber conditions to determine seed bank properties. As mentioned elsewhere in this volume, this method may underestimate both species richness and population sizes in the seed bank.

Table 1
Size and diversity of seed banks at xeric tundra sites

Location	Latitude (°N)	Elevation (m)	Vegetation type	Seed bank size (seeds m^{-2})	Number of species	n	Area sampled (m²)	Core depth (cm)	Reference
Alexandra Fjord, Ellesmere Island, Northwest Territories	78° 53'	<50	*Dryas* fell-field	130	5[a]	4	1.0	2	Freedman et al. (1982)
Alexandra Fjord, Ellesmere Island, Northwest Territories	78° 53'	<50	*Cassiope* heath	55	9[a]	4	1.0	2	Freedman et al. (1982)
Eagle Summit, Alaska	65° 30'	950–1100	*Dryas* and heath	85–3142[b]	8[c]	26	0.1809	5	Fox (1983)
Rankin Inlet, Northwest Territories	62° 48'	<100	Lichen heath	1769	5[a,d]	36	0.2037	10	Archibold (1984)
Churchill, Manitoba	58° 45'	<100	*Dryas* heath	0	0[a]	36	0.2037	10	Archibold (1984)
Plateau Mountain, Alberta	50° 13'	~2400	Dry alpine meadow	521	4	36	0.2037	10	Archibold (1984)

[a] Additional species were identified in the cores, but they were vegetative emergents.

[b] Range for five areas within three sites.

[c] Some graminoid and dicot emergents were not identified to species. Numbers shown are minimum estimates based only on identified species (J.F. Fox, personal communication).

[d] Nongerminating seeds of additional species were found, but are not included.

Table 2

Size and diversity of seed banks at mesic tundra sites

Location	Latitude (°N)	Elevation (m)	Vegetation type	Seed bank size (seeds m^{-2})	Number of species	n	Area sampled (m^2)	Core depth (cm)	Reference
Eagle Creek, Alaska	65° 26'	750	Cotton grass tussock tundra	3367	5	40	0.0785	40	McGraw (1980)
Kuparuk Ridge, Alaska	68° 40'	975	Cotton grass tussock tundra	1173	>5[a]	65	0.2091	25–40	Gartner et al. (1983)
Atigun River, Alaska	68° 20'	790	Carex/Salix tundra	779	13	16	0.1963	20	Roach (1983)
Eagle Summit, Alaska	65° 30'	950–1100	Snowbed vegetation	310–2717[b]	13[c]	70	1.0904	5	Fox (1983)
Rankin Inlet, Northwest Territories	62° 48'	<100	Sedge-dominated tussock tundra	78	2	36	0.2037	10	Archibold (1984)
Alexandra Fjord, Ellesmere Island, Northwest Territories	78° 53'	<50	Carex fen	1	1	4	1.00	2	Freedman et al. (1982)

[a] Two species reported in Gartner et al. (1983); three others were identified as distinct species (B.L. Gartner, personal communication).

[b] Range for 15 areas within three sites.

[c] Some graminoid and dicot emergents were not identified to species. Numbers shown are minimum estimates based only on identified species (J.F. Fox, personal communication).

Table 3
Size and diversity of seed banks at hydric tundra sites

Location	Latitude (°N)	Elevation (m)	Vegetation type	Seed bank size (seeds m^{-2})	Number of species	n	Area sampled (m^2)	Core depth (cm)	Reference
Barrow, Alaska	71° 18'	~4	Sedge meadow	205	1	36	0.2830	16	Leck (1980)
Atigun River, Alaska	68° 20'	790	Sedge meadow	533	5	16	0.1963	20	Roach (1983)
Eagle Summit, Alaska	65° 26'	950–1100	*Carex* moss	80–2802[a]	2[b]	23	0.2614	5	Fox (1983)
Rankin Inlet, Northwest Territories	62° 48'	<100	Sedge grass	10	1	36	0.2037	10	Archibold (1984)
Churchill, Manitoba	58° 45'	<100	Sedge grass	0	0	36	0.2037	10	Archibold (1984)
Plateau Mountain, Alberta	50° 13'	~2400	Grass sedge (seepage fen)	462	5	36	0.2037	10	Archibold (1984)

[a]Range for four areas within three sites.

[b]Some graminoid and dicot emergents were not identified to species. Numbers shown are minimum estimates based only on identified species (J.F. Fox, personal communication).

Tables 1–3 summarize seed bank properties of the 18 field sites studied to date. All were in North America and only two were located in areas with distinct alpine floristic affinities (Plateau Mountain, Alberta, Canada; Archibold, 1984). Xeric sites (Table 1) had thin mineral soils and were generally located on rocky, exposed ridges. Their seed banks were variable in size (0–3142 seeds m^{-2}) and often they were diverse. Mesic sites (Table 2) were often dominated by sedges, with considerable buildup of organic matter. Seed bank size was again variable (1–3367 seeds m^{-2}), but tended to be larger than those of xeric sites. Generally diversity was low except in a mesic site near the boundary between two sites where 13 species contributed to the seed bank (Roach, 1983). In hydric sites (Table 3), densities were 0–2802 seeds m^{-2} and diversity was low. None of the data shows clear trends in buried seed density or diversity with latitude or elevation. However, Fox (1983) found a correlation of total soil seed densities with productivity in arctic tundra of central Alaskan mountains.

III. Seed Bank Composition

To summarize the most common arctic and alpine seed bank taxa, the three families having the most abundant seeds in each study were tabulated and a composite list was assembled (Table 4). Six common families

Table 4
Relative abundance of tundra seed bank taxa

Relative abundance	Family	Genera
Common	Cyperaceae	*Carex, Eriophorum*
	Saxifragaceae	*Saxifraga, Chrysosplenium*
	Juncaceae	*Luzula*
	Poaceae	*Poa, Puccinellia, Trisetum, Festuca*
	Polygonaceae	*Polygonum, Oxyria*
	Rosaceae	*Dryas, Thalictrum, Potentilla*
Less frequent	Ericaceae	*Ledum, Empetrum*
	Caryophyllaceae	*Stellaria*
	Salicaceae	*Salix*
	Papaveraceae	*Papaver*
	Brassicaceae	*Cardamine, Draba*
	Liliaceae	*Tofieldia*
	Ranunculaceae	*Anemone*
	Asteraceae	*Antennaria*

were identified by this procedure. Eight other families were represented at lower abundances in the seed banks. The common families included typical tundra genera (e.g., *Eriophorum, Saxifraga, Oxyria,* and *Dryas*) as well as more cosmopolitan genera (e.g., *Carex, Poa, Polygonum,* and *Thalictrum*). Three of six common families were monocots, while only one of the other eight families was monocotyledonous. Undoubtedly, other genera and families exist in arctic and alpine seed banks, but they were not detected in the communities sampled.

Tundra seed banks contain a variety of growth forms, including woody dwarf shrubs (e.g., *Dryas octopetala, Saxifraga oppositifolia,* and *Salix* spp.), graminoids (e.g., *Carex bigelowii, Eriophorum vaginatum,* and *Poa lanata*), and herbaceous dicots (e.g., *Chrysosplenium tetrandrum, Oxyria digyna,* and *Thalictrum alpinum*). Monocot seeds were slightly less abundant than dicot seeds (40 versus 60% of all seeds found). All were seeds of perennial species, reflecting the rarity of true annuals or biennials in tundra. Annual species found in such environments should have seed banks as a buffer against the occasional (or perhaps frequent) failure of reproduction. Indeed, Reynolds (1984) did find sizeable seed banks (600–2240 seeds m^{-2}) for *Koenigia islandica, Polygonum confertiflorum,* and *Polygonum douglasii,* three alpine annuals of the Beartooth Mountains of Montana and Wyoming.

Composition of the seed bank and the vegetation could be dissimilar. Graminoids tend to be overrepresented in some sites of interior Alaska (McGraw, 1980; Fox, 1983). Fox (1983) noted a marked discordance between the aboveground community and the seed bank. In addition, the only seed bank species found by Leck (1980) in coastal wet meadow tundra was inconspicuous in the aboveground community. In contrast, Roach (1983) found only eight instances of significant differences (of 70 comparisons) between composition of the seed pool and of the aboveground vegetation. Moreover, she also found that seed bank richness (number of species) varied in concert with aboveground species richness on a transect across an ecotone between wet and mesic communities.

For three studies, it was possible to compare the presence/absence of species in the standing vegetation and in the seed bank [McGraw, 1980; Freedman *et al.,* 1982; Fox 1983 (dicots only)]. Based on this small sample, half of the total tundra species encountered were found both in the seed bank and the vegetation (mean + SE; 50.5 ± 6.3%). A large proportion (43.6 ± 4.4%) were found only in the vegetation, while a few were present only in the seed bank (Fox 1983; 5.9 ± 5.9%). The latter two categories may be overestimates because absence either in the vegetation or in the seed bank could reflect inadequate sampling, or in the case of the buried seed pool, inappropriate germination assay conditions, improper timing of sampling (e.g., if seed banks are temporarily depleted), and/or dormancy that was not broken by the assay conditions. A majority of tundra species, then, appear to have seed banks.

IV. Selection for Delayed Germination in Arctic and Alpine Environments

The timing of reproduction is a key life history trait in terms of its influence on population growth rate (Cole, 1954) and hence on fitness. Delayed germination effectively shifts the reproductive schedule of an individual so that offspring are recruited into the population later than they are produced. Individuals with delayed reproduction may therefore be at a considerable selective disadvantage. For selection to favor delayed germination, there must be an offsetting advantage. Two factors may compensate for the disadvantages of delayed germination: (1) the high probability of reemergence associated with soil disruption, and (2) the greater relative success of seedling establishment in disturbed (vis-à-vis undisturbed) environments.

Despite the popular concept that tundra plant communities are fragile (see Ives, 1970, and Billings, 1973, for discussion of this issue), natural disturbances are diverse and ubiquitous (Table 5). Disturbances associated with freeze/thaw activity disrupt the soil extensively (Billings and Mooney, 1968). Freezing and thawing of soil causes disturbances of widely variable size, from very small (e.g., needle ice) to very large (e.g., mud/iceflows). Permanently frozen soil (permafrost) is widespread in the arctic and occasional in alpine tundra. With seasonal freezing and thawing of the active layer of soil above the permafrost, local disruption is common, especially in association with patterned ground. Thermokarst erosion, caused by local melting of the permafrost, is less common, but where it occurs, long-buried organic soil is often exposed. In steep alpine areas, disturbance is frequently a result of the combined effects of freeze/thaw activity, mass wasting, and erosion.

Disturbances caused by humans are increasing as exploitation of arctic and alpine environments accelerates (Table 5). These disturbances, often locally intense, may destroy or remove the top layers of soil. Large-scale activities of man in these environments require extraordinary measures to prevent permanent damage, but when the effort is made, as with the Alyeska Pipeline project, impacts can be minimized (Alexander and Van Cleve, 1983).

In the cold soils of arctic and alpine tundra, nutrients often limit community productivity because of low rates of decomposition (Shaver and Chapin, 1980). Changes in soil structure and albedo in response to disturbance result in relatively fertile conditions in disturbed areas. In

Table 5
Some disturbances that can disrupt seed-containing soil in Arctic and Alpine environments

Natural disturbances

1. Avalanche
2. Desiccation cracking
3. Fire
4. Flooding
5. Frost creep
6. Glaciers
7. Landslides
8. Mass wasting
9. Mud flows
10. Needle ice
11. Animal activity (foraging, migration, and burrowing)

12. Patterned ground and associated disturbances (ice-wedge polygons, stone nets, frost boils, pingoes and ice mounds, and string bogs)
13. Rock falls
14. Rock glaciers
15. Sheet wash
16. Solifluction
17. Thermokarst erosion
18. Wind

Human disturbances

1. Construction sites/camps
2. Grazing by domestic livestock
3. Mineral exploration
4. Mining
5. Pipelines
6. Power lines
7. Fire

8. Recreational uses (including hiking trails, campsites, trampling, skiing, and off-road vehicles)
9. Reservoirs
10. Roads

addition, reduced competition due to destruction of the standing vegetation can increase levels of other limiting resources. As a result, disturbed arctic and alpine sites of any scale are frequently better environments for seedling establishment and growth than undisturbed sites or microsites.

The differential success of seedlings in disturbed and undisturbed environments has been documented most extensively in mesic cotton grass tussock tundra. A thorough search for seedlings showed that seedling populations existed for seven of eight species in the undisturbed community (253 ± 40 seedlings m^{-2} in the most well-studied community, ranging from 81 to 2270 seedlings m^{-2} in other undisturbed sites; McGraw and Shaver, 1982). However, seedling age structures, and branching patterns indicated that survival and growth to adult size from seed was occurring for only two of the eight species in undisturbed tundra and that establishment was rare and slow for those two (McGraw and Shaver, 1982).

Seedlings were also found on sites disturbed by fire, blading (with a caterpillar tractor), or thermokarst in tussock tundra at densities comparable to undisturbed sites (range, 67–3376 seedlings m^{-2}). In contrast, on sites that were disturbed by removal of standing vegetation, survival of seedlings was much higher and growth rates far greater than on undisturbed sites (Chapin and Chapin, 1980; McGraw and Shaver, 1982; Chester and Shaver, 1982a; Gartner et al., 1983, 1986). The sedges *Eriophorum vaginatum* and *Carex bigelowii* most successfully colonize disturbances, and the presence of a buried seed pool, more than any other factor, accounts for their initial invasion (McGraw, 1980; Chester and Shaver, 1982a; Gartner et al., 1983). Based on these findings, several studies have recommended stockpiling organic soil layers for later revegetation efforts in areas disturbed by humans (McGraw, 1980; Chapin and Chapin, 1980; Gartner et al., 1983; Shaver et al., 1983).

Rapid growth after germination on cleared plots was found to reestablish complete vegetational cover within a decade in tussock tundra (Chapin and Chapin, 1980). During 7 yr after clearing, total community productivity was four times greater than in adjacent undisturbed tundra, indicating that resource availability remains high long enough for establishment (Chapin and Chapin, 1980). The nutrient limitation of productivity appeared to be removed by the perturbation because fertilizer addition stimulated no increase in productivity in the cleared plots, but prompted a large increase in productivity in adjacent, undisturbed tundra (Chapin and Chapin, 1980; Shaver and Chapin, 1980). Successful establishment of sedge species from the seed bank has also been noted in undisturbed tussock tundra, but upon close examination, this success was localized in small natural disturbances associated with frost boil activity (Hopkins and Sigafoos, 1951; Gartner et al., 1986).

Despite the many examples of improved establishment with disturbance, many arctic and alpine species do not establish readily in disturbed areas. In tussock tundra, and probably other mesic or hydric sites, species lacking seed banks are absent or rare following disturbance. For example, grass species are rare in the emerging community following blading in tussock tundra, and they are absent in the seed bank. Although reestablishment does occur for such species (J. B. McGraw, personal observation, B. L. Gartner, personal communication), the process may take several decades or more, particularly on xeric sites. In rocky, exposed fell-field sites, a nurse plant phenomenon has been observed, where establishment appears to succeed only within the canopy of an adult plant. Griggs (1956) described how several species of fell-field plants typically become established within cushions of *Silene acaulis* in Rocky Mountain National Park. For such species, disturbances may have higher resource availability, but the risks associated with exposure to snowblast and desiccating winds outweigh considerations of resource availability (Urbanska and Schutz, 1986). Seed banks are found in exposed sites, as noted earlier, but their role is not clear. Perhaps buried seeds in such sites are sensitive to the microenvironment within the canopy so that germination still occurs only when the chance for successful establishment is high.

V. Seed Bank Dynamics and Persistence

For arctic and alpine species, there have been no complete accounts of the processes regulating gains (seed rain and immigration) and losses (germination, emigration, or mortality due to predation, decay, pathogens, or natural aging) from a seed bank. Nevertheless, these processes are of interest because they are likely to occur at different rates in arctic and alpine sites than in other environments. Total seed rain, for example, may be low in tundra communities because productivity is low (Billings and Mooney, 1968; Fox, 1983) and a relatively low proportion of the energy captured is allocated to reproduction (Chester and Shaver, 1982b). Both immigration and emigration of seeds may occur due to movement by wind or water, or in the guts or on surfaces of animals (Bonde, 1969; Ryvarden, 1971, 1975; Elven and Ryvarden, 1975; Grulke and Bliss, 1983). Virtually nothing is known about rates of predation of seeds in tundra seed banks. Natural decay or aging would be expected to be low since soil temperatures are low and decomposition rates are slow. Tundra plants have high temperature optima for germination (20–30°C; Sayers and Ward, 1966; Billings and Mooney, 1968; Bliss, 1971); as a result germination rates may be low under prevailing field conditions even in midsummer (Bell and Bliss, 1980). This temperature response may have selective value in the present environment, ensuring germination and rapid early growth under warm conditions. Alternatively, it may result from the fact that many tundra species evolved from species of warmer climates and the trait is a genetic holdover from the ancestral species.

The rates of processes controlling change in seed bank size will vary depending on the environment, species, and probably even population and genotype. Nevertheless, available information suggests that tundra seed banks are characterized by low turnover rates relative to seed banks of other plant communities. This prediction is consistent with a partial description of seed bank dynamics for *Eriophorum vaginatum* based on empirical data gleaned from three separate studies (McGraw, 1980; Chester and Shaver, 1982a; McGraw and Shaver, 1982) carried out in tussock tundra at Eagle Creek, Alaska (Fig. 1). In tundra sites with accumulating organic soils, an active and inactive seed bank can be identified based on proximity to the soil surface. Seed rain and germination interact only with the active seed bank. Predation, migration, and

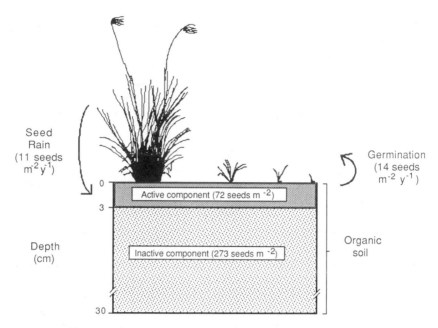

Figure 1. Conceptual model of seed bank for *Eriophorum vaginatum* at Eagle Creek, Alaska. The seed rain and germination values are annual rates. Pool sizes are shown for the active (0–3 cm) and inactive (3–30 cm) components (McGraw, 1980). Data are for 1 yr.

mortality within the seed bank were not measured, but assuming the rates of these processes are slow, the data clearly indicate a seed bank with low turnover.

In the accumulating organic soils of many tundra communities, seeds that are progressively buried by deposition of litter from above may become layered by age. Potentially such layering provides a record of past communities, populations, and genotypes at a site. Van der Valk and Davis (1979) successfully used the vertical buried seed profile to infer a cyclical pattern of wetland community change. No such attempts have been made for tundra seed banks.

A population matrix model was devised to account for depth distributions in age-layered organic soils and to determine whether population history (abundance aboveground) could be inferred from the observed depth distribution for a particular species (McGraw, 1987). It was concluded that depth distributions with peaks below the soil surface could be produced with no historical change in the seed rain; historical effects were confounded with the effects of soil compression, soil decay, and seed survival. Vertical movement in the profile, if it occurs, would further blur the historical record. Although attempts to model buried seed populations will undoubtedly bear fruit in the future, more empirical data on seed burial dynamics and age-specific survival are necessary before it will be possible to infer population history as can be done with pollen. At present, it does seems clear from depth profiles that graminoids persist longer than dicots, at least in tussock tundra (McGraw, 1980; Roach, 1983).

The question of seed persistence in tundra soils is sure to provoke debate (see Priestley, 1986, for an excellent general review), but the inevitable conclusion of the debate at present is that no one knows precisely how long seeds can survive under natural conditions. Porsild *et al.*'s (1967) claim of >10,000-yr-old seeds for *Lupinus arcticus*, while provocative and exciting, is unfortunately based on relatively weak evidence (Godwin, 1968). The seeds were extracted from collared lemming burrows that were buried in permanently frozen silt. By analogy with radiocarbon-dated burrows in central Alaska, supplemented with arguments about the biogeography of collared lemmings, Porsild *et al.* arrived at an age of >10,000 yr for the seeds. As Godwin (1968) indicated, however, the age would be unequivocal only if the seeds themselves were radiocarbon dated. Despite Porsild *et al*'s (1967) explicit challenge at the conclusion of their paper, there has been only one published attempt to discover ancient buried seeds in the last two decades. Kjoller and Ødum (1971) searched for viable seeds of three species commonly found in placer mine areas in Alaska. Most notable among these was *Descurainia sophioides*, known as Mammoth flower among the local miners because it is commonly associated with deposits containing Pleistocene woolly mammoth fossils. Kjoller and Ødum failed to find viable buried seeds, although they did find viable spores of an actinomycete. As the authors readily admitted, failure to find seeds does not rule out their existence.

Seeds of arctic and alpine plants would be expected to be longer lived than those of any other biome. As pointed out by Billings and Mooney (1968), respiratory rates of arctic plants drop quickly at low temperatures (Scholander *et al.*, 1953) so that stored reserves in seeds should be adequate to maintain embryos indefinitely in cold soils that are frozen for much of the year. In addition, a high degree of seed longevity is expected because of the low rates of decay in tundra soils that result from cold temperatures and often low pH conditions. If seeds remain viable and are buried deeply enough to enter the permafrost zone, not only do they experience cold, dry conditions ideal for preservation (see Roberts, 1960), but they cannot be eaten or leave the seed pool via emigration or germination. They are effectively trapped until the overlying soil is disrupted and the permafrost melts. The possibility of occasional long-term dormancy on the order of hundreds or thousands of years has even fueled speculation that seeds might persist through glacial periods, trapped in or under the ice (Billings and Mooney, 1968). Billings and Mooney (1968) suggested that peculiar, heretofore inexplicable, biogeographic patterns of arctic plants might be explained in this way. In fact, the presence of viable seeds in glacier snow and ice has been documented, although no ages have been reported for those seeds (Bonde, 1969; Ryvarden, 1971, 1975). In a study of a plant community entombed by glacial ice, Bergsma *et al.* (1984) recovered seed-containing fruits of several species. None of the seeds germinated after a 3-wk incubation at 20°C.

The existence of a seed bank containing long dormant individuals has important genetic, as well as ecological, consequences. The seed pool represents a genetic "memory" of the selective effects of past environments. With mathematical models, Templeton and Levin (1979) showed that traits unrelated to dormancy would be influenced by germination from a seed bank. In general, genetic change was retarded. This retardation was enhanced as the mean age of germinating seeds increased. One result could be that seed bank species track the environment less effectively. In responding to long-term directional environmental change, introduction of genotypes from the seed bank would lower mean population fitness and act as a genetic load on the population. In later models, Brown and Venable (1986) included dormancy as a variable, as well as other genetic traits. Their conclusions magnified some predictions of Templeton and Levin (1979) in some cases and tempered others, but qualitatively the results were the same. In tundra, where seeds may germinate hundreds or thousands of years after dispersal, environmental changes could be large enough that differences in fitness between old and young seeds would be expected. This idea has not been tested, but if correct, this effect of genetic load could be important. If disturbances caused by humans release old seeds from dormancy at a high rate, such seeds could have a significant genetic impact on tundra species. Thus, artificial disturbances could have important genetic as well as ecological impacts in arctic and alpine environments. The genetic impact would not be confined to the disturbance because gene flow could occur by pollen or seed dispersal.

Seed banks containing long-lived seeds may represent a transect backward in time that could be exploited by plant population biologists. Unlike pollen profiles, buried seed profiles contain a living genetic history. Such a transect in arctic and alpine environments may represent a particularly long time span, and therefore has the potential for answering questions about microevolution within populations. This could be done by direct examination of genotypes from different levels in the soil profile and in the vegetation.

Additional descriptive studies of tundra seed banks at the community level are necessary, particularly in alpine areas. At present, comparisons between arctic and alpine sites, or comparisons of tundra with other biomes, are based on too few data sets. In addition, quantitative studies of buried viable seeds found in snow and glacial ice are needed. Experimental studies of seed bank population dynamics are needed to determine rates of processes controlling seed bank size in undisturbed communities. Particularly enlightening would be studies of closely related species with contrasting seed bank properties or seed characteristics. Methods will be needed to allow marking of seeds for release and recapture in the field. Studies of seed profile age structure and seed persistence are needed, including attempts to resolve the question of maximum seed longevity with direct radiocarbon dating of viable seeds. The physiology of old seeds requires further study to determine how viability is retained and whether there are important genetic components to long-term viability. Once a buried seed profile is dated, it may then be possible to use it to answer questions about the genetic history of populations and the ecological and genetic consequences of seed banks based on both empirical and theoretical studies. It is in these latter areas that the study of arctic and alpine seed banks may make a unique contribution to biology.

VI. Summary and Conclusions

A decade ago little was known about the size, composition, or even the existence of seed banks in arctic or alpine vegetation. Based on scanty data, it had been suggested that there was a decline in seed bank size with increasing latitude and, presumably, elevation. This hypothesis has since been refuted with the publication of seven studies that have reported substantial seed banks in many arctic and alpine sites. Arctic and alpine seed banks are generally smaller than those found in wetlands and grasslands, but comparable to those of temperate forests. Species abundance in the seed bank frequently does not mirror that of the aboveground vegetation. However, the seed banks contain a wide variety of plant growth forms and taxa, often including community dominants. More than half of all tundra species have seed banks. We suggest that the prevalence of seed banks is favored by the high frequency of natural disturbance and the relative success of seedling establishment in those disturbances. In addition, the cold, often frozen, tundra soils favor long-term seed preservation. Little is known of seed bank dynamics of tundra species, but probably they have a low turnover rate, and seeds have correspondingly long persistence times. The maximum longevity of seeds is not known, but it is potentially great. Seed banks of arctic and alpine communities thus may provide ecologists with a historical transect that may provide answers to questions concerning the genetic consequences of seed banks and microevolution within populations.

CHAPTER 7

Seed Banks and Vegetation Processes in Coniferous Forests

O. W. Archibold

Department of Geography
University of Saskatchewan
Saskatoon, Saskatchewan, Canada

I. Introduction

Persistent seed banks represent a mechanism of survival for species growing in regions subject to periodic disturbance. Throughout the coniferous forests, fire is the major natural hazard, but the plants are well adapted and recovery even on the most intensely burned sites starts rapidly. Often the first plants to establish are opportunistic species which produce numerous, readily transported seeds. Many plants will develop vegetatively. Few species are dependent on seed reserves that have accumulated in the soil between disturbances, and the seed bank is characteristically small in these regions. The size of the buried seed pool reflects the type, intensity, and frequency of disturbance. The manner in which these factors influence seed production, storage, and subsequent plant establishment is discussed below.

II. The Environmental Setting

The coniferous forest formation of North America consists of two main divisions. Most extensive is the boreal forest, which stretches as an unbroken belt across the cool temperate regions of the continent. Floristically distinct are the western montane forests, which extend southward along the western cordillera from Alaska to California. Climatically the boreal forest is characterized by long, cold winters and short, warm summers, with conditions generally becoming milder and wetter toward the south and east. Permafrost is common throughout the region. Though highly variable as a result of different glacial processes, soils are typically shallow brunisols or luvisols with fibrisols often associated with the poorly drained sites. General descriptions of the vegetation of the boreal forest are given in Rowe (1972). The principal species are *Picea mariana, Picea glauca, Abies balsamea,* and *Pinus banksiana,* with *Betula papyrifera* and *Populus tremuloides* increasing in importance in the more southerly boreal mixed woods and in disturbed sites. In the northern open boreal forest regions, *Picea glauca, Picea mariana,* and *Larix laricina* are often the sole dominants.

Ground cover in the boreal forest is restricted to species that have adapted to the rigorous climate, the shade cast by the evergreen canopy, and the poor seedbed and rooting medium resulting from a thick mat of slowly decaying needles. In the northernmost districts the ground cover is dominated on the drier sites by a variety of lichens (e.g., *Cladonia mitis, Cetraria nivalis,* and *Stereocaulon paschale*), with mosses such as *Pleurozium schreberi, Hylocomium splendens,* and *Ptilium crista-castrensis* associated with damper sites or closed-canopy forest. Shrubs such as *Vac-*

cinium vitis-idaea, Vaccinium myrtilloides, and *Arctostaphylos uva-ursi* become increasingly common in the more southerly districts, with *Ledum groenlandicum* and species of *Sphagnum* and *Carex* found in the muskegs.

The climatic and physiographic setting of the western coniferous forests is quite distinct from that of the northern forests, and a wide variety of habitat conditions are encountered (Krajina, 1969). In the coastal zone the principal species in the north are *Picea sitchensis, Tsuga heterophylla,* and *Thuja plicata. Pseudotsuga menziesii* is important in southern British Columbia and Washington, with *Sequoia sempervirens* gaining importance in more southerly locations. There is an abundance of shrubs and herbaceous species in the mild, moist coastal regions (Franklin and Dyrness, 1973). In the interior montane regions the forest types are closely related to altitudinal gradients. Comparatively open shrub–herb understories occur throughout this region, but in the drier parts grass-dominated parkland habitats are found (Pfister *et al.,* 1977).

III. Seed Bank Studies

The importance of buried seed in the regeneration of coniferous stands was discussed by Hofmann as early as 1917. In many burned and logged sites in northwestern Idaho, Washington, and Oregon, the density of regrowth stands showed little relationship to the distribution of the seed trees that remained after the disturbance. Attention was turned to the duff layer on the assumption that "the duff must be the storage medium of the seed, and that seed must have been produced and stored in the forest floor before the fire and have retained its viability through the fire" (Hofmann, 1917, p. 12). Seeds of *Pseudotsuga menziesii* and *Pinus monticola* were reported at 19 seeds m^{-2} from duff samples.

Despite this early study, relatively little research has been conducted on buried seed reserves in coniferous forests, although seeding habits, seedling establishment, and the effects of fire in commercial stands have been discussed extensively by sylviculturalists. However, over the past 20 yr an increasing literature on buried seeds has developed, focusing on regional differences within the coniferous forests, the distribution of seeds within the soil profile, and the role of buried seeds in regeneration following disturbance.

A. Latitudinal and Altitudinal Responses in Seed Banks

A general decrease in buried seed reserves in higher latitudes was postulated by Johnson (1975) based on studies of soil cores from various stands of *Picea mariana, Pinus banksiana,* and *Picea glauca* in the subarctic. Although germination trials failed to produce any seedlings, some seeds of *Empetrum nigrum, Picea* spp; *Betula* spp; and *Vaccinium vitis-idaea* proved to be viable using the tetrazolium test. The paucity of viable seed is perhaps related to the harshness of the environment in these latitudes. Johnson (1975) suggested that rapid germination of seeds or vegetative reproduction proves to be more reliable in areas where plants must establish during a brief growing season. A small complement of buried seed was also reported by Whipple (1978) for mountain sites in Colorado. He found 52.8 seeds m^{-2} in an old-growth *Abies lasiocarpa* and *Picea engelmannii* stand and only 3.2 seeds m^{-2} in a drier high-elevation site dominated by *Pinus contorta.*

In a comparative study of arctic and alpine sites, Archibold (1984) reported buried seed reserves of 393 seeds m^{-2} and 707 seeds m^{-2} for two tree line forest stands in Alberta, but none for a subarctic tree line site near Churchill, Manitoba. Similarly, Elliott (1979) found an absence of buried viable seed from northern tree species in the forest–tundra ecotone. Both Elliott (1979) and Nichols (1976) suggested that seed production is restricted to periods of climatic amelioration, although Wein (1974) recorded the establishment of *Picea mariana* in a forested site near Inuvik. Payette *et al.* (1982), studying seed populations of *Larix laricina* and *Picea mariana* in tree line sites in northern Quebec, found only four viable tree seeds in the 5000 seeds extracted from soil cores. Stand population studies suggested that seedling establishment was restricted by climatic conditions. In the absence of a seed bank they concluded that the species were maintained by rapid germination during less severe climatic periods. Temperatures above 15°C are reported for germination of *Picea mariana* seeds collected from northern sites (Black and Bliss, 1980).

B. Distribution of Buried Seeds within the Soil Profile

The depth distribution of buried seeds beneath coniferous forests was first reported by Kellman (1970b). He counted emergents in soil cores collected from a 100-yr-old *Pseudotsuga menziesii* and *Tsuga heterophylla* forest site in coastal British Columbia. These yielded >1000 seeds m^{-2} representing 19 species. Most (845 seeds m^{-2}) emerged from the upper 5 cm; 171 seeds m^{-2} came from a depth of 5–10 cm. Similarly, Strickler and Edgerton (1976) noted a depth-related decrease in soil cores taken from three *Abies grandis* sites in Oregon. Seedling emergence from the litter averaged 766 seeds m^{-2} compared to 475 seeds m^{-2} from the upper 2 cm of mineral soil and 197 seeds m^{-2} for the 2- to 4-cm soil layer. Seedling emergence in conifer-dominated sites in New Brunswick

(Moore and Wein, 1977) ranged from 580 seedlings m^{-2} for a *Picea mariana* and *Pinus strobus* stand, to 180 seedlings m^{-2} for a *Larix laricina* bog. Total seedling emergence was highest in deciduous-dominated stands with as many as 3400 seedlings m^{-2} reported on a clear-cut *Betula papyrifera* and *Fagus grandifolia* site. In all sites the highest percentage of seedlings emerged from the 0- to 2-cm layer of organic soil, the layer most susceptible to destruction by fire. Recently, Kramer and Johnson (1987) reported on the distribution of seed beneath mature coniferous forest in Idaho. Buried viable seed densities averaged 1065 m^{-2}; the upper 5 cm of the soil, consisting of compacted litter, contained 67% of the seed, while the remainder were present in the 5- to 10-cm mineral layer.

C. Species Representation in the Seed Bank

Species which are excluded from the mature coniferous forest because of unfavorable growing conditions are often well represented in the buried seed pool. Conversely, seed of the dominant species is typically sparse or absent. Only 3 of the 19 species in the soil cores studied by Kellman (1970b) were growing in the vegetation. Approximately 70% of this buried seed was *Alnus rubra*, an early, shade-intolerant regrowth species. No conifer seedlings germinated in the samples. The lack of correspondence between the buried seed pool and the established cover is consistent with other studies. It is, however, evidence of the mechanism whereby many short-lived, early successional species are maintained in a region which is prone to periodic disturbance.

In a more extensive study, Kellman (1974a) monitored the production, movement, and storage of seed in a 100-yr-old mixed conifer forest and an adjacent regrowth stand that was logged and burned about 7 yr prior to the study. Annual seed input ranged from 14.5 to 46.4 seeds m^{-2} in the mature forest over a 3-yr period. In the regrowth stand there was a progressive decline in seed input from 641 to 249 seeds m^{-2}. Kellman attributed this to reduced seed production by two weedy species, *Hypochaeris radicata* and *Lactuca muralis*, possibly because of successional changes in community structure. The buried seed complement of the top 10 cm of soil totaled 206 seeds m^{-2} representing 16 species for the mature forest site. Although 6 (38%) of these species were present in the vegetation cover, only *Gaultheria shallon*, *Tolmiea menziesii*, and *Vaccinium parvifolium* were actively contributing to the seed rain. No tree species were represented. In the regrowth stand, 2612 seeds m^{-2} representing 16 species were recorded; 11 (69%) of these were present in the vegetation while 13 (81%) were collected in the seed rain.

The high seed input and storage noted within the regrowth community decline noticeably at the forest boundary (Fig. 1). Different re-

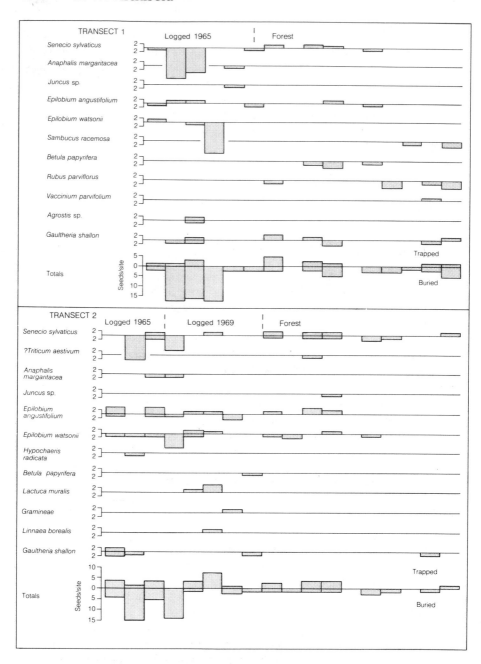

Figure 1. Seed inputs and storages for two transects across the boundary between logged areas and mature forest in coastal British Columbia (After Kellman, 1974a). For each species and transect totals, values (seeds/site) above the line represent seed rain (trapped) and those below the line line represent buried seeds.

productive mechanisms were postulated by Kellman (1974a) for the two sites. In the mature forest continuous seed production and seedling storage would be the norm here, because tree seeds quickly lose their viability. In the regrowth stand, establishment following disturbance requires rapid germination of the resident seed pool and rapid population growth as soon as conditions are favorable. The paucity of seed reserves of primary species in the secondary community could reflect low production, poor dissemination, or short viability. Consequently, their reestablishment may ultimately depend on residual plants within the region or chance seedling establishment.

The buried seed complement and subsequent seed rain in a burned mixed-wood stand in northern Saskatchewan was monitored by Archibold (1979, 1980). Of the 426 seeds m^{-2} counted in soil cores collected immediately after the fire, 87% came from the seed bank. Species of *Carex* were the most abundant, representing 32.6% of the germinants. Tree species were also well represented, with *Betula papyrifera* seedlings comprising 22.6% and *Picea glauca* comprising 14.1%. Herbs totaled 15.6% of the germinants and included *Epilobium angustifolium*, *Cornus canadensis*, *Mentha arvensis*, and *Viola palustris*. Of the eight shrub species, only *Prunus pensylvanica*, *Rubus strigosus*, and *Vaccinium vitis-idaea* appeared to have originated from buried seed and accounted for 4% of the seedlings. A total of 1698 seeds (898 seeds m^{-2}) were collected in seed traps set out in the first year following the fire (Archibold, 1980). Herbaceous species accounted for 63% of the seed rain, tree species 34%, graminoids 2%, and shrubs 1%. In the second year the seed rain dropped to 651 seeds (344 seeds m^{-2}), of which graminoids comprised 52%, tree species 32%, herbs 13%, and shrubs 3%. The number of seeds entering the site in the first year exceeded the residual buried seed complement by a factor of 5; in the second season seed input was about double the seed reserves in the soil. The potential therefore exists for the rapid restoration of buried seed reserves, but only six of the species found in the seed traps were present as buried seed. Because of predation or loss of viability, several years will elapse before the preburn buried seed complement is restored in the area.

IV. Seed Bank Dynamics

Harper (1977) provided a general model of seed bank dynamics in which the size of the seed bank represented a balance between the seed rain and losses through germination, predation, and death. The role that each of these components plays in the seed bank of coniferous forests is outlined below.

A. Seed Production by Conifers

Seed production characteristics for the common coniferous forest species are presented in Table 1. In the boreal forest, *Abies balsamea* and *Picea mariana* stands produce comparatively light seed crops, whereas stands of *Larix laricina*, *Picea glauca*, and *Betula papyrifera* can have much heavier seedfall. However, annual variation in seed crops can be considerable. Frank and Safford (1970) for example, found that seedfall over 6 yr in Maine ranged from 5200 to 450,000 seeds ha^{-1} for *Picea glauca* and 400 to 550,000 seeds ha^{-1} for *Abies balsamea*. Likewise, Reukema (1982) reported seedfalls ranging from 0 to 3 million seeds ha^{-1} over 29 yr for a young (39 yr at start of experiment) stand of *Pseudotsuga menziesii*. Viability of the seed is also variable and is generally low for the boreal species (Table 1). Average seed viability is 25% for *Picea glauca* and about 40% for *Abies balsamea*. In extreme sites, viability of *Larix laricina* seed has been reported as low as 3% (Payette *et al.*, 1982). Seed viability in the western and montane species is normally between 50 and 70%. However, seed of *Pinus monticola* estimated at between 8000 and 32,000 seeds ha^{-1} in northern Idaho showed between 2 and 20% germination, giving between 160 and 6400 viable seeds ha^{-1} (Hofmann, 1917).

B. Seed Losses in Coniferous Forests

Seed losses can occur in many conifers as a result of insect attack. Detailed descriptions of various insect feeding groups are provided by Keen (1958), Hedlin (1974), Furniss and Carolin (1977), and Martineau (1984). Extensive seed loss in species of *Picea* may result from infestations of cone insects such as *Laspeyresia youngana* and *Hylemya anthracina*, and 100% seed loss has been reported in *Pinus monticola* caused by *Conophthorus monticolae*. In *Pinus ponderosa*, seed losses of 47% have been caused by *Laspeyresia piperana*; even greater losses have been reported from *Dioryctria auranticella*. *Pseudotsuga menziesii* suffers seed loss from numerous insect species; the larvae of *Barbara colfaxiana*, *Contarinia oregonensis*, and *Dioryctria abietivorella* are particularly damaging. Indirect seed losses can also occur following outbreaks of defoliating insects. Notorious among these are *Gilpinia hercyniae*, *Lambdina fiscellaria*, and *Choristoneura fumiferana*. Further seed losses may result from predation by birds and rodents (Martell, 1979). Approximately 95% of *Abies lasciocarpa* seed placed in experimental plots was consumed by rodents in 2 wk (Fowells, 1965). High losses are characteristic of *Pinus ponderosa*, *Pseudotsuga menziesii*, and *Larix laricina*. *Thuja plicata* and *Picea englemannii* are less prone to predation. While cone serotiny in *Pinus contorta*, *Pinus banksiana*, and *Picea mariana* may afford some protection, squirrels commonly remove cone bearing branches.

Table 1
Sylvicultural data for selected coniferous tree species[a]

Species	Frequency of good seed crops (yr)	Dispersal distance (m)	Viability (%)	Average seeding rates (ha⁻¹)	Seed longevity (yr)
Northern boreal species					
Abies balsamea	2–4	30–45	24–72	2400 (increases to 800,000 after fire)	5
Betula papyrifera	Most years	Close to parent tree	Rapidly lose viability	14,400,000	1.5
Pinus banksiana	3–4	60–90	Varies with age	500,000–1,700,000	5
Picea mariana	1	25–50	Good	80,000	5
Picea glauca	2–6	600	25	5,000,000	10
Larix laricina	5–14	120	27	2,000,000	3
Populus tremuloides	4–5	several km	Rapidly lose viability	—	0.5
Western and montane species					
Abies lasiocarpa	3	—	38	—	1
Thuja plicata	2–3	120	73	600,000	2
Pseudotsuga menziesii	3–7	100–200	Good	3500	4
Picea englemannii	2–6	200	69	40,000	3
Tsuga heterophylla	3–4	600	56	3,200,000	2
Pinus contorta	1–3	60	64	40,000	5
Pinus ponderosa	2	50	Good	100,000	3
Sequoia sempervirens	1	60–90	Varies with age	1,200,000	1
Picea sitchensis	3–4	400	60	Low	—
Pinus monticola	3–4	120	40	16,000	2

[a]After Fowells (1965) and U.S. Department of Agriculture (1948).

C. Viability of Conifer Seed

The size of the forest seed bank is also affected by gradual loss of viability. Except for *Picea glauca*, viability is typically lost within 5 yr (Table 1) for cleaned dried seed held under optimal conditions—normally low temperature in sealed containers. However, under natural conditions this period could be much shorter. Estimates of the duration of viability were made by Hofmann (1917) based on the establishment of seedlings several years following fire events. Germination could be delayed up to 10 yr for *Taxus brevifolia*; 6 yr for *Pseudotsuga menziesii*, *Tsuga heterophylla*, and *Pinus monticola*; 5 yr for *Abies amabalis*; and 3 yr for *Abies procera*. The viability of *Pinus monticola* seed stored in the duff was reported to be 40% after 1 yr, 25% after 2 yr, and below 1% after 3 yr (Fowells, 1965). Seeds of *Picea mariana* may retain their viability for only 10–16 months (Fraser, 1976). Isaac (1935) reported viabilities ranging from 4 to 74% for *Pseudotsuga menziesii* seed after the first year, but after the second year viability averaged only 0.7%. Germination might be delayed until the second postfire season in *Pinus banksiana*, *Picea mariana*, and *Pinus strobus*, but no delayed emergence was noted for *Abies balsamea* (Thomas and Wein, 1985b). For *Betula pubescens* and *Betula verrucosa* seeds in Swedish forest soils, the viable seed complement declined to 6% of the original population after the first year, and to 3% during the second year (Granström and Fries, 1985).

D. Seedbed Requirements for Seedling Establishment

Successful establishment of forest plants is closely tied to the suitability of the seedbed. Daniel and Schmidt (1972) investigated the effect of litter type on subsequent germination of associated species. They concluded that litter from *Picea engelmannii* provided an unsuitable seedbed for its own seed (3.5% germination), and also for seed of *Abies lasiocarpa* (1.3%), *Pseudotsuga menziesii* (9.5%), and *Pinus contorta* (11.9%). In contrast, *Abies lasiocarpa* and *Pseudotsuga menziesii* litter did not seriously impair germination of other species, but was autoinhibitory. *Pinus contorta* litter was neutral to all species. Others similarly report the need for seedbed preparation to promote good seedling establishment (e.g., Griffith, 1931; Smith and Clark, 1960). In the coniferous forests such preparation invariably means a reduction in thickness of the surface organic layers. This occurs naturally with fire.

Van Wagner (1983) concluded that most of the common boreal forest species require burned or bared surfaces for good germination. Carleton (1982) identified this as the key factor for successful establishment of *Picea mariana* seedlings in northern Ontario, where, in the absence of fire, the rapid growth of *Pleurozium schreberi* can overtop the slow-growing tree seedlings. Thomas and Wein (1985a), using artificially shaded plots, demonstrated a differential response in seedling establishment in four coniferous species. *Pinus banksiana* was little affected by shade conditions, but *Pinus strobus*, *Abies balsamea*, and *Picea mariana* were more successful under increased shade. Delayed emergence over two seasons has been noted for buried seed of *Pinus banksiana*, *Pinus strobus*, and *Picea mariana* (Thomas and Wein, 1985b). This delay could increase seedling survival in severely burned sites once conditions have been ameliorated by a rapidly expanding herbaceous cover.

V. Fire as an Agent of Disturbance

Plant response to fire is dependent on fire frequency, fire intensity, and depth of burn (Van Wagner, 1983). Fire frequency may be <10 yr in stands of *Pinus ponderosa* growing in the western United States, increasing to >200 yr in *Picea glauca* forests growing in floodplain sites in Alaska (Chandler *et al.*, 1983). For the boreal forests of Canada, it is estimated at between 50 and 100 yr (Heinselman, 1973; Van Wagner, 1978; Zackrisson, 1977). If adequate material accumulates, high-intensity crown fires result, readily killing most northern thin-barked species (Lutz, 1956). The depth of burn is largely controlled by the moisture content of the organic surface layer. This both modifies the seedbed and determines the fate of the buried propagules.

A. Plant Strategies in Fire-Prone Environments

The comprehensive model of vegetation regrowth following fire proposed by Noble and Slatyer (1977) has been adapted to northern coniferous forests by Rowe (1983). Five survival strategies are recognized (Fig. 2). The shade-intolerant invaders produce large numbers of wind dispersed propagules which quickly establish on burned sites. The evaders typically store seed in the canopy, duff, or mineral soil. Shade-intolerant evaders are characterized by rapid germination, while seeds of late successional perennial evaders steadily accumulate in the soil. The evaders are best represented in the seed bank, particularly in forests subject to short or intermediate fire cycles. The avoiders tend to be late successional species that establish from dispersed seed only if suitable environmental conditions have developed at a site. Shade-intolerant species whose adult stage can survive low-severity fires are termed resisters. Species that resprout following fire are classed as endurers. The endurers may represent 75% of the flora in northern forests having

Short fire cycle	Intermediate fire cycle	Short, intermediate, or long fire cycle	Very long fire cycle
Endurers VI* species	Resisters WI species	Invaders DI species	Avoiders DT species
Populus tremuloides	Pinus banksiana	Ceratodon purpureus	Abies balsamea
Apocynum androsaemifolium	Eriophorum vaginatum	Polytrichum piliferum	Picea glauca
Arctostaphylos uva-ursi	Evaders CI species	Epilobium angustifolium	Linnaea borealis
Maianthemum canadense	Pinus banksiana	Calamagrostis canadensis	Mitella nuda
	Picea mariana		
Evaders SI species	Evaders ST and SI species	Endurers VT species	Avoiders DR species
Corydalis sempervirens	Rubus strigosus	Alnus crispa	Goodyera repens
		Aralia nudicaulis	Circaea alpina
Geranium bicknellii	Comandra livida	Cornus canadensis	Hylocomium splendens
Aralia hispida	Prunus pensylvanica	Equisetum sylvaticum	Peltigera aphthosa
		Pteridium aquilinum	

Dry   ⟶   Moisture gradient   ⟶   Moist

* *Mode of regeneration and reproduction*

Vegetative-based:
V species - able to resprout if burned in the juvenile stage
W species - able to resist fire in the adult stage and to continue extension growth after it (though fire kills juveniles)

Disseminule-based:
D species - with highly dispersed propagules
S species - storing long-lived propagules in the soil
C species - storing propagules in the canopy

Communal relationships

T species - tolerants that can establish immediately after a fire and can persist indefinitely thereafter without further perturbations
R species - tolerants that cannot establish immediately after fire but must wait until some requirement has been met (e.g., for shade)
I species - intolerants that can only establish immediately after a fire. Rapid-growth pioneers, they tend to die out without recurrent disturbances

Figure 2. Survival strategies of selected coniferous forest species in environments with varying fire cycles. The two evader groups with SI and ST species have strategies based on buried seed banks (after Rowe, 1983).

longer fire cycles, although success is largely determined by fire intensity. With very long fire cycles the reserve of buried viable seeds will
probably decline, and regeneration from seed may be restricted to species with highly dispersive propagules or from the seedlings, as suggested by Kellman (1974a).

The underground organs of boreal species occur at variable depths
(Flinn and Wein, 1977). McLean (1969), working in *Pseudotsuga menziesii*
stands in the interior of British Columbia, demonstrated a significant
increase in fire resistance in species rooted at least 5 cm below the
surface of the mineral soil. The deeper rooting species are rarely destroyed by fire. These are the resisters (Rowe, 1983) and they are well
represented in the regrowth cover. Buried seeds, however, are typically
found in the shallow surface layers of the soil and duff. For example,
84% of the viable seed reported by Kellman (1970b) was found in
the upper 5 cm of the soil, where many seeds will be destroyed during a
fire. Successful establishment of those that survive or are disseminated
into the area may be further restricted by the availability of favorable
microsites.

B. The Role of Buried Seeds in Postfire Recovery

The mechanism of plant establishment following fire in a mixed-wood
site in Nova Scotia was investigated by Martin (1955) using open and
covered plots. He concluded that all of the herbaceous and shrub species
that developed during the first 2 yr survived the burn and generally
reestablished through vegetative processess. Variation in recovery patterns reflected the intensity of the burn; any seedlings present tended to
be associated with patches of bare mineral soil where all humus was
destroyed. Hofmann (1917) noted that grasses, herbaceous plants, and
shrubs became established in areas where the duff was burned, while
tree seedlings were restricted to areas not reached by ground fire.

Ahlgren (1960) identified three plant associations at fire sites in Minnesota (Table 2). Some species, which tended to occur infrequently in
the region, were found only on unburned shady sites. These were characteristically perennials that reproduced vegetatively. Other species
were restricted to burned sites. These were annuals that typically reproduced by seed. The third group consisted of fire-tolerant species that
germinated or arose vegetatively in burned sites and became important
components of the forest understory. Only 20% of the species associated
with burned sites established solely from seed; seeding might supplement vegetative reproduction in a further 36% of the species. Although
Ahlgren (1960) recognized that some seedlings could have arisen from
buried seed reserves, no specific data were given.

Table 2
Species distribution and method of reproduction in fire sites in Minnesota[a,b]

Class A—Species found only on unburned land		Class C—Species found on burned and unburned sites	
Cypripedium acaule		Cornus canadensis	V (S)
Cinnia latifolia		Maianthemum canadense	V (S)
Goodyera repens		Vaccinium angustifolium	V (S)
Polypodium virginiatum		Clintonia borealis	V (S)
		Lycopodium obscurum	V
		Aster macrophyllus	V (S)
		Rubus idaeus	V–S
Class B—Species found only on		Rubus pubescens	V–S
burned land		Viola spp.	S (V)
Geranium bicknellii	S	Alnus crispa	V
Epilobium angustifolium	S	Ribes glandulosum	S
Epilobium glandulosum	Spores	Pinus banksiana	S
Marchantia polymorpha	S	Betula papyrifera	V–S
Corydalis sempervirens	V	Fragaria vesca	V–S
Dryopteris spinulosa	V (S)	Pyrus americana	V–S
Carex disperma	S (V)	Aralia nudicaulis	V (S)
Viola incognita	V	Gaultheria hispidula	V–S
Lonicera canadensis	S (V)	Populus tremuloides	V–S
Polygonum cilinode	S	Dryopteris thelypteris	V
Anaphalis margaritacea	V–S	Picea mariana	S
Aralia hispida	V–S	Apocynum androsaemifolium	V–S
Aster ciliolatus	V–S	Linnea borealis	V
Anemone quinquefolia	S	Oryzopsis asperifolia	V–S
Comptonia peregrina	V (S)	Amelanchier sp.	V
Carex sp.		Corylus cornuta	V (S)
		Diervilla lonicera	V (S)
		Trientalis borealis	V–S
		Rosa acicularis	V (S)
		Calamagrostis canadensis	V (S)

[a]From Ahlgren, C.E. (1960), *Ecology* **41**:431–445. Copyright © 1960 by the Ecological Society of America. Reprinted by permission.

[b]Methods of reproduction: S, seed origin; V, vegetation origin; less frequent method shown in parentheses.

Hall (1955) reported on vegetation changes in a mixed-wood stand dominated by *Abies balsamea* and *Picea rubens* in New Brunswick over a 4-yr period following clearing and annual burning for blueberry production. Common boreal species in the ground cover (e.g., *Maianthemum canadense, Cornus canadensis, Lycopodium obscurum, Lycopodium annotinum, Trientalis borealis, Vaccinium myrtilloides,* and *Aralia nudicaulis*) propagated from underground organs. Evidence of seedling establishment was noted for *Aster lateriflorus, Lactuca biennis, Hieracium aurantiacum, Rubus strigosus, Prunus pensylvanica, Aralia hispida*, and *Sambucus pubens*. However, it was not clear if these arose from previously buried seed or from newly transported material. Seeds of *Vaccinium angustifolium* and *Betula populifolia* were extracted from soil samples by coarse sieving.

Lutz (1956) provided a review of plant response to fire in Alaska. *Picea glauca* regeneration occurred primarily through seed input from adjacent unburned areas; regeneration from buried seed reserves was considered unimportant. For *Picea mariana*, restocking of burned sites was largely from seeds present in unopened cones at the time of the fire. Although sprouting from the root collar was common in young *Betula papyrifera* stands, initial establishment of pioneer stands and restocking beneath older stands was normally through seedling reproduction. However, most of the seed was stored in the canopy rather than buried in the soil. Fire-damaged stands of *Populus tremuloides* and *Populus balsamifera* typically reproduced from root suckers, but Lutz (1956) noted that seedling establishment from wind-transported seed could be extensive on exposed mineral soil. Understory species, including *Arctostaphylos uva-ursi* and *Empetrum nigrum*, originated from buried seed, but most shrubs arose vegetatively. Seeds of herbaceous plants typically invaded burned sites by wind dispersal. Short-term seed storage was noted in Alaska for *Alnus crispa, Salix* spp., and *Vaccinium* spp. (Viereck, 1973), but the thin, soft seed coats of these species appeared to afford little protection to fire injury. Plants with thick, hard seed coats, including *Viburnum* spp., *Rosa* spp., *Cornus* spp., *Geocaulon* spp., *Corydalis* spp., and *Shepherdia* spp., were considered persistent, fire-dependent residents of the seed pool.

Lyon and Stickney (1976), described three wildfire sites in the Rocky Mountains. About 80% of the postfire successional vegetation cover was derived from materials which survived the burn. Survival through seeds and fruits accounted for 11–31% of community composition, but some of this seed may have been stored in the canopy. Species in the persistent seed bank, known to germinate following fire, include *Ceanothus velutinus, Geranium bicknellii, Dracocephalum parviflorum*, and *Iliamna rivularis*.

Although many coniferous forest species reproduce vegetatively, seed production is essential for others. Postfire regeneration is closely related to the nature of the burn. Successful seedling establishment is linked to disturbance regimes that provide suitable seedbed and growing conditions. Fire is required by many conifers to reduce litter, by herbaceous annuals to remove competitors, and by shrubs to break seed dormancy. Some species depend on serotiny or sprouting for survival.

Others must survive the burn. For these, regeneration requires continual release of seed by mature individuals that survive in areas of patchy or low-intensity fires.

The subordinate species are best represented in the buried seed reserves, although they are also dependent on variable fire regimes. Rowe (1983) suggests that frequent fires favor short-lived herbs that reproduce quickly. Seeds incorporated into the mineral soil may be afforded protection from fire by accumulating layers of litter. These species will succeed if the litter layer is frequently reduced by low-severity fires. The longer lived shrubs may contribute to the seed bank for many seasons, distributing their seed throughout the litter layer. Short or intermediate fire cycles of low severity are most favorable for these species. As the period between fires lengthens, the differential viability of the buried seed would increasingly determine the nature of the postfire community. Where the dominant species are absent from the seed bank of a climax conifer stand, the likelihood increases that they will be replaced by hardwood species capable of vegetative regeneration following an intense burn.

VI. Summary and Conclusions

Buried seed reserves are not extensive in coniferous forests. Often seeds are not viable or quickly lose viability. Many are lost to predators. The dominant species are rarely characterized by abundant buried seed reserves, but the early successional species are well represented. Severity of the environment and disturbance events determine plant regeneration strategies in the coniferous forests. As climatic conditions deteriorate with increasing latitude, plant survival is more likely assured through vegetative reproduction, with a corresponding general decline in the buried seed population. Historically, fire has been the principal agent of disturbance in coniferous forests. Regeneration is inexorably linked to fire disturbance. Several adaptive strategies, including reliance on seed banks, serotiny, sprouting, and dispersal, interact with climatic severity and fire intensity to dictate the vegetation patterns observed in coniferous forests.

CHAPTER 8

Seed Bank Dynamics in Temperate Deciduous Forest

S. T. A. Pickett and
M. J. McDonnell

Institute of Ecosystem Studies
The New York Botanical Garden
Mary Flagler Cary Arboretum
Millbrook, New York

I. Introduction

The notion that seeds buried in the soil can be important in the dynamics and structure of plant communities apparently was applied first to meadows and arable lands (Duvel, 1902; Brenchley, 1918; Brenchley and Warington, 1930). The notion was soon applied to temperate deciduous forests and successions (Oosting and Humphreys, 1940). Starting with this early work, the guiding question has been and still is, to what extent does the presence and composition of seed banks affect the structure of deciduous forest communities? Because seed banks are composed of species having different ecological strategies, the answer requires that establishment from the seed bank be determined under a wide variety of ecological conditions. In addition, the opportunities and constraints for success in deciduous forest seed banks must be put in an adaptive perspective.

We, therefore, examine the generalizations that have emerged so far from studies of persistent seed banks in deciduous forests. The focus will be on community-level questions, rather than on the seed bank dynamics of individual species. The existence of syndromes of species behaviors relative to seed banks will be considered, and differences in seed bank dynamics resulting from community or environmental differences will be examined. Several specific questions will be addressed throughout this chapter: (1) What are the dynamics of seed banks in deciduous forests? (2) What physical conditions govern the behavior of the seed bank? (3) Do seed bank dynamics diverge within the biome? (4) Are there patterns in the seed bank within a community? (5) Are species strategies apparent in the seed bank? (6) Do species of particular successional positions have similar seed bank dynamics? (7) Are the patterns explainable in terms of natural selection? (8) How do physiological aspects of seed bank dynamics affect dynamics at the community level? We will be able provisionally to answer several of these questions, but others remain unanswered, leading us to conclude the chapter with suggestions of critical avenues for research on seed banks in deciduous forests.

II. Tentative Generalizations about Seed Banks

We begin by reviewing the generalizations that are currently supported in the literature on seed banks in deciduous forests. We label such generalizations "tentative" because any empirically (inductively) derived generalization can be considered a hypothesis to be tested in additional systems or scrutinized with improved methods. Most of the patterns and processes considered below are far from definitive.

A. Differences between Standing Vegetation and Seed Banks

Perhaps the soundest generalization is that seed banks diverge from the standing vegetation at the site. For example, Olmstead and Curtis (1947) found that 15 species in the seed banks of deciduous and coniferous stands in Maine lacked reproductive adults in those same stands. Seed banks of successional communities in Chiba, Japan contained species of earlier and later "stages," but did not contain all species of the current community (Numata *et al.*, 1964). Such differences are represented by a mean similarity of 45% (Sorensen's index) across a sere in southwestern Japan (Nakagoshi, 1985a). The conclusion that standing vegetation and seed banks differ markedly is echoed throughout the literature on deciduous forest seed banks (Oosting and Humphreys, 1940; Livingston and Allessio, 1968; Brown and Oosterhuis, 1981; Hill and Stevens, 1981; Fenner, 1985; Nakagoshi, 1985a). Only in repeatedly disturbed arable fields does seed bank composition sometimes coincide with that of standing vegetation (Jensen, 1969; R. G. Wilson *et al.*, 1985).

B. Relation to Succession and Gap Dynamics

The concern with seed banks during succession is motivated by the important role of buried seed in vegetation dynamics. The significance of the seed bank for the establishment of early successional communities has been experimentally demonstrated by Marks and Mohler (1985). Sterilization of the soil to kill seeds reduced the contribution of colonizing annuals to the postagricultural succession and speeded the dominance of herbaceous perennials.

The richness and density of seed banks often decrease through old field succession. On the North Carolina Piedmont, seed banks have the highest density (13,181 seeds m^{-2}) in first-year old fields, with an irregular decline to 1180 seeds m^{-2} in old-growth oak–hickory (Oosting and Humphreys, 1940), although species richness shows slight increases in the fifth year and in the old growth. In early old-field succession in Massachusetts, total species richness of the seed bank fluctuates, but the number of species first appearing in the seed bank of a field decreases with field age (Livingston and Allessio, 1968). In southwest Ohio, seed banks decreased from a high of 43 species in second-year fields to a low of 14 species in old-growth beech–maple forest (Roberts *et al.*, 1984; G. Snyder and J. Vankat, unpublished data). The succession at Chiba, Ja-

Table 1
Species richness and seed density[a,b]

Age (yr)		Richness (0.4 m^{-2})		Density (m^{-2})	
1979	1983	1979	1983	1979	1983
0	4	26	30	628	1078
5	9	30	29	1613	1435
75	79	38	27	3180	3098
~200	—	33	28	850	923

[a]Species were taken from the seed banks of four communities representing a sere on Mt. Futatani, Kobe, Japan. Each stand was sampled in December 1979, and again in December 1983.

[b]Data compiled from Nakagoshi (1981, 1985b).

pan, shows consistent decline in both species richness and total seed density (Numata *et al.*, 1964). In the Hiba Mountains of Japan, species richness is lowest in herbaceous and shrub-dominated old fields, greatest in the oak-dominated forest, and lower again in the dense beech forest (Nakagoshi, 1984a). Seed densities were higher in the herbaceous and shrub communities (Nakagoshi, 1984a). Seed banks from a different chronosequence in southwest Japan, which ran from successional *Mallotus–Aralia* to old-growth *Castanopsis cuspidata*, had less variation in species richness, but great variability in total seed density (Table 1; Nakagoshi, 1981).

Nakagoshi's (1985b) study of the *Mallotus–Aralia* to *Castanopsis* chronosequence represents the only direct long-term study of seed bank dynamics in deciduous forest of which we are aware. The seed richness and density in the first-year field increased over the 4-yr period of the study, the fifth-year field remained roughly constant over that same period, and the seventy-fifth-year stand declined in seed bank richness over 4 yr (Table 1). The early increase confirms patterns inferred from other chronosequences discussed above. The decline in richness between 75 and 79 yr may correspond to the decline inferred in other studies at roughly the same age (e.g., Livingston and Allessio, 1968; Marquis, 1975). The large change in the old-growth stand may represent cycling through time in fruit production or predation prior to entering the seed bank, but no sound interpretation is possible.

The use of chronosequences to infer the temporal pattern of seed banks has also been applied to post-clear-cutting successions in forest. Species richness was independent of age in the sequence of 14 northern hardwoods stands, ranging between 1 and >100 yr old, in and near the

Hubbard Brook Experimental Forest, New Hampshire. Although the number of species was not statistically related to age of stand, the density of common species (*Sambucus, Rubus,* and *Prunus pensylvanica*) in the seed bank fluctuated, peaking in 15- to 35-yr-old stands (Bicknell, 1979). In secondary cherry–maple stands of the Allegheny Plateau of Pennsylvania, five stands, ranging in age from 35 to >100 yr since cutting, were sampled to determine seed banks of the tree species. Densities of tree seeds were high through 70 yr, but declined in the oldest stand. The density of seeds of individual species did not change consistently through the sequence, except for the decline in density of *Prunus pensylvanica* in stands older than 53 yr. Declines in density and richness have been found in intermediate-aged pine plantations in the Harvard Forest, Massachusetts (Livingston and Allessio, 1968), and in *Picea sitchensis* plantations older than about 30 yr in Britain (Hill and Stevens, 1981).

The increased opening of gaps in the canopy as forests age is one of the most conspicuous features of forest succession (Watt, 1947; Raup, 1957; Bormann and Likens, 1979; Oliver, 1981), and studies often find increased richness of seed banks in old-growth stands compared to late secondary forest. For example, Bicknell (1979) found *Sambucus* and *Rubus* in seed traps within an intact 60-yr forest near a clear-cut. Input of new seeds in gaps has been hypothesized to increase in the oldest stands (Oosting and Humphreys, 1940; Marquis, 1975; Mladenoff, 1985). Increased removal rates of fleshy fruits appear in gaps compared to old-growth mesic forest in Illinois, suggesting that birds would also bring seeds into gaps preferentially (Thompson and Willson, 1978; Moore and Willson, 1982). Bird-disseminated shrub and herb species are important in the seed bank of a large old-growth maple–hemlock–birch forest in Michigan, especially in gaps (Mladenoff, 1985).

The available data indicate that the dissimilarity of seed banks and the standing vegetation clearly changes during the course of succession. Quantification of the differences is rare, however, and many papers invite the reader to inspect tables listing species in the seed bank and standing vegetation. The publication of such raw data is commendable because relationships, patterns, and tests that the authors of the studies may not have envisioned can be extracted later, but statistical comparison and summarization are too infrequent in the seed bank literature on deciduous forests. Nakagoshi's (1984a, 1985a) quantitative comparison of seed bank and standing vegetation stands out, although some caution is called for because the sequence he presents includes two plantations and communities encompassing some site variation. He found that coefficients of similarity were most variable in the youngest fields, and averaged around 40%. Similarity peaked at slightly over 50%

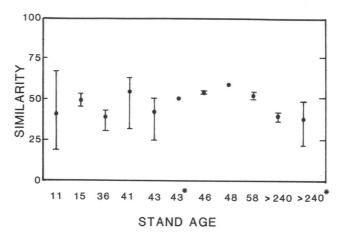

STAND AGE

Figure 1. Similarity between the standing vegetation and composition of the seed banks across a chronosequence in the Hiba Mountains, Japan. Mean and range of values are shown. The communities and their ages are as follows: 11 yr, *Rubus–Aralia*; 15 yr, *Corylus–Sasa*; 36 yr, *Weigela–Shortia* with *Gaultheria*; 41 yr, *Weigela–Shortia* with *Quercus*; 43 yr, *Chamaecyperus–Lindera* (plantation) on mesic slope; 43* yr, *Chamaecyparis–Lindera* (plantation) on moist flats; 46 yr, *Quercus–Castanea* with *Ilex*, lower elevation; 46 yr, *Quercus–Castanea* with *Oplismencus*, higher elevation; 58 yr, *Quercus–Fagus*; >240 yr, *Fagus–Lindera*, typical site; >240* yr, *Fagus–Lindera* in hollows. Redrawn from Nakagoshi (1984a).

in the 41-yr-old shrub community and in the 48-yr-old *Quercus–Castanopsis* community. Lowest similarities, averaging ~30%, appeared in the shrub communities and the oldest *Fagus* forest (Fig. 1).

Indeed, low correspondence of the vegetation and seed bank floras has been noted by many investigators in old-growth forests (Piroznikow, 1983). Perhaps these trends in heterogeneity of the seed bank reflect joint changes in the canopy integrity, input of seeds from within and outside the stand, as well as decay of initial seed banks.

Seed banks in old-growth forests appear to be compositionally and numerically more constant across years than seed banks of grasslands (Piroznikow, 1983; cf. Thompson and Grime, 1979). Species of old-growth forests have persistent seed banks (Nakagoshi, 1984b) composed of higher proportions of native species than those of other communities (Nakagoshi, 1985a). The proportion of native species in the standing vegetation also generally increases in old-field seres (Bard, 1952; Beckwith, 1954). However, the seed banks of some old-growth forests in highly managed or modified landscapes can have large numbers of exotic species (>80% of total seed density, M. J. McDonnell, unpublished data).

C. Relation to Species Strategies

Forest herbs are very rare or completely lacking in forest seed banks (Nakagoshi, 1985a; Brown and Oosterhuis, 1981; Petrov and Belyaeva, 1981). Many forest species were absent from the seed bank in the old-growth forest at Bialowieza, Poland (Piroznikow, 1983). Virtually all woodland species were absent from the seed banks of a wide variety of forest, grassland, and woodland stands in Britain (Donelan and Thompson, 1980). Mladenoff (1985) did find a variety of forest residents, including *Erythronium americanum* and *Viola* spp., which were the most common species in his old-growth seed banks. Burial by ants, the dispersal agents of these two genera, may be responsible for their appearance in the seed bank. Because many of the herbs Mladenoff (1985) found in the seed bank were present in the stand, their seed could be relatively short-lived in the soil. He did not find increased herb richness in the seed bank of gaps 0–8 yr old. Furthermore, creation of small gaps, which mimicked snapping of standing trees, did not stimulate establishment of *Erythronium* or other herbaceous species characteristic of closed canopies in the Allegheny hardwood forest in Pennsylvania. However, *Aster acuminatus* did invade where soil was disturbed by small mammals (Collins and Pickett, 1987).

Several generalizations result from the classification of floristic information on seed banks by reproductive and dispersal strategy, and by physiological requirements. The species most often found in seed banks in deciduous forests are those described as "intolerant," "early successional," or "colonizing." Regardless of the terminology used, and any assumptions about where such species evolved, it is clear that such species characteristically occupy open sites. Often these species are not at all characteristic of the forest (e.g., *Ranunculus*, *Hypericum*, and *Plantago* in a *Tilia–Carpinus* forest near Moscow; Petrov and Belyaeva, 1981). Otherwise, genera such as *Prunus*, *Betula*, *Rubus*, *Carex*, and *Juncus* recur in lists from throughout the north temperate regions, and often account for the greatest densities of viable seeds in the bank (Oosting and Humphreys, 1940; Marquis, 1975; Moore and Wein, 1977; Hill and Stevens, 1981). The majority of seeds in persistent soil banks are small (Piroznikow, 1983), which may relate to ease of penetrating litter or to burial by worms (Thompson, 1987). Large seeds, exceeding 32 mg, do appear in occasional early successional seed banks, but are consistently present only later in the sere (Fig. 2; Nakagoshi, 1984a). Tolerance of dense shade (e.g., Salisbury, 1942; Grime, 1977) and the ability of seedlings to emerge through thick litter appear to be important selection pressures for large-seededness later in succession (Thompson, 1987). The rarity of very large seeds in forest seed banks may relate to the selection value of masting and predator satiation, which are inconsistent with seed longevity (Silvertown, 1980b).

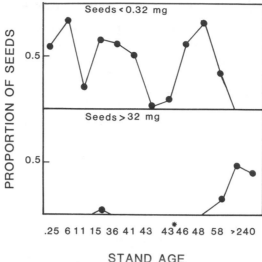

Figure 2.   Proportion of total seeds in the seed bank representing contrasting size classes across a chronosequence on the Hiba Mountains, Japan. The upper panel shows the seeds <0.32 mg, and the lower panel shows seeds >32 mg. Intermediate size classes shown in the original are omitted for simplicity here. The communities ≥11 yr of age are the same as in Fig. 1. The 0.25- and 6-yr communities are *Rubus–Aralia* types. Redrawn from Nakagoshi (1984b).

The syndromes for dispersal differ through succession. Based on species composition of the seed bank in autumn, most common species are animal dispersed throughout the entire succession, with the next major group representing wind dispersal (Nakagoshi, 1984a). Studies of the standing vegetation often report a compensatory shift in the proportions of wind- and animal-dispersed species through time. This shift is echoed in the samples of seed banks in spring, which better represent the persistent component of the seed bank. Likewise, actual densities of seeds of each type in the seed bank show a more pronounced shift from wind to animal dispersal later in succession. Animal-dispersed seeds comprise >50% of the seed bank density only in Nakagoshi's (1984a) three late-successional communities. Autochorous (self-dispersed) species are important in the persistent seed bank, and are common only early in succession (Fig. 3; Nakagoshi, 1984a). Because seed bank studies rarely consider dispersal, it is difficult to evaluate the dispersal strategies of seed bank species in deciduous forests.

A common correlate with the lack of persistent seed banks in forest species is the presence of alternative regeneration strategies. Tree species lacking a persistent seed bank are those that sprout (e.g., *Fagus*) or possess seedling banks (e.g., *Prunus serotina*, *Acer saccharum*, *Tsuga*; Marquis, 1975). The forest herbs also possess alternative modes of propagation (e.g., stolons and bulbs). The capacity of these species to survive the more common disturbances in mesic deciduous forests is great (Collins *et al.*, 1985).

The idea that seed banks and other tactics for persistence in an area are compensatory has been evaluated using schemes to recognize reproductive options. The first classifications of seed bank patterns were based on herbaceous communities of Britain (Thompson and Grime, 1979; Grime, 1981). Nakagoshi (1985a) has used a similar scheme to show in detail how the seed bank behaviors in various forests in Japan relate to other modes of reproduction, and how both relate to the successional status of the forest and its disturbance regime. Species that do not possess persistent seed banks in deciduous forests often have one or more of these characteristics: wind dispersal; broad germination tolerances; bulbs, corms, or perennating rhizomes; shade tolerance of early life cycle stages; ability to penetrate litter via long roots; scatter-hoarded seeds; slow growth; tolerance to herbivory; and long vegetative phases. Not all of these tactics may appear in a single species; some species possessing persistent seed banks may also exhibit some of these characteristics. In general, the relationships among seed bank and other reproductive and dispersal characteristics need to be rigorously and comprehensively examined.

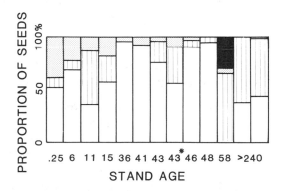

Figure 3.   Proportion of total seeds in the seed bank representing different dispersal syndromes in the Hiba Mountains chronosequence. For community types, see Figs. 1 and 2. The open portion of the bars indicates anemochorous seed (wind dispersed), the hatched portion indicates endozoochorous seed (animal dispersed), the stippled portion indicates autochorous seed (self dispersed), and the solid portion represents clitochorous seed (acorns dispersed close to parent trees). Modified from Nakagoshi (1984a)

D. Seed Banks and Ecosystem Processes

The observation that forest herbs and tolerant woody species do not store seeds in the soil might suggest that seed banks are important only to the persistence of early-successional herbaceous plants and woody species requiring open sites. However, the dynamics of seed banks are important aspects of the dynamics of vegetation and ecosystems. The well-known role of species arising from the seed bank is shown by the revegetation of deforested watersheds in the Hubbard Brook experiments (Bormann and Likens, 1979). The dominant species of the initial successional vegetation, *Prunus pensylvanica*, *Betula*, and *Rubus*, are seed bank species (Marks, 1974). These genera, and others arising from seed banks, are responsible for the rapid revegetation of deforested sites throughout the temperate regions (Marquis, 1975; Hill and Stevens, 1981; Brown and Oosterhuis, 1981). The nutrient uptake that these seed bank species accomplish on disturbed sites is a major component of ecosystem recovery (Marks and Bormann, 1972; Marks, 1974; Bormann and Likens, 1979). Seed banks play a similar role in the immediate re-vegetation of postagricultural old fields. *Ambrosia artemisiifolia* is a paradigm for seed longevity, complex dormancy and specific germination requirements (Pickett and Baskin, 1973), and high resource demand (Vitousek, 1983).

Despite their importance for certain life history groups and for ecosystem processes, it is worth emphasizing that seed banks will not be important conservation tools for species of late-successional, closed habitats. These are the very species that generally do not form persistent seed banks (Brown and Oosterhuis, 1981).

E. Environmental Correlates of Seed Bank Size and Composition

There is great interest in the environmental relationships and mechanisms that control seed bank size and composition. Dormancy and germination requirements, the standing vegetation, soil depth and other characteristics, and controls of decay rates are widely recognized factors influencing seed bank dynamics. However, relatively few such mechanisms have been studied directly and most information is correlative. The environmental control of dormancy and germination is covered by Baskin and Baskin (Chapter 4, this volume). The remaining factors are less well known. We present information on environmental factors and relationships beginning with fine-scale or local factors, and ending with the influence of the landscape. Some generalizations are at least tentatively supported by existing studies.

First, vegetation type appears to be related to the composition and density of the seed bank. Deciduous stands support richer and denser seed banks than do coniferous stands in Maine (Moore and Wein, 1977), where canopy composition is more important than stand age. Bogs in the same area had no, or depauperate, seed banks (Moore and Wein, 1977). On a coarser scale, communities containing more therophytes are usually found to have larger seed banks than are found in communities rich in other life forms (Piroznikow, 1983).

Another environmental factor frequently mentioned as controlling seed bank composition is soil type. Of course, this factor will often be confounded with vegetation type. Soils with low pH support smaller seed banks in certain cases (Brown and Oosterhuis, 1981), and organic soils support smaller seed banks than do mineral soils (Moore and Wein, 1977). The smaller seed banks of peat soils may be a result of poor permeability of such soils to seeds (Hill and Stevens, 1981).

Environmental control of decay in seed banks may be important influences on density and composition. Decay rates of seeds in soil have been investigated primarily in agricultural and fallow soils (Cavers and Benoit, Chapter 14, this volume), and information on decay rates in deciduous forest soils is sparse. Lower decay rates appear at greater depth in field soils (Stoller and Wax, 1974; Fenner, 1985). Steady rates of decay have been found for weed seed banks in pasture (K. Thompson, 1978) and for *Prunus pensylvanica* in closed forest (Marquis, 1975). Steady declines in seed banks of intolerant species suggest that seed banks of such species are not replenished in closed communities.

Decline in the number of seeds in the seed bank with soil depth is a widely recognized phenomenon. The largest number of seeds is found in shallow horizons of mineral soil, with rapid decline with depth (e.g., Numata *et al.*, 1964; Moore and Wein, 1977; Nakagoshi, 1981; Hill and Stevens, 1981). Litter may contain large numbers of seed. The inconsistent sampling of organic horizons among various studies weakens this generalization, however. Litter layers are not always sampled because it is thought that they would harbor only recently produced seed. In at least one situation, litter and humus layers contain fewer seeds than are found in mineral soil (Hill and Stevens, 1981).

The factors discussed above are those that act within a site. Sensitivity of seed banks to surrounding features of the landscape has been mentioned in the literature, but rarely has it been assessed. One clear case of the importance of the biogeography of a site exists. Nakagoshi (1985a) notes that the seed bank of a forest on an island in Japan's Inland Sea is depauperate. Edge effects have been recognized as potentially important in the composition of seed banks through dispersal patterns (Cavers and Benoit, Chapter 14, this volume), but seed banks have not been studied relative to edges and in contrasting landscapes in deciduous forests.

III. Seed Banks and
Adaptation to Disturbance

A. Gradients of Disturbance

Characteristics which enable a plant species to store seed in the soil for extended periods of time (seed banking) may be adaptive in environments subject to frequent disturbance. This hypothesis can be tested by comparing the seed banks at sites of increasing age since the time of the disturbance, assuming a constant decay of seed adapted to disturbance or its immediate aftermath (Donelan and Thompson, 1980), or at sites having different disturbance regimes. While the second test would yield the strongest evidence, only the first has been attempted so far, although correlation with a current disturbance regime has been found for single species (e.g., *Mentha pulegum*; Panetta, 1985). Based on considerations of the contrasting characteristics of seeds of stress-tolerant species versus seeds of species that can survive disturbance and persist in the soil, K. Thompson (1978) theorized that seed banks should correlate positively with some measure of disturbance. Assuming that a vegetational gradient mimicking a succession represents a gradient of receding impact of disturbance, Donelan and Thompson (1980) located nearby sites in County Durham, England, ranging from a highly disturbed site (a quarry), through pastures having different grazing pressures, to a small, old woodlot representing a long-undisturbed site (see Donelan and Thompson, 1980, in Table 2). The total number of seeds fluctuated across the sites and did not reflect the presumed disturbance gradient. The gradient includes some sites that are questionable parts of the sere (the youngest and oldest), and the disturbances represent rather different agents and frequencies. Thus, there is difficulty in interpreting the study as a test of K. Thompson's (1978) hypothesis.

Seed banks in abandoned coppice woods in England show the different ways that disturbance impinges on seed banks (Brown and Oosterhuis, 1981). Coppicing, which involves thinning the canopy, represents a relatively mild disturbance to a stand. Without such disturbance, species richness declines, and shade-intolerant species of active coppice are absent. The intolerant species that dominate the ground layer after thinning of the coppice are the major components of the seed bank. Thus, different components of the disturbance regime of a complete coppice cycle favor different regeneration and seed bank strategies.

Using different disturbance treatments, Marquis (1975) found variation in species responses in experimental plots in the Allegheny hardwoods forest. Clear-cutting favored immediate germination, while partial

Table 2

Studies of seed banks of temperate deciduous forest communities and successions[a]

Vegetation type	Seed density	Species richness[b]	Reference
North America			
Maine			Olmstead and Curtis (1947)
24-yr red pine plantation	All: 2.2×10^6 (acre^{-1}) Viable: 2.1×10^4	8	
30-yr spruce–fir	All: 11.5×10^6 Viable: 2.0×10^5	13	
70-yr white pine	All: 4.06×10^6 Viable: 6.5×10^5	11	
80-yr white pine	All: 1.3×10^6 Viable: 0	5	
50-yr beech–birch–maple	All: 8.7×10^5 Viable: 1.5×10^5	10	
110-yr beech–yellow birch–sugar maple	All: 4.0×10^6 Viable: 3.7×10^5	12	
150-yr sugar maple	All: 4.9×10^5 Viable: 4.4×10^4	8	
Massachusetts			Livingston and Allessio (1968)
1-yr horseweed field	232 (ft^{-2})	17	
2-yr goldenrod field	424	23	
8-yr *Andropogon* field	117	19	
5-yr white pine stand	419	25	
7-yr white pine stand	196	21	
15-yr red pine plantation	453	21	
21-yr red pine plantation	147	21	

(continued)

Table 2 (*Continued*)

Vegetation type	Seed density	Species richness[b]	Reference
25-yr white pine plantation	466	21	
35-yr red spruce plantation	164	15	
36-yr red pine plantation	116	24	
37-yr white pine plantation	130	21	
41-yr red pine plantation	224	21	
42-yr white pine plantation	145	14	
47-yr red pine plantation	315	16	
47-yr white pine plantation	125	20	
80-yr white pine stand	311	20	Marquis (1975)
Pennsylvania			
35-yr Allegheny hardwoods	1.5×10^6 (acre^{-1})	6	
53-yr Allegheny hardwoods	1.9×10^6	7	
65-yr Allegheny hardwoods	2.5×10^6	6	
70-yr Allegheny hardwoods	1.2×10^6	6	
>100-yr Allegheny hardwoods	0.4×10^6	6	
New Brunswick			
Betula–Fagus	3400 ± 970 (m^{-2})	~5 taxa	Moore and Wein (1977)
Acer–Fagus	1950 ± 620	—	
Acer clear-cut	1390 ± 260	—	
Acer–Abies	1230 ± 260	—	
Picea–Pinus	580 ± 90	—	
Picea	370 ± 140	—	
Larix bog	320 ± 100	—	
Picea clear-cut	180 ± 80	—	
Bog	0	—	

Location	Density	Species richness	Reference
New Hampshire			Bicknell (1979)
1 to >100 yr stands	117 (m^{-2})	3 taxa	
Michigan			Mladenoff (1985)
Gap	421.9 (0.01 ha^{-1})	18 taxa (1.4 m^{-2})	
Forest	190.5	20 taxa	
North Carolina			Oosting and Humphreys (1940); cited in Nakagoshi (1984b)
85-yr old-field pine	3318 (m^{-2})	—	
112-yr old-field pine	1470	—	
>200-yr oak–hickory forest	1181	—	
Ohio			Roberts *et al.* (1984)
2-yr *Trifolium*	—	41	
10-yr *Solidago*	—	33	
50-yr *Solidago–Fraxinus–Prunus*	—	39	
90-yr *Acer–Ulmus*	—	21	
Europe			
England, County Durham			Donelan and Thompson (1980)
15-yr abandoned quarry	1077 (m^{-2})	20 (0.73 m^{-2})	
Arable field	468	23	
Grazed grassland	1329	24	
Ungrazed grassland	377	28	
Gorse scrub	290	25	
Hawthorn scrub	181	22	
>80-yr ash with hawthorn	369	14	
~100-yr ash with sycamore	758	15	
Oak woodlot	27	6	

(continued)

Table 2 (*Continued*)

Vegetation type	Seed density	Species richness[b]	Reference
England, County Essex and County Sussex			Brown and Oosterhuis (1981)
Combined coppice woods	610 (m^{-2})	11.6 (0.45 m^{-2})	
Scotland/Wales			Hill and Stevens (1981)
Brown earths	1000–5000 (m^{-2})	—	
Peaty gleys	500–2500	—	
Deep peat	50–250	—	
Poland			Piroznikow (1983)
Spring	Seed 6075 ± 1223 (m^{-2})	32	
	Seedlings 190 ± 32		
Fall	Seed 8115 ± 1964	—	
	Seedlings 24 ± 7	—	
USSR			Petrov and Belyaeva (1981)
Querceto–Tilietum	1500 (m^{-2})	24	
Japan			Numata *et al.* (1964)
Erigeron old fields	2607 (400 m^{-2} soil)	25	
Imperata old fields	571	22	
Pinus old fields	178	32	

138

	Seed density[a]	Species richness[b]	Reference
Castanopsis forest	5574 (m⁻²)	—	Hayashi (1977, cited in
Pinus densiflora	13,650	—	Nakagoshi, 1984b)
Pinus thunbergii	2575	—	
0-yr *Mallotus–Aralia*	628 (m⁻²)	26 (0.4 m⁻²)	Nakagoshi (1981)
5-yr *Mallotus–Aralia*	1613	30	
75-yr *Rhododendron–Pinetum*	3180	38	
Photinio–Castanopsietum (climax)	850	33	
Weigela–Shortia (*Quercus*)[c]	4900 (m⁻²)	25	Nakagoshi (1984a)
Chamaecyparis–Lindera (moist)	10,000	29	
Quercus–Castanopsis (*Ilex*)	2000	34	
Quercus–Castanopsis (*Oplismensus*)	2700	34	
Fagus–Lindera	800	25	
Pinus–Quercus, stand 14	400 (m⁻²)	17	Nakagoshi (1984b)
Pinus–Quercus, stand 15	560	13	
Tsuga–Neolitsia, stand 16	795	15	
Tsuga–Neolitsia, stand 17	490	18	
Sciadophytis–Chamaecyparis, stand 18	1325	8	
Sciadophytis–Chamaecyparis, stand 19	515	10	

[a]Seed densities are reported in the units of the original study; the units are given in parentheses following the first data entry of each study and are identical for all other data entries for a given study. Note that due to differences in methodologies, the data are not necessarily readily comparable among different studies.

[b]Species richness, unless otherwise noted, represents the total species encountered.

[c]Only communities having either the fewest or most species, or the lowest and highest seed densities, are shown. The values are from graphs and are only approximate.

cutting resulted in delayed germination of some species from the seed bank (e.g., *Prunus pensylvanica*). Over the 2 yr of the study, germination of *Prunus serotina* was favored by canopy closure, while germination of *Acer rubrum* fluctuated with canopy opening. In the first year, this species exhibited a tendency for delayed germination under a full canopy but germinated more rapidly as canopy density decreased. During the second year, germination of *Acer rubrum* was high under the full canopy.

A variety of species responses in seed banks is apparent. Nakagoshi (1985a) combines this observation with his classification of the variety of possible regeneration behaviors of coexisting species to suggest that the disturbance regime is a major correlate of regenerative variety in forest species. On the basis of seed bank and other reproductive tactics, he shows how species can be partitioned coarsely by the kinds of disturbed sites commonly present in forests. Tests of this hypothesis must compare sites having disturbance regimes that differ in frequency of gap formation, spectrum of gap size, mode of gap formation (e.g., whether trees are uprooted or snapped off), and anthropogenic versus natural disturbance. Seed banks responding to different disturbance regimes should differ in density, richness, and type of seed bank (*sensu* Thompson and Grime, 1979; Nakagoshi, 1985a). Indeed, the spectrum of dispersal types and other reproductive syndromes in both the seed bank and standing vegetation should differ among disturbance regimes. Understanding the adaptive nature of seed banks (not necessarily selected for *in situ*) requires knowledge of the reproductive options available to the species and the complete disturbance regime. Such information is currently lacking.

Disturbance regimes dominated by natural and anthropogenic agents should not necessarily generate divergent seed banks. However, seed banks may shift in composition and size, because anthropogenic disturbances are often more intense, frequent, and larger compared to the natural ones they replace or complement. Seed banks reflecting human-dominated disturbance regimes are likely to contain species evolved under similar natural disturbance regimes or to have originated in similar human-dominated landscapes elsewhere (e.g., introduced crop weeds). For example, *Ambrosia artemisiifolia* belongs to a taxon that diversified in habitats subject to frequent disturbance (Payne, 1970).

Many of the colonizing plants in North American old fields are derived from permanently or fluctuating open habitats, rather than forest gaps (P. L. Marks, 1983). In such marginal, open habitats seed banks may reflect evolution to avoid physical hazards or fluctuating biotic interactions rather than the periodic closing or disappearance of the site. The ability to determine confidently the selective pressures behind various attributes of seed banks is limited. The wide migration of most plant taxa in geological time obscures selective history. The long history of natural environmental fluctuations, the large role of natural disturbances, the potential influence of preindustrial peoples, and the documented long history of environmental modification by many cultures may all have played a selective role. These factors must be considered in generating and evaluating adaptive interpretations of seed banks.

B. Problems with Adaptive Interpretations

The hypothesis that seed banks are adaptive under certain disturbance regimes cannot be evaluated without addressing three fundamental questions related to how seed banks operate. These involve how seeds enter the seed bank, whether the source of seeds is local or distant, and the nature of heterogeneity of seed banks within a stand.

The early assumption that seeds in the soil were present from the onset of succession and entered via soil formation above them (Oosting and Humphreys, 1940) is now disfavored. Early-successional incorporation due to cultivation, or freeze–thaw cycles, or continued incorporation due to animal activity seems more likely (Livingston and Allessio, 1968; Brown and Oosterhuis, 1981). The role of earthworms appears to be especially great, as seeds of such a large number of species in seed banks are small and smooth (Thompson, 1987). Furthermore, continuous entry of many seeds into seed banks seems likely, because seeds are found in litter layers (Hill and Stevens, 1981), and large proportions of seeds in the soil are wind and animal dispersed (Nakagoshi, 1981).

The locale of origin of seeds in soil is also problematical. Brown and Oosterhuis (1981) found that the seed bank did not correlate with distance from glades or rides (open riding trails) in abandoned coppice woods, suggesting that such open areas were not the source of seeds of shade-intolerant species in the soil. The failure of species from adjacent intact areas to occur in cleared forest areas suggests that dispersal across stand boundaries is not an important source of seed in soil (Hill and Stevens, 1981). However, this inference is weakened by the instantaneous nature of dispersal at the time of disturbance compared to the long time that seeds may have been entering the seed bank and the difference in environmental factors influencing germination. Marquis (1975) also found that the seed bank of particular stands in the Allegheny hardwood forest correlated best with the standing vegetation rather than with age of the stands. However, his characterization of the seed bank includes many relatively short-lived tree species, and did not include herbaceous species.

The statements concerning the sources of seeds in seed banks are, on the whole, derived not from direct measurement of the dispersal process, nor even from direct correlative study of the question of source.

The specific mechanism of incorporation of seeds into the soil is not addressed in any study from the deciduous forests. Rather, ideas about source and incorporation of seeds are inferred in the course of addressing other questions. The growing literature on factors affecting dispersal patterns of seeds in deciduous forests suggests that factors such as distance to seed source, height of seed source, and availability of dispersal vectors are important to the recruitment of seeds into plant communities (McDonnell and Stiles, 1983; Holthuijzen and Sharik, 1985; Murray, 1987a; McDonnell, 1986). It appears, though, that both wind- and bird-disseminated plants are dispersed over a distance of only tens of meters (A. J. Smith, 1975; Johnsen *et al.*, 1981; Holthuijzen and Sharik, 1985). However, dispersal studies and seed bank studies should be integrated to test this conclusion.

IV. Limitations of the Data Base and Opportunities for Future Research

There is a firm empirical basis of studies of seed banks in deciduous forests, but the open questions illustrate the great opportunities and needs for further study. Four areas in particular require attention: (1) methodological improvements, (2) integration of seed bank patterns with relevant processes, (3) expanded comparison among and within sites, and (4) the study of seed banks in an evolutionary context. The first area, methodological improvements, is discussed in Chapter 1 (this volume). Thus, our discussion will begin with the need to integrate processes.

A. Integration with Relevant Processes

A conspicuous lack in the study of deciduous forest seed banks is the failure to include processes that may actually control the dynamics of the seed banks. The concept of seed budgets (Fig. 4; Kellman, 1974a)

Figure 4. Model of the soil seed bank budget. Although the seed budget presented here consists of the same basic phenomena as presented by Harper (1977), additional mechanistic and spatial detail is suggested. (A) Origins of seeds that may be incorporated into the seed bank. Global refers to the flux of seed from distant sources, and local refers to origin in adjacent landscape elements. (B) Modes of incorporation of seeds into the seed bank. The box represents a particular horizon of the soil. The arrows above the box indicate entry into a horizon and the arrows below, representing the same processes, indicate potential exit from that horizon. (C) Fates of seed once incorporated into the bank or a particular horizon.

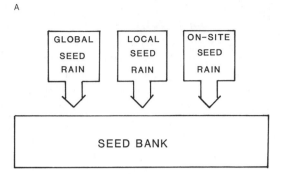

needs to be applied rigorously to seed banks in deciduous forests. For instance, entry into the seed bank has only been inferred from the observation that seed banks in the vast majority of communities are composed of species that are important in earlier successional communities (Oosting and Humphreys, 1940). Contemporaneous studies of the origin and magnitude of the seed rain would be the strongest inference (Fig. 4), but such joint studies are almost nonexistent (but see Bicknell, 1979). Indeed, exactly how seeds enter the soil bank once they reach a site must also be determined (Fig. 4). Percolation within the soil and the role of animals as transporters or predators in seed banks need to be determined. This information is necessary to answer questions raised about adaptiveness of seed banks. The patterns of seed banks through succession and along other environmental gradients suggest that the controls on seed budgets may change over these gradients. Such mechanistic changes deserve serious, focused study.

B. Critical Comparisons

The studies of seed banks in deciduous forests have been conducted in remarkably few locations within the broad variety of communities present in the biome (Table 2). In the United States, for example, the northern hardwoods forests (including the transitional hardwoods of New England and the Allegheny hardwoods), beech maple forests, and transitional conifer–mixed hardwoods, all occupying the northern reaches of the biome, have received the most attention. Mixed mesophytic, oak–hickory, and pine-dominated regions of the biome have been essentially ignored. The possibility that climatic, disturbance, and animal effects on seed banks differ within the biome in ways that alter seed bank composition, densities, and dynamics cannot be evaluated without a broader base of studies for comparison.

The question of whether seed banks in deciduous forest reflect species adaptation to disturbance must await further comparisons of well-documented gradients of disturbance intensity or other aspects of disturbance regime. Determination of seed banks in stands having a variety of disturbance histories is needed. Likewise, the potential for correlation of various terms of the seed budget (Fig. 4) with disturbance regime and particular disturbances is an intriguing question. Most studies of seed banks have been conducted in highly modified landscapes, where exotic species and edges abound. The role of these factors can best be determined by comparison with landscapes having more continuous cover of closed forest and lower frequencies of introduced taxa. Thus, studies in large, old-growth forests (cf. Mladenoff, 1985), including the various components of the forest regeneration cycle (Watt, 1947), require investigation.

C. Evolutionary Context

An evolutionary interpretation requires the ability to explain adaptively the characteristics of the species comprising seed banks. We have already noted the poor likelihood of determining the habitat or disturbance regime under which seed characteristics actually evolved. Knowing where species originated, the nature of disturbance regimes (human and natural) which have influenced them in the past, and the seed bank dynamics of related species would enhance the ability to pose reasonable evolutionary hypotheses about species in seed banks.

An adaptive interpretation of seed banks, on the other hand, does not necessarily require knowing the evolutionary origin of the species or their traits. Adaptation involves the degree of correspondence of seed characteristics and behaviors to a range of conditions. Thus, mechanistic richness is the basis for an adaptive interpretation of seed banks. The detailed flow model of seed banks (Fig. 4) incorporates the mechanisms that must be understood. For deciduous forests, few of the mechanisms are well known for many species or for contrasting habitats.

V. Summary and Conclusions

We will summarize by briefly answering the eight questions posed in the Introduction of this chapter.

What are the dynamics of seed banks in deciduous forests? Several tentative generalizations concerning seed bank dynamics have been derived from the literature, and these serve as a foundation for other studies in stands similar to those studied so far, or as hypotheses to test in new situations or with improved methods. Seed banks differ from the standing vegetation, except in large, recently disturbed gaps or abandoned fields where banks and vegetation are more, though not completely, similar. Seed banks are usually most diverse and dense early in succession, and decline with time. Such trends are not necessarily monotonic; the largest seed banks do not always appear in the youngest successional communities. Seed banks are important factors in the regeneration of disturbed communities and ecosystems, even though not all species of a succession reside in seed banks.

What physical conditions govern the behavior of the seed bank? Seed banks decline with depth in the soil, and are smaller in organic soils and at low pH. These correlations are based on very few data from deciduous forests. Furthermore, the mechanisms underlying these trends are uncertain. Penetration into the soil, invertebrate and fungal predation, failed germination, and physiological death must be evaluated.

Do seed bank dynamics diverge within the biome? The existing studies of seed banks represent only a few of the many associations of the deciduous forest biome. In North America, northern mesic forests have received the most attention, leaving the central, southern, western, and more xeric associations of deciduous forests uninvestigated. Gradients of disturbance intensity and other important aspects of disturbance regime cannot be effectively compared on the basis of existing data.

Are there patterns in the seed bank within a community? The level of replication within communities and the relationship of samples to physical and biotic heterogeneity within a stand are inadequate to answer this question. In addition, the relationship of the seed bank to the landscape in which a community is embedded, and the role of local versus adjacent seed sources, cannot yet be addressed.

Are species strategies apparent in the seed bank? Seed banks, especially their most persistent components, are largely composed of shade-intolerant or early-successional species. Very few large-seeded or shade-tolerant species reside for long in the seed bank of deciduous forests. Few studies systematically or quantitatively address additional aspects of species strategies, so that community summaries are not now possible.

Do species of particular successional positions have similar seed bank dynamics? The great variety of reproductive tactics apparent in plant assemblages also appears in the variety of seed bank patterns. Not all pioneer species, for example, rely on seed banks for establishment, although some of the more widespread dominants do. Likewise, a few late-successional species exploit seed banks, but alternative reproductive tactics, including vegetative spread, a bank of persistent, shade-tolerant seedlings, or high fecundity and rapid turnover of seedlings, are found in late- or intermediate-successional species. The variety of reproductive strategies and degrees of reliance on seed banks may relate to the mosaic nature of forests and the resultant variety of regeneration opportunities.

Are the patterns explainable in terms of natural selection? The tendency of species relying on seed banks to be those shade-intolerant invaders of disturbed sites suggests that disturbance may be a powerful selective force in the evolution of seed banks. However, because many other pressures (e.g., predation and relation to seed rain) on seed banks have not been explicitly examined and brought to bear on seed budgets, they cannot be ruled out or ranked relative to disturbance. This is one of the major tasks for future studies of seed banks in deciduous forests.

How do physiological aspects of seed bank dynamics affect the dynamics at the community level? Much is known about physiology of dormancy and germination (Baskin and Baskin, Chapter 4, this volume).

However, the compilation of this information for entire seed banks in contrasting situations would be required to answer adequately this question. The compilations that do exist, and which have interesting implications for community organization and evolution of seed banks, apply to grassland or meadow communities.

Additional conclusions concerning the overall state of seed bank science in deciduous forests emerge from the above summary of answers. Three broad themes appear, concerning methods, integration of factors, and comparison at various scales. First, increased methodological sophistication is required in seed bank studies in deciduous forests. Increased replication and attention to important sources of ecological heterogeneity must characterize seed bank studies in the future. Second, the concept of seed budget should be used to integrate important processes that can influence seed bank composition and density. Seldom have factors such as seed input, movement in a soil profile, and predation been considered along with the basic information about seed banks. Many interesting inferences have been drawn in the literature, but only explicit measurement and manipulation of the processes that govern seed banks can be used to evaluate the predictions. Finally, to assess the significance of seed banks in deciduous forests, comparisons must be made at a variety of scales. Measurement of seed banks within communities, at replicate sites within associations, and in neglected associations of deciduous forests are all required. With this information in hand, the questions we have addressed in a necessarily preliminary way can be unambiguously answered, and new hypotheses may emerge to generate a rigorous science of seed banks in deciduous forest.

Acknowledgments

Dave Mladenoff and Susan Bicknell kindly provided access to important unpublished material. Mary Leck also provided assistance with bibliographic references. Contribution to the program of the Institute of Ecosystem Studies with support from the Mary Flagler Cary Charitable Trust.

CHAPTER 9

Tropical Soil Seed Banks: A Review

Nancy C. Garwood

Department of Botany
Field Museum of Natural History
Chicago, Illinois
and
Smithsonian Tropical Research Institute
Balboa, Panama

I. Introduction

Since Symington (1933) reported seeds in Malayan forest soil, studies of tropical soil seed banks have proliferated. The crucial and controversial question, whether soil seed banks play an important role in regeneration of tropical forests, has been difficult to resolve because regeneration from dormant seeds that have accumulated in the soil through time has seldom been distinguished from regeneration from seeds that have recently dispersed into a site (Denslow, 1980; Sarukhán, 1980; Whitmore, 1978, 1983, 1984; Brokaw, 1985a). To evaluate the role of the soil seed bank, its contribution must be compared with that of other regeneration pathways over the entire range of natural and man-made disturbances found in the tropics. If the seed bank is important, we need to understand the dynamics of seeds in soil and the causes of spatial and temporal heterogeneity, including changes during succession.

In this review, I briefly summarize the regeneration strategies of tropical plants, then discuss the dynamics of seeds in soil, tropical seed bank strategies, the role of the seed bank in regeneration, changes in the seed bank during succession, and nonsuccessional patterns of spatial and temporal heterogeneity. The seed banks of a variety of tropical vegetation types are summarized; this review focuses on the most frequently sampled tropical vegetation, lowland forest, secondary regrowth, and farmland. Although a few lower montane and waterlogged caatinga forests have been studied, little is known of seed banks in nutrient-poor, swamp, and elfin forests, in alpine páramo (but see Guariguata and Azocar, 1988), or in grasslands and savannas. [See Richards (1952), Whitmore (1984), and Walter (1985) for descriptions of tropical vegetation.]

II. Regeneration Strategies of Tropical Species

Tropical species can be grouped into regeneration strategies based on similar regeneration requirements (see Baur, 1968; Gómez-Pompa and Vázquez-Yanes, 1974; Whitmore, 1984; Hartshorn, 1978; Denslow, 1980; Alexandre, 1982; Brokaw, 1985a; Clark and Clark, 1987; Swaine and Whitmore, in press). While no simple classification adequately categorizes the diversity of tropical species, these regeneration strategies provide a useful first-order approach. Tropical forest species regenerate from one or more pathways (Alexandre, 1982; Brokaw, 1985a): seed rain (recently dispersed seeds), the soil seed bank (dormant seeds in the soil), the seedling bank (established, suppressed seedlings in the understory), advance regeneration (suppressed saplings in the understory), coppice (root or shoot sprouts of damaged individuals), or canopy in-growth (lateral growth of the canopy).

Primary species germinate and establish in the shaded understory of undisturbed forest. They are large-seeded, shade-tolerant, slow-growing, and long-lived species. Species with a persistent seedling bank eventually benefit from a treefall gap, but understory species may continue to grow slowly in the shade. In contrast, **short-lived pioneer species** germinate and establish only in large forest gaps or man-made clearings. They are small-seeded, shade-intolerant, and fast-growing species. **Long-lived pioneer species** (i.e., late secondary species) germinate in sun or shade, but grow only in forest gaps. They are intermediate between short-lived pioneer and primary species. They dominate late secondary forests but are also components of mature forest canopy (Hartshorn, 1980). **Weedy species**, usually herbs, shrubs, and vines, have pioneer-like traits, but are never components of the tropical forest (Hopkins and Graham, 1984a). They are common and extremely important in man-made environments. Weedy species and short-lived pioneers reproduce annually, sometimes continuously (Gómez-Pompa and Vázquez-Yanes, 1974; Whitmore, 1983; M. K. Marks, 1983a,b). Long-lived pioneers and primary species reproduce annually or supraannually. Successional (or secondary) species collectively refer to pioneer and weedy species.

III. Dynamics of Seeds in Soil

Soil seed banks include all seeds buried in the soil and those on the soil surface. Seeds in the litter/humus layer are often included (see Section VIII, Appendix). The length of time seeds reside in the soil seed bank will be determined by their physiological properties, including germination, dormancy, and viability; by the environmental conditions where they land and subsequent changes; and by the presence of seed predators and pathogens. This section focuses on the changes within seeds as well as in their surroundings.

A. Ecophysiology of Germination

1. Dormancy in Tropical Species

Extensive surveys of germination of fresh seeds sown under suitable conditions (Marrero, 1949; Taylor, 1960; de la Mensbruge, 1966; Koebernik, 1971; Macedo, 1977; Ng, 1978, 1980; Alexandre, 1980; Garwood, 1982, 1983) indicate that rapid germination is the most common response. Dormancy, however, is frequent in some regeneration strategies and vegetation types (Vázquez-Yanes and Orozco Segovia, 1984).

Seeds of many rapidly germinating species have short-term viability (a few months at most) and **no dormancy**. They occur in all regeneration strategies (Table 1). Many primary and late secondary species, including most economically important timber and fruit trees, have seeds that are nondormant, large, moist, and recalcitrant (i.e., impossible to dry without killing; Chin and Roberts, 1980). The small seeds of pioneer *Eucalyptus* species also lack dormancy (Baur, 1968).

Seeds of many rapidly germinating short-lived pioneer and weedy species remain facultatively dormant when conditions are unsuitable and may have long-term (≥ 2 yr) viability (Vázquez-Yanes and Orozco Segovia, 1984). **Facultative dormancy** is regulated by at least two physiological mechanisms that detect environmental changes associated with forest disturbance (Vázquez-Yanes and Orozco Segovia, 1984). The first, mediated by phytochrome, detects increases in the red/far-red ratio reaching the soil after the leafy canopy is removed. The second, often associated with breaking hard seed coats, detects increasing temperature fluctuations in the soil following exposure or burning. These mechanisms have been extensively studied in Neotropical and African pioneer species (see Vázquez-Yanes and Orozco Segovia, 1984; Fenner, 1980b; Marks and Nwachuku, 1986), but in few Asian or Australian species (Aminuddin and Ng, 1982).

Seeds of two other groups of species do not germinate rapidly under suitable conditions. In the first group, seeds remain seasonally dormant

Table 1

The expected distribution of tropical seed bank strategies among regeneration strategies[a]

Seed bank strategy	Regeneration strategy[b]			
	Weedy species	Short-lived pioneers	Long-lived pioneers	Primary species
Transient	R + A	R + A	R + A/I	R + A/I
Transient with seedling bank	—	—	—	R + A/I
Pseudo-persistent	R + C	R + C	—	—
Delayed-transient	—	—	D + A/I	D + A/I
Seasonal-transient	S + C/A	S + C/A	S + A/I	S + A/I
Persistent	F + C/A	F + C/A	—	—

[a]Based on the distribution of germination behaviors and frequency of dispersal.

[b]Germination behavior + dispersal frequency. Germination behavior denoted by R, rapid germination and short-term viability; S, seasonal dormancy and intermediate-term viability; D, delayed germination and intermediate-term viability; and F, facultative dormancy and long-term viability. Dispersal frequency denoted by C, continuous; A, annual; and I, supraannual (intermittent).

through an environmentally adverse period, usually a dry season. **Seasonal dormancy** is found in all regeneration strategies (Garwood, 1983; Vázquez-Yanes and Orozco Segovia, 1984; Table 1). The physiological basis is complex and little studied (Longman, 1969; Marks and Nwachuku, 1986). In the second group, germination occurs after several months (or years), often asynchronously, and does not follow seasonally adverse periods. **Delayed germination** is most frequent in large-seeded late secondary and primary species (S. A. Foster, 1986; Hopkins and Graham, 1987), especially those with hard or fibrous seed coats (Kiew, 1982; Vázquez-Yanes and Orozco Segovia, 1984; Hopkins and Graham, 1987). Dormancy is probably mechanically imposed. Seed longevity in both groups is intermediate between species with no dormancy and those with facultative dormancy.

2. Changes in Physiology during Dispersal

At the time of dispersal, seeds of each species have ecophysiological properties that can be modified before seeds hit the soil. Seed-handling behaviors of tropical dispersers, including passage through the digestive tract, sometimes alter the percentage or speed of germination (Vázquez-Yanes and Orozco Segovia, 1984; Howe, 1987). The effects are species specific for both dispersers and plants (Janzen, 1982b; Lieberman and Lieberman, 1986). Passage can also kill larvae of seed predators (Lamprey *et al.*, 1974) or change the physiological mechanisms controlling dormancy. Seeds of *Cecropia obtusifolia* lose far-red light inhibition of germination when passed through spider monkeys and tayras (a mustelid), but not bats (Vázquez-Yanes and Orozco Segovia, 1986). Because plants generally have a suite of dispersers, not all seeds arriving at the soil surface will have the same potential for persisting in the soil.

3. Changes in Physiology after Dispersal

Physiological properties can change while seeds are buried, causing germination, loss of viability, or change in germinability without loss of viability. Of 13 Nigerian weed species, dormancy cycled seasonally in five, changed permanently after burial in two, but remained constant in the rest (Marks and Nwachuku, 1986). In four Mexican pioneer *Piper* species, sensitivity of germination to length of photoperiod and quality of light changed during burial, but the timing and extent of changes were species specific (Orozco Segovia, 1986; Vázquez-Yanes and Orozco-Segovia, 1987).

The soil environment must be at least partially responsible for the physiological changes affecting germination because changes occur faster in seeds buried in the soil than in seeds imbibed in the dark on agar or stored dry (Orozco Segovia, 1986). Leaching may remove inhibitors from the seed or surrounding soil. Seeds of one *Piper* species buried 16 months had significantly higher germination when thoroughly rinsed before sowing (52%) than when sown directly from the soil without rinsing (20%); although germinability declined during burial in four *Piper* species, it increased (tripled in two species) in the month when the soil was wettest (Orozco Segovia, 1986). Responses of tropical seeds to abiotic soil factors, including oxygen concentration (Ponce de León, 1982), nitrate concentration (Uhl and Clark, 1983; Orozco Segovia, 1986), and flooding (Cole, 1977; Uhl and Clark, 1983; Orozco Segovia, 1986), are variable. Allelopathic inhibition of germination by chemicals leached from pioneer vegetation (e.g., Anaya Lang and Rovalo, 1976) and microbial enhancement of germination through degradation of hard seed coats (Edmisten, 1970) have been noted. Determining the causal mechanisms for change in physiological properties of seeds in the soil is critical to understanding dynamics of tropical soil seed banks.

B. Incorporation of Seeds into Soil

How seeds become incorporated into the soil is not well known, but is essential to understanding both spatial heterogeneity of seed banks and seed longevities. Scatter-hoarding rodents bury large seeds down to 8 cm deep (Smythe, 1978; Hopkins and Graham, 1983; Hallwachs, 1986), ants bury smaller seeds in nests and refuse piles (Roberts and Heithaus, 1986), and dung beetles bury seeds trapped in dung balls down to 12 cm deep (Estrada and Coates-Estrada, 1986). Seeds might become trapped and buried in soil spaces created by biotic or abiotic factors: large terrestrial animals (e.g., armadillos and coatis) dig holes, smaller burrowing animals (e.g., arthropods, terrestrial crabs, and caecilians) and dying roots create tunnels (Young, 1985; Hopkins and Graham, 1983), uprooted trees leave pits >1 m deep (Putz, 1983), and drying soil develops cracks 1–80 cm deep (Smela, 1987; J. Cavelier, personal communication). In some tropical regions, earthworms overturn enormous amounts of soil by burrowing and casting activity (0.2–39.0 mm yr^{-1}) and termites and ants overturn smaller quantities (Lal, 1987). In addition, rain may wash small seeds into coarse-textured soil (Hopkins and Graham, 1983) and plowing will bury seeds at farmed sites. These processes are spatially heterogeneous. Accordingly, the depth profiles should be extremely variable over small distances.

Seed density and number of species decreased with depth in 13 of 15 soil profiles in forest, savanna, regrowth, and farmland (Kellman, 1978; Cheke *et al.*, 1979; Holthuijzen and Boerboom, 1982; Hopkins and Graham, 1983; Putz, 1983; Young, 1985; Enright, 1985; Young *et al.*, 1987; Putz and Appanah, 1987; Sabiiti and Wein, 1987), but remained nearly constant in one profile (Young *et al.*, 1987), and was greatest at intermediate depth in another (Kellman, 1978). Seeds were found to all

Table 2
Mean seed densities and number of species in soil seed banks of mature tropical vegetation

Location and Site[a]	Ref. no.[a]	Vegetation type[b]	Altitude[b] (m)	Edge[c]	Soil depth[a] (cm)	Seed density m^{-2} Mean[e,g]	CV	[N]	Number of species[f,g] (area)
Neotropics									
Florida (USA)									
Everglades NP	[16]	Subtropical hammock forest	<10	—	3	81[m,n]	—	[3]	21 (7500)
Everglades NP	[17]	Hammocks and herbaceous glades	<10	—	—	<200	—	[—]	—
Puerto Rico									
El Verde	[6]	Forest	424	C	~12	587[l]	—	[3]	—
El Verde	[6]	Forest creek bed	424	C	~12	196[l]	—	[1]	—
Mexico									
Los Tuxtlas	[7]	Evergreen forest	120	~C	12	344–862 S	36	[16 × 8]	4–14 (640)
Los Tuxtlas	[7]	Evergreen forest	131	~C	12	175–689 S	40	[16 × 8]	2–7 (640)
Costa Rica									
Monteverde FR	[40]	Elfin forest	1500	—	10	406	49	[5]	—
La Selva	[34]	Lowland evergreen forest	<100	E	5	2340[r]	94	[—]	—
				CD	5	4700–2000[l] T	46–70	[—]	—
Panama									
Barro Colorado NM	[35]	Semideciduous forest	25	—	2	318[i]	—	[240]	54 (75,000)
Barro Colorado IS	[25]	Semideciduous forest	25	Fx	10 (20)	742 (827)	35	[10] [3]	48 (8550)
San Blas	[42]	Premontane forest	—	F	5	937	39	[10]	—
Venezuela									
San Carlos	[26]	Caatinga forest	120–170	—	5	200	88	[30]	14 (12,000)
San Carlos	[26]	Caatinga forest	120–170	—	—	270	—	[—]	—
San Carlos	[22, 26]	Tierra firme forest	120–170	—	5	177	59	[13]	13 (5200)

(continued)

Table 2 (*Continued*)

Location and Site[a]	Ref. no.[a]	Vegetation type[b]	Altitude[b] (m)	Edge[c]	Soil depth[d] (cm)	Seed density m^{-2} Mean[e,g]	CV	[N]	Number of species[f,g] (area)
San Carlos	[21]	Tierra firme forest	120–170	A	5	752	—	[17]	7 (6800)
Las Cruces	[41]	Páramo (*Espeletia* sp.)	4200	—	5	9–46 S	—	[60 × 2]	—
Surinam									
Mapane	[23]	Forest (*Cecropia* spp.)	<100	F	2	33–127 S	—	[5 × 6]	—
					2	73	—	[5]	—
					(20)	(152)	—	[5]	—
French Guiana									
St. Elie[v]	[20]	Evergreen forest	<50	—	15	60	—	[20]	—
St. Elie	[33]	Evergreen forest	<50	C	2	331	40	[16]	—
St. Elie	[33]	Evergreen forest	<50	EF	2	160–205 T	31–86	[16 × 4]	—
St. Elie	[33]	Evergreen forest, under the pioneer adult	<50	DEy	2	372–2008 R	8–41	[4 × 3]	—
St. Elie	[33]	Evergreen forest, away from pioneer adult	<50	DEz	2	156–500 R	24–78	[4 × 3]	—
Africa									
Ivory Coast									
Divo & Adio-podoumé	[13]	Wet forest (*Trema* sp.)	—	—	5	300	—	[—]	—
La Mamba	[2]	Forest (*Musanga* sp.)	—	C	5	124	—	[1]	—
Ghana									
Shai	[18]	Forest outlier in grassland	120	B	4	107	—	[2]	22[h] (20,000)
Akosombo	[18]	Southern marginal forest	340	8	4	384	—	[2]	38[h] (20,000)
Obdoben	[18]	Dry semideciduous forest	120	8	4	696	—	[2]	43[h] (20,000)
Kade	[18]	Moist semideciduous forest	180	D	4	623[m]	—	[2]	29[h] (20,000)

Kade [36]	Moist semideciduous forest	150	D	5	900[l]	22	[25]	—
Neung [18]	Wet evergreen forest	60	F	4	163	—	[2]	22[h] (20,000)
Atewa [18]	Upland evergreen forest	780	F	4	45	—	[2]	17[h] (20,000)
Nigeria								
Usonigbe FR [3]	Lowland forest	—	—	6	182[k,n]	—	[1]	11 (1871)
Omo FR [3]	Lowland forest	—	—	6	186[k,n]	—	[1]	12 (3871)
Uganda								
Queen Elizabeth NP [37]	Unburned savanna (*Acacia* sp.)	—	—	2 (4)	576 (821)	29 (24)	[10] [10]	— —
Queen Elizabeth NP [37]	Burned savanna (*Acacia* sp.)	—	—	2 (4)	520 (768)	12 (28)	[10] [10]	— —
Asia								
Thailand								
Doi Suthep Mt. [15]	Lower montane forest [yr 1]q	1350	C	5	243[j,n]	—	[1]	27 (7500)
Doi Suthep Mt. [15]	Lower montane forest [yr 2]q	1350	CD C	5 5 (25)	2–49 T 17 (48)	—	[6] [1]	2–9 (60,000) —
Doi Suthep Mt. [15]	Lower montane forest [yr 1]	1000	D	5	161[j,n]	—	[1]	22 (7500)
Kasetsart Univ. [15]	Dry dipterocarp forest [yr 1]	550	—	5	137[j,n]	—	[1]	16 (7500)
Malaysia								
Pasoh FR [39]	Lowland dipterocarp forest	—	Fx	10 (20)	131 (148)	35 —	[8] [6]	30 (5656) —
Pasoh FR [14]	Lowland dipterocarp forest	—	DE A	—	55	54	[3]	—
Sabah [9]	Lowland wet forest	—	—	15	62 58[m,o]	15 26	[2] [5]	29 (202,295)
Australia–New Guinea								
Papua New Guinea								
Mt. Susu NR [31]	*Araucaria* rain forest	900	D	8 (3)	1325 (525)	7 —	[4] [4]	18 (1600) —
Gogol Valley [43]	Lowland forest	<100	B	5	398[p]	—	[2]	—

(continued)

157

Table 2 (Continued)

Location and Site[a]	Ref. no.[a]	Vegetation type[b]	Edge[c]	Altitude[b] (m)	Soil depth[d] (cm)	Seed density m^{-2} Mean[e,g]	CV	[N]	Number of species[f,g] (area)
Australia Queensland									
	[24]	Evergreen forest (L,C)	—	100	5	588	41	[12]	64 (30,000)
	[24]	Evergreen forest (L)	—	60	5	519	78	[12]	64 (30,000)
	[24]	Semideciduous forest (L)	—	80	5	1069	42	[12]	79 (30,000)
					5	1120^l	—	[1]	—
					(50)	(3616^l)	—	[1]	—
	[24]	Evergreen forest (L)	—	250	5	594	41	[12]	60 (30,000)
					5	480^l	—	[1]	—
					(50)	(1040^l)	—	[1]	—
	[28]	Evergreen forest	—	<70	5	592	12	[4]	30 (5000)
	[29]	Semideciduous forest (L)	—	80	3	823	14	[8]	58 (20,000)

[a]References as in Section VIII, Appendix. FR, Forest reserve; NP, national park; NR, nature reserve; IS, island; NM, nature monument.

[b]Vegetation type and altitude are from references or inferred from general sources. All forests are mature and undisturbed by recent human activity, except several that were selectively logged >15 yr ago (L) or were cyclone damaged ~25 yr ago (C). For population studies, the species is given in parentheses.

[c]Distance of sample or forest from nearest gap, successional forest, or agricultural land: A, adjacent; B, surrounded (applies to ≤5-ha plots); C, ≤50 m; D, 100–200 m; E = 201–500 m; F = >>500 m; x, ≥20 m from treefall gap; y, directly under old pioneer tree (>20 m tall and >25 yr old); z, 30 m from old pioneer tree.

[d]Depth of sample from surface; parentheses denote depth of a deeper sample.

[e]Mean density (mean number m^{-2}) is based on seeds of all species germinated under high light conditions (unless otherwise indicated). Densities of deeper samples given in parentheses were calculated by adding densities from all depths; this was approximate where sample sizes differed between depths (e.g., Refs. 25 and 39 in the Appendix). Ranges represent variation among seasons (S), along transects (T), or among replicates (R). CV, Coefficient of variation (SD × 100/mean); N, number of samples per replicate; a second number indicates replication among seasons, positions along transects, etc.

[f]Number of species is the number in the total area sampled (number of replicates × surface area of sample). This is a minimum value, because some taxa represented groups of species that were hard to distinguish. Total area sampled (cm^2) is given in parentheses.

[g]Whenever ferns and contaminating greenhouse weeds have been separately enumerated in a reference, they have been excluded from densities, CV, and number of species.

158

[h]Number of species based on both sun and shade samples; densities based on sun-germinated samples only.

[i]Density and number of species based on both sun- and shade-germinated samples.

[j]Density was based on the assumption that values in Table 1 (see Ref. 15 in the Appendix) were total seeds in the 7500-cm² sample (not seeds m⁻²).

[k]There are typographical errors in Keay's (see Ref. 3 in the Appendix) Box II(a) table: the total number of *Discoglypremna* should be 2 (not 0), the number of *Fagara macrophylla* below it should be 2 (not 22), the total is reduced from 80 to 62 (excluding 27 contaminants), and density is 160 seedlings m⁻². Also, a 3 is entered twice for Box I(a).

[l]Mean and CV estimated from graph.

[m]Ferns excluded.

[n]Probable local or greenhouse contaminants excluded.

[o]May include some plants establishing from rhizomes.

[p]Reference received after statistical summaries in text were completed.

[q]Densities at the upper montane site were measured in two consecutive years using different sampling designs.

[r]Densities in interior of the same forest may be lower (306 seeds m⁻², Lieberman *et al.*, unpublished data, cited in [34]).

Table 3

Mean seed densities and number of species in soil seed banks of tropical secondary regrowth, highly disturbed vegetation, and agricultural land

Location and site[a]	Ref. no.[a]	Vegetation type[a] (disturbed area,[b] ha)	Altitude[a] (m)	History[c]	Age[d] (yr)	Edge[e]	Soil depth[a] (cm)	Seed density m^{-2} Mean[a]	CV	[N]	Number of species[a] (Area)
Neotropics											
Florida, USA											
Everglades NP	[16]	Farms/regrowth	<10	FA/C,CB	1–3	—	3	4253–29,974[k,l]	—	[3 × 3]	31–40 (7500)
	[16]	Woody regrowth	<10	—	17–35	—	3	328–1297[k,l]	—	[3 × 2]	27–35 (7500)
	[17]	Young regrowth	<10	—	—	—	—	3161	—	[—]	—
	[17]	Older regrowth	<10	—	—	—	—	1569	—	[—]	—
Puerto Rico											
El Verde	[6]	Clearing edge (0.03)	424	C	1	C	~12	310[f]	—	[2]	—
Mexico											
Los Tuxtlas	[7]	Regrowth	~100	C	<1	C	12	862–2672 S	37	[16 × 8]	4–11 (640)
	[7]	Regrowth	~100	CB,FA	7	C	12	1982–4051 S	25	[16 × 8]	4–12 (640)
Belize											
Central Farm	[11]	Pastures and fields	—	F	~4–20	—	4	6488	—	[78]	54 (2280)
	[11]	Cultivated fields	—	F	~4–20	—	4	7623	—	[~41]	—
	[11]	Pastures	—	F	~4–20	—	4	5226	—	[~37]	—
	[12]	Crop field	—	F	26	—	3 (10)	6760 (9800)	62	[20] [20]	23 (500)
	[12]	Pasture	—	F	>28	—	3 (10)	6760 (13,000)	21	[20] [20]	33 (500)
Costa Rica											
Florencia	[19]	Regrowth (0.33)[m]	650	CU	8–9[j]	—	4	7786[i]	63[f]	[10]	67 (1140)
Norte	[19]	Preburn slash (0.33)[m]	650	C	11 wk	—	4	6000[i]	29[f]	[10]	51 (1140)

160

[Ref]										
[19]	Postburn field (0.33)^m	650	CB	2 day	—	4	3000^i	68^f	[10]	37 (1140)
[38]	Regrowth	650	C	3	—	4 (20)	3340^f (6775)	— (51)	[12] [12]	— 40 (684)
[38]	Regrowth	650	CU	11	—	4 (20)	1889–3201 S (9516)	34–16 83	[12] [12]	— 47 (684)
[38]	Regrowth forest	650	—	75	C,D	4	1532 (6990)	(16) 47	[3] [3]	21 (171) 21 (660)
[32]	Regrowth forest	650	—	~70	~D	20 (40)	1812 (2205)	—	[6] [6]	— —
[34]	Regrowth	650	—	<15	C	5	18,900	60	[—]	—
Venezuela San Carlos										
[21]	Burned slash (0.25)	~120	CB	—	A	5	157	—	[20]	8 (8000)
[26]	Farm	~120	CB	2	—	5	1055	—	[15]	—
[22,26]	Farm (0.25)	~120	CB	3	A	5	581	109	[1]	9 (6000)
[22]	Farm, under fruit trees	~120	CB,FA	4	—	5	932^n	—	[1]	—
[22]	Farm, under slash	~120	CB,FA	4	—	5	126^n	—	[1]	—
[22]	Farm, bare soil	~120	CB,FA	4	—	5	74^n	—	[1]	—
[26]	Degraded pasture	~120	F	6	—	5	1250	44	[10]	16 (4000)
[26]	Cow droppings	~120	F	—	—	—	12,100	45	[8]	—
French Guiana, St. Elie										
[33]	Regrowth (1.0)	<50	CB^g	2	B	2	204–1552	38–72	[4 × 2]	—
[33]	Regrowth (0.16)	<50	C	4	B	2	196	56	[8]	—
[33]	Regrowth (25)	<50	CpB	6	E	2	228–2460	19–64	[4 × 4]	—
[33]	Regrowth (20-m strip)	<50	C	9	C	2	892	26	[8]	—
Africa Senegal, Fann, Dakar										
[4]	Crop fields	—	F	~3	—	—	5029–7183 (7982–14,084)^t	66–109	[6]	36–38 (60,000)

(continued)

161

Table 3 (Continued)

Location and site[a]	Ref. no.[a]	Vegetation type[a] (disturbed area,[b] ha)	Altitude[a] (m)	History[c]	Age[d] (yr)	Edge[e]	Soil depth[a] (cm)	Seed density m^{-2} Mean[a]	CV	[N]	Number of species[a] (Area)
Ghana, Kade											
	[36]	Regrowth	150	—	20	D	5	10,348	21[f]	[25]	—
	[36]	Burned forest	150	B	—	E	5	350[f]	20[f]	[25]	—
	[36]	Crop fields	150	F	1	E	5	4900[f]	57[f]	[25]	—
	[36]	Forest, open canopy	150	—	—	D,E	5	1650[f]	30	[25]	—
Nigeria, Calabas											
	[27]	Crop fields	—	—	—	—	4	2240[s]	—	[10]	27 (52,000)
Asia											
Thailand, Kasetsart Univ.											
	[15]	Wasteland	350	—	—	—	5	59[h,k]	—	[1]	11 (7500)
	[15]	Relict teak copse	350	—	—	—	5	65[h,l]	—	[1]	10 (7500)
Malaysia, Pasoh FR											
	[39]	Treefall gaps	—	—	2–5	B,C	10	64	48	[4]	~8 (2828)
	[14]	Regrowth (0.40)	—	C	4	B,C	—	48	67	[2]	—
Australia-New Guinea											
Papua New Guinea, Gogal Valley											
	[30]	Just clear-cut[p]	<100	CL	0	—	5	4100[r]	—	[5]	43 (12,500)
	[43]	Just clear-cut[p]	<100	CL	0	—	5	793[o,r]	—	[5]	—
	[43]	Regrowth[p]	<100	C	0.5–2.0	—	5	3848–9231[o]	—	[5]	—
	[43]	Regrowth, 2 sites	<100	C	1.5–2.5	—	5	4762–7824[o]	—	[2]	—
	[43]	Regrowth, 8 sites	<100	C	3.5–11	—	5	630–2526[o]	—	[2]	—
	[43]	Regrowth forest (<1)	<100	B	~55	B	5	757[o]	—	[20]	43[o,q] (50,000)
Australia, Queensland											
	[28]	Regrowth forest	<70	CB	>30	—	5	1302–1440	6–12	[4 × 2]	35–58 (5000)
	[28]	Regrowth forest	<70	CBm	>30	—	5	334	26	[4]	26 (5000)

[28]	*Eucalyptus* re-growth	<70	CBa	>50	—	5	244	30	[4]	21 (5000)
[28]	*Imperata* grass-lands	<70	CBa	>30	—	5	370	22	[4]	18 (5000)
[28]	Grass/shrub-land	<70	CBe	6	—	5	2060	26	[4]	31 (5000)
[28]	Pine plantation	<70	CBe	~20	—	5	3116	20	[4]	24 (5000)

[a]References are in the Appendix; headings and terms as in Table 2.

[b]Size of clearing, farm, or disturbed vegetation from which samples were collected.

[c]Disturbance history: F, farm under cultivation/pasture for ≥5 yr; FA, farm recently abandoned (≤1 yr); FO, farm abandoned ≥5 yr; C, cut once but not burned; CB, cut and burned once; CpB, cut and partially burned once; CL, cut and sampled during logging; CBm, cut and burned several to many times in past; CBa, cut then burned annually; CBe, cut and burned but fire excluded for >5 yr; CU, cut once, burning history not specified; B, burned once but not cut.

[d]Age of vegetation since time of last disturbance.

[e]Distance of disturbed site from forest: A, adjacent; B, surrounded (for plots ≤5 ha); C, ≤25 m; D, 26–50 m; E, 100–500 m; F, ≥500 m.

[f]Value estimated from graph or histogram.

[g]Cutting history from de Foresta *et al.* (1984).

[h]Density based on the assumption that the values in Table 1 (see Ref. 15 in Appendix) were seeds in 7500-cm² sample (not seeds m⁻²).

[i]Sequential samples at the same site.

[j]Includes patches of relict ~70-yr-old forest.

[k]Ferns excluded.

[l]Probable local or greenhouse contaminants excluded.

[m]Three replicates, each 0.33 ha.

[n]Seeds of woody species only.

[o]Reference received after statistical summaries in the text were completed.

[p]The same site was sampled five times in 2 yr; the discrepancy in mean density between Refs. 30 and 43 for the just cut site is not discussed in Ref. 43.

[q]Tree species only.

[r]Because germination had not yet begun in the clear-cut site, these values may represent predisturbance densities in old regrowth.

[s]Approximate mean density per sample, based on total number of seedlings (dicots, Ref. 27a; monocots, Ref. 27b) in 13 samples (0.40 m²) during the year.

[t]Estimated density based on subsamples monitored another 6 wk after the rainy season began.

depths sampled (to 50 cm; Tables 2 and 3). Within the same soil profile, some species are more abundant near the surface, others in deeper layers, while some show a nearly constant distribution over the sampled depths (Kellman, 1978; Young, 1985; Young *et al.*, 1987). Seed density in the litter/humus layer above the soil is extremely variable. Density, expressed as the percentage of the seed bank in the upper ≈3 cm of soil, was 22–33% in forests in Surinam and New Guinea (*Cecropia* only, Holthuijzen and Boerboom, 1982; Enright, 1985), 106% in a cornfield in Belize, but only 9% in a nearby pasture (Kellman, 1978). Differences in the depth profile of the seed bank within and between sites have been attributed to differences in forest soil texture (Hopkins and Graham, 1983), cultivation history of field and pastures (Kellman, 1978), and successional changes in seed rain (Young *et al.*, 1987), but differences within a given profile must reflect variation among species in seed rain, rates of incorporation, and/or seed longevity.

Whether deeply buried seeds contribute to regeneration depends on the depth from which they can germinate and the rate at which seeds in deep soil are brought to the surface. Ability of seeds to germinate generally decreases with depth and is influenced by soil type, seed size, and species (Juliano, 1940; Pandya and Baghela, 1973; Schwerzel, 1976; Schwerzel and Thomas, 1979; Shukla and Ramakrishanan, 1982; Bliss and Smith, 1985; Fatubarin, 1987). The processes that bury seeds can also bring them to the surface, but I doubt whether they account for much regeneration following gap creation except on the bare soil around uprooted trees (Putz, 1983; Riera, 1985).

C. Longevity of Seeds in Soil

Potential longevity of seeds in the soil is determined by the inherent viability of seeds and dormancy mechanisms that prevent germination under the canopy. Soil conditions, environmental changes that trigger germination, and predation by pathogens and animals reduce longevity.

1. Artificial Longevity

Studies based largely on tree species indicate that seeds of tropical species collectively have shorter artificial longevity (shelf life) than those of temperate species (de la Mensbruge, 1966; Moreno Casasola, 1976a, 1977; Vázquez-Yanes and Orozco Segovia, 1984; S. A. Foster, 1986). Shelf life of tropical seeds is not a good predictor of potential longevity in the soil. After 2 yr of storage, 81% of the seeds of *Pterocarpus rohrii* are still viable; however, fresh seeds germinate immediately on the soil surface, but rot if buried (Moreno Casasola, 1976b, 1977). The shelf life of *Piper* species is much higher than their longevity in soil (Orozco

Segovia, 1986). In addition, the shelf life of seeds contained in dried forest soil exceeds 3 yr (Castro Acuña and Guevara Sada, 1976), but this is not an estimate of seed longevity in the soil although frequently so cited.

2. Indirect Estimates of Longevity

Ng (1980) suggested that potential longevity in the soil might be estimated from extensive germination studies by graphing the number of species with viable but ungerminated seeds against time since planting. For Malaysian woody species, number of species decreases exponentially; after 6 wk, only half have viable seeds (Ng, 1980). Unfortunately, Ng's (1980) viability curve underestimates seed longevity of pioneer species in forest shade, because seeds were sown in light shade just below the soil surface, a condition that promotes germination of pioneer species (Bell, 1970; Hall and Swaine, 1980; Aminuddin and Ng, 1982). Thus, these data cannot indicate whether pioneer species persist in the soil longer than more shade-tolerant species (e.g., S. A. Foster, 1986). Because Ng's graph is often reproduced (Ng, 1983; Vázquez-Yanes and Orozco Segovia, 1984; Whitmore, 1984), it must be interpreted carefully.

3. Direct Estimates of Longevity

Burying seeds in the soil directly measures longevity (Table 2). These are biased estimates of longevity of naturally dispersed seeds because burial of large numbers of seeds in small containers may exclude many seed predators, change the microenvironment, and affect pathogen infection (Vázquez-Yanes and Smith, 1982).

In the most extensive study of tropical seed longevity (Hopkins and Graham, 1987), seeds of primary species had much shorter longevity than those of pioneer and late secondary species (Table 4). This confirmed a long-held but little-tested notion based on artificial longevity of commercial species. Seeds of most primary species and a few late secondary species germinated as readily when deeply buried as they did when on the surface, indicating short longevity and lack of facultative dormancy; in contrast, buried seeds of most pioneer and late secondary species remained viable and dormant (Hopkins and Graham, 1987). There appears to be more variation in longevity among successional species at other sites compared to Australia (Table 4). Agricultural weeds are a very diverse group (Table 4), making weed eradication by hand cultivation difficult. At a well-studied site, Los Tuxtlas, there is also considerable unexplained variation within species (e.g., *Cecropia obtusifolia, Piper auritum,* and *Piper hispidum*), perhaps arising from environmental heterogeneity or differences in methods.

Some variability might be explained by depth of burial (Table 4).

Table 4
Viability of seeds of tropical successional and primary species buried in the soil

Location	Methods of burial[a] and testing[b]	Depth of burial[c] (cm)	Species[d]	Length of burial[e] (yr)	Relative percentage of viable seeds remaining in the soil after burial[f]		References[g]
					Successional species (N)	Primary species (N)	
Australia (Queensland)	M; IG	50	Rainforest trees, shrubs, and climbers	0.1	79 (26)[u]	32 (16)[u]	Hopkins and Graham (1987)
				0.5	68 (31)[u]	11 (15)[u]	Hopkins and Graham (1987)
				1.0	63 (31)[u]	2 (16)[u]	Hopkins and Graham (1987)
				2.0	60 (27)[u]	0 (11)[u]	Hopkins and Graham (1987)
Mexico (Los Tuxtlas)	W; G	10–20	Pterocarpus rohrii	0.1	—	0 (1)[i,k]	Moreno Casasola (1976b)
	W; G	10–20	Nectandra ambigens	0.4	—	3 (1)[i,m]	Moreno Casasola (1976b)
	W; G	10–20	Calophyllum brasiliense	0.4	—	70 (1)[i]	Moreno Casasola (1976b, 1977)
	W; G	10–20	Chamaedorea tepijolete	0.4	—	73 (1)[i]	Moreno Casasola (1976b, 1977)
	M; G	5	Heliocarpus donnell-smithii	1.0	70 (1)	—	Vázquez-Yanes and Orozco Segovia (1982a)
	M; G	5	Cecropia obtusifolia	1.0	12–82 (1)[n]	—	Vázquez-Yanes and Smith (1982)
	W; G	—	Cecropia obtusifolia	0.6	<5 (1)[s]	—	Martínez-Ramos and Alvarez-Buylla (1986)
	M; G	5	Piper hispidum	1.0	2–22 (1)[n]	—	Vázquez-Yanes and Orozco-Segovia (1982b)
	M; G	5	Piper hispidum	1.0/1.9	32/3 (1)	—	Pérez-Nasser (1985)
	M; G	5	Piper aff hispidum	0.5/1.0	71/0 (1)	—	Orozco Segovia (1986)

Location		No.	Species						Reference
	M; G	5	*Piper auritum*	1.0	74–95		(1)[n]	—	Vázquez-Yanes and Smith (1982)
	M; G	5	*Piper auritum*	0.5/1.0	11/4		(1)		Orozco Segovia (1986)
	M; G	5	*Piper aequale*	1.0/1.5	49/15		(1)		Orozco Segovia (1986)
	M; G	5	*Piper umbellatum*	1.0/1.5	41/31		(1)		Orozco Segovia (1986)
	M; G	5	*Trema micrantha*	1.0/1.9	112/56		(1)		Pérez-Nasser (1985)[r]
Mexico	W; GT	~5	*Calathea ovandensis* In gap	1.0	24		(1)[j,m]	—	C.C. Horvitz and D.W. Schemske (unpublished data)
			In shaded understory	1.0	83		(1)[j]	—	C.C. Horvitz and D.W. Schemske (unpublished data)
Mexico	—	—[t]	*Cordia elaegnoides*	0.4	1		(1)[s]	—	Sarukhán (1978)
Puerto Rico	T; G	—	*Palicourea riparia*	3.0	+		(1)	—	Lebrón (1979)
Costa Rica	MV; G	15	*Phytolacca rivinoides*, *Witheringia* spp.	2.2	100		(3)		K.G. Murray (1986, 1988)
Venezuela	M	2–3	*Espeletia timotensis*[v] (Andean páramo)	1.0	—	55	(1)		Guariguata and Azocar (1988)
Surinam	P; G	25	*Cecropia sciadophylla*	4.5	100		(1)	—	Holthuijzen and Boerboom (1982)
			Cecropia obtusa	3.3	96		(1)	—	Holthuijzen and Boerboom (1982)
Malaysia	T; G	—	*Pinus caribaea*	0.2	0		(1)[p]	—	Aminuddin and Ng (1982)
	T; G	—	*Gmelina arborea*	0.5	37		(1)[p]	—	Aminuddin and Ng (1982)
	T; G	—	*Vitex pinnata*	0.5	58		(1)[p]	—	Aminuddin and Ng (1982)
	T; G	—	*Sapium baccatum*	0.5	93		(1)[p]	—	Aminuddin and Ng (1982)
	—		*Macaranga hypoleuca*	0.1/1.0	50/0		(1)		Taylor (1982)[q]
Venezuela	C; G	+50	Herbs, shrubs, and trees	1.0	≥25		(8)[j]	—	Uhl and Clark (1983)

(continued)

167

Table 4 (Continued)

Location	Methods of burial[a] and testing[b]	Depth of burial[c] (cm)	Species[d]	Length of burial[e] (yr)	Relative percentage of viable seeds remaining in the soil after burial[f]		References[g]
					Successional species (N)	Primary species (N)	
Nigeria	M; GI	5	Weedy agricultural herbs	1.2	0 (10)	—	Marks and Nwachuku (1986)
				1.2	≥50 (5)	—	Marks and Nwachuku (1986)
Zimbabwe	M; G	3–30	Weedy agricultural herbs	5.0	2–10 (3)[h]	—	Schwerzel (1976)
Philippines	G; G	15	Weedy agricultural herbs	≤1.0	0 (3)	—	Juliano (1940)
				2.0–4.0	0 (9)	—	Juliano (1940)
				6.0–6.5	4–139 (11)	—	Juliano (1940)

[a]Seeds were buried as follows: G, open glass vials filled with sand; P, unglazed clay pots with lids, filled with forest soil; C, aluminum cans with drainage holes, filled with sterilized soil, covered with mesh; W, cages of 3 × 3-mm mesh filled with soil; M, nylon bags of fine mesh filled with soil or seeds only; V, mesh-covered vials; and T, soil-filled trays.

[b]Methods of testing viability included G, germination; T, tetrazolium test; and I, insepction (seeds not decomposed or rotten were considered viable but dormant).

[c]Depth that containers were buried in the forest soil; a plus (+) indicates that containers were placed above soil.

[d]All are species found in mature forest and forest gaps except where noted.

[e]Length of time seeds were buried in containers in the soil. In most studies, several containers were buried and retrieved over a 1- to 2-yr period. Most studies had only one replicate for each time period, making it difficult to separate spatial and temporal variation in loss of viability.

[f]The percentage of viable, ungerminated seeds in the sample after burial for the specified time divided by the initial percentage of viable seeds in the sample before burial; N, number of species in each group; +, viable seeds present but percentage unknown. Seeds that germinated during burial were excluded from the remaining viable fraction. (When initial viability is measured by germination and a large proportion of the seeds are initially dormant, the relative percentage of viable seeds can be >100% after seeds break dormancy). Species in each study were grouped by maximum relative percentage of viable seed over the study period, except in Hopkins and Graham (1987), where means of species examined are given. Successional and primary species groups are as defined in the text.

[g]Other studies note or infer that pioneer seeds remain viable for several years, but give no quantitative data (e.g., Aikman, 1955, for Ochroma lagopus).

[h]Mean of six depths.

[i] Percentage germination from seeds planted on soil surface at time of burial was used as initial viability.

[j] Final percentage germination or viability; no initial values given.

[k] Most seeds decomposed while buried in the soil.

[m] Most seeds germinated while buried in the soil.

[n] Because of the high nondirectional variability, the range over 330–390 days is given. Maximum percentage germination of seeds imbibed in the dark then germinated in light was used as initial viability. All values estimated from graphs.

[p] Maximum percentage germination of seeds planted in full sun or light shade was used as initial viability. All values estimated from graphs.

[q] Cited in Putz and Appanah (1987).

[r] Cited in Vázquez-Yanes and Orozco-Segovia (1987).

[s] Most seeds were destroyed by predators or pathogens.

[t] The two depths used were not reported in this review.

[u] Mean of all species.

[v] The dominant species in Andean páramo.

Longevity of agricultural weeds buried very near the surface is short because seeds germinate (Juliano, 1940; Pandya and Baghela, 1973; Schwerzel, 1976). Away from the surface, there is no overall correlation between depth and survival of agricultural weeds from the Philippines and Zimbabwe, although some species have higher survival at particular depths (Juliano, 1940; Schwerzel, 1976). Cultivation brings facultatively dormant seeds nearer the surface, triggering germination (Schwerzel and Thomas, 1979).

4. Causes of Reduced Longevity

Longevity in the soil is probably not limited by seed energy reserves. Germinability and viability of imbibed seeds stored in the dark >1 yr remained high in four *Piper* and seven pioneer species of other genera (Vázquez-Yanes, 1976b; Vázquez-Yanes and Smith, 1982; Orozco Segovia, 1986). Large dead seeds often have ample reserves remaining (Baker, Chapter 2, this volume).

Postdispersal seed predation and pathogen attack may reduce longevity. Remains of seed coats implicate predators or pathogens in the rapid disappearance of buried seeds of *Cecropia obtusifolia* (Martínez-Ramos and Alvarez-Buylla, 1986). Rodents remove 99% of buried *Cordia elaeagnoides* seeds (Sarukhán, 1978). In contrast, burial of two large-seeded late secondary species, *Hymenaea courbaril* and *Gustavia superba*, reduces seed predation (Hallwachs, 1986; Sork, 1987). It should be noted, however, that most tropical species studied are relatively large-seeded primary and late secondary species (see Augspurger, 1983, 1984a,b; Howe and Smallwood, 1982; Clark and Clark, 1987; Schupp, 1988) and are not expected to have extended longevity in the soil.

Environmental factors not associated with gap formation, such as unpredictable rainfall during the dry season, also cause seed death. Under experimental conditions, seed survival of three seasonally dormant Panamanian species varied with species and depended on the length, continuity, and position of dry and wet periods (Fig. 1). Species with short-term seed viability are also sensitive to variable rainfall (Augspurger, 1979). Greater temperature fluctuations and lower moisture levels in savanna inhibit germination of some forest species (Ponce de León, 1982).

Fires in savanna and adjacent forest and in slash-and-burn agriculture fields kill seeds of many species, although seeds of a few species are unaffected or stimulated to germinate by the high temperatures (Vázquez-Yanes, 1974, 1976a; López-Quiles and Vázquez-Yanes, 1976; Alexandre, 1978; Ewel *et al.*, 1981; Ponce de León, 1982; Gillon, 1983; Uhl and Clark, 1983; Hopkins and Graham, 1984a; Sabiiti and Wein, 1987). The effect of fire decreases with depth of seed burial (Brinkmann and Vieira, 1971).

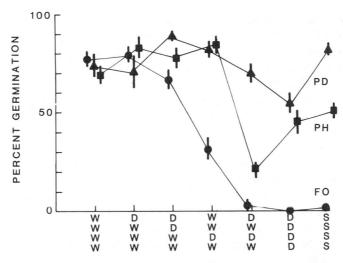

Figure 1. Germination (%; $\bar{x} \pm$ SE) of seeds of three Panamanian species with seasonal dormancy using seven experimental dry season watering regimes. Each 4-month-long watering regime is described by the sequence of the following codes. W, Watered daily for 1 month; D, not watered for 1 month; and S, watered only on the first 3 days of each month. Seeds were watered daily for an additional 5 months. Total percentage germination after 9 months, scored weekly, is given. Species means are connected for clarity, although treatment is not a continuous variable. Species, all in the Rubiaceae, were *Faramea occidentalis* (FO; 12 replicates, 5 seeds each), *Psychotria deflexa* (PD; 6 replicates, 35 seeds each), and *Psychotria horizontalis* (PH; 6 replicates, 24 seeds each). (N. C. Garwood, unpublished data.)

D. Fate of Seeds during Disturbance

Hopkins and Graham (1984b) followed the fate of seeds in forest soil placed in small to large gaps and in the shaded forest understory. Over a 14-wk period, 24–33% of the seed bank germinated in medium ($\approx$175 m^2) and large (40,000 m^2) gaps and 1% in a small gap (30 m^2) and the understory. More seeds remained dormant in the understory and small gap (32–44%) than in the medium and large gaps (3–9%). However, the process of disturbing the soil caused high mortality of seeds at all sites (56–71%). Other studies also note that soil disturbance (apart from changes in light and temperature) stimulates germination (Bell, 1970; Liew, 1973; Alexandre, 1978). Shallowly buried seeds of the late secondary herb *Calathea ovandensis* also germinate in gaps but most remain dormant in deep and intermediate shade (Table 4). Whether seed mortality is high in undisturbed soil following gap creation must yet be tested.

The enormous environmental heterogeneity within gaps should affect the fate of seeds. In treefall gaps, the pits and mounds of the root zone are clear of roots, litter, and shade, while the crown area is usually covered by leaf litter and branches (Oldeman, 1978; Denslow, 1980; Hartshorn, 1980; Orians, 1982; Putz, 1983; Riera, 1985). Seedlings of pioneer species are relatively more abundant on bare soil than elsewhere in the gap (Putz, 1983; Riera, 1985; Lawton and Putz, 1988; Murray, 1988); seeds germinate better on bare mounds than under gap litter (Riera, 1985). Under experimental gap conditions, deep litter lowers survival of small-seeded pioneers, has no inhibitory effect on large-seeded pioneers, but enhances germination of very large-seeded shade-tolerant species (Molofsky and Augspurger, 1988). Burial enhances germination in a large-seeded late secondary species in gaps but not in the understory, but only in young, rather than old, forests (Sork, 1985). Schulz (1960a,b) also notes that covering seeds with litter or soil enhances germination in 30 canopy species.

IV. Tropical Soil Seed Bank Strategies

Potentially more soil seed bank strategies exist in the tropics than in temperate zones (Thompson and Grime, 1979) because reproduction occurs throughout the year. I describe five basic seed bank strategies and several variations, based on germination behaviors and temporal patterns of seed dispersal. Their distribution among the four regeneration strategies is summarized in Table 1.

Transient seed banks are composed of short-lived, nondormant seeds that are dispersed for short periods during the year (Fig. 2A). They may occur in any of the regeneration strategies (Table 1). In some primary species, long-lived seedling banks are formed (Fig. 2B; Table 1). **Persistent** seed banks, expected in weedy and short-lived pioneer species (Table 1), are composed of long-lived seeds having facultative dormancy that are dispersed for short or long periods (Fig. 2C). Seasonal dormancy can be imposed on otherwise persistent seed banks (Fig. 2D). **Pseudo-persistent** seed banks, also expected in weedy and short-lived pioneer species (Table 1), are composed of short-lived, nondormant seeds that are dispersed continuously throughout the year. If dispersal is frequent but not continuous, seed bank size will fluctuate (Fig. 2E). **Seasonal-transient** seed banks are composed of seasonally dormant seeds with intermediate longevity that are dispersed for short or long periods (Fig. 2F). They may occur in any regeneration strategy (Table 1). **Delayed-transient** seed banks, expected primarily in late secondary or primary species (Table 1), are composed of seeds with delayed germination (often asynchronous) not associated with seasonally adverse condi-

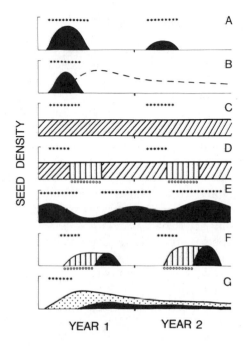

Figure 2. Tropical soil seed bank strategies. (A) Transient seed bank. (B) Transient seed bank replaced by seedling bank. (C) Persistent seed bank. (D) Persistent seed bank with period of seasonal dormancy. (E) Pseudo-persistent seed bank of fluctuating size. (F) Seasonal-transient seed bank. (G) Delayed-transient seed bank. Fruiting periods are denoted by asterisks, dry seasons by small open circles, seedling banks by dashes, germinable seeds without dormancy that must germinate or die by black areas, seeds with seasonal dormancy by vertical lines, seeds with facultative dormancy under forest canopy by slanted lines, and seeds with delayed germination by stippling.

tions (Fig. 2G). Some seeds may remain in the seed bank for as long as 1–2 yr.

Because the community soil seed bank is composed of the seed banks of individual species, all seed bank strategies may co-occur in the soil. One strategy may predominate, but dominance may change seasonally. To evaluate the role of the persistent seed bank in regeneration, it is necessary to identify what fraction of the seeds or species in the soil seed bank are in the persistent seed bank, rather than the various transient seed banks. At least two of the following criteria must be known to identify seed bank strategies: seasonal changes in the seed bank, seasonal timing of dispersal, or the presence of seasonal or facultative dormancy. Most studies document only one criterion.

1. Seasonal-Transient Seed Banks

Although a few equatorial rain forests are truly aseasonal, most tropical regions show seasonality in rainfall (Whitmore, 1984; Walter, 1985) and the timing of seed dispersal (Howe and Smallwood, 1982; Sabatier, 1985). Seasonal-transient seed banks predominate in agricultural fields in seasonally dry West African climates. In Senegal, Ghana, and Nigeria, the most abundant weedy species are dispersed primarily during the rainy season. Dormancy in all species is enforced or innate during the dry season, with germinability of seeds peaking at the beginning of the rainy season (Miège and Tchoumé, 1963; J. Jenik, unpublished data, cited in Longman, 1969; M. K. Marks, 1983a,b).

Seasonal-transient seed banks are also found in up to 60% of the species in a seasonally dry Panamanian forest (Garwood, 1983). Approximately 20% of all species studied are dispersed late in the rainy season and are seasonally dormant through the 4-month-long dry season (Garwood, 1983). Another 40% are dispersed during the dry season, but do not germinate until the beginning of the rainy season. Several pioneer species with abundant seeds in the soil (Putz, 1983) are also dormant through the dry season (Brokaw, 1986), suggesting seasonally dormant persistent seed banks (Fig. 2D). Ewel et al. (1979) report large seasonal-transient seed banks in subtropical Florida. The phenology of many savanna species indicates that seasonal-transient seed banks are not uncommon (Sarmiento and Monasterio, 1983). In wet regions that lack pronounced dry seasons, seasonal-transient seed banks may occur in drier habitats or drier periods. At a wet, relatively aseasonal Venezuelan site, little germination occurs during the drier season on exposed slash-and-burn farmland (Uhl et al., 1981). In contrast, Aiyar (1932a,b) reports that seeds are seasonally dormant through the wettest season in a very wet Indian forest.

If seasonal dormancy is under strong selective pressure, widespread species should exhibit geographical variation in dormancy. Variation in seasonal dormancy in the African savanna grass *Digitaria milanjiana* is correlated with rainfall (Hacker, 1984). Ecotypes from low-rainfall regions have dormant seeds, but those from high-rainfall regions do not.

2. Persistent Seed Banks

Many of the seeds in the soil seed bank appear to be facultatively dormant because more seedlings and species germinate from samples put in sunny than in shady locations (Keay, 1960; Bell, 1970; Hall and Swaine, 1980; C. C. Horvitz and D. W. Schemske, personal communication). However, no studies have simultaneously documented two of the three criteria stated above needed to distinguish persistent from various transient seed banks.

The existence of a persistent seed bank can be inferred if the average

Table 5

Residence times of seeds in the soil and comparison of size of soil seed bank with seed rain onto the soil

Method[a]	Ref. no.[b]	Location	Vegetation[b] (age of regrowth)	Comparison	Soil depth[c] (cm)	Seed density m^{-2}[d] [Mean ± SD (N)]		Sampling interval[e] (months)	Residence time[f] (years)
						Bank + rain	Rain		
A	[39]	Malaysia	Mature forest	All species	10°	34 ± 14 (17)	5 ± 3 (17)*	9	4.3
A	[40]	Costa Rica	Mature forest	All species	10°	153 ± 33 (8)	21 ± 8 (8)*	12	6.1
B	[21]	Venezuela	Farm, post-burn (0 months)	Woody plants	—^{p}	1.0 (6)	0.05 (6)*	4	6.3
				Grasses	—^{p}	0.5 (6)	1.50 (6)^{ns}	4	0.0^i
						Bank	Rain + bank		
C	[38]	Costa Rica	Clearing (0 months)	All species	—^{n}	1021 ± 617 (6)	912 ± 529 (6)*	2	Long
D	[20]	French Guiana	Mature forest	All species	15	67 (10)	54 (10)	5	Long
						Bank	Rain		
E	[26]	Venezuela	Mature forest	All species	5	180 (13)	50 (10)	12	3.6
				Secondary species	5	>118 (13)	8 (10)	12	>14.8
				Primary species	5	<62 (13)	42 (10)	12	<1.5
				Cecropia sp.	5	80–126^h (13)	5 (10)	12	16.0–25.2^h
E	[38]	Costa Rica	Forest (75 yr)	All species	4	1532 (12)	1185/490^i (6)	12	1.3/3.1^i
				All species	20	6990 (12)	1185/490^i (6)	12	5.6/13.6^i
			Regrowth (1–3 yr)	All species	4	3340 (12)	3194/3215^i (6)	12	1.0/1.0^i
				All species	20	6755 (12)	3194/3215^i (6)	12	2.1/2.1^i
E	[17]	Florida^q	Young regrowth	All species	—	3161 (—)	788 (—)	1^m	0.3

(continued)

Table 5 (Continued)

Method[a]	Ref. no.[b]	Location	Vegetation[b] (age of regrowth)	Comparison	Soil depth[c] (cm)	Seed density m^{-2d} [Mean ± SD (N)]		Sampling interval[e] (months)	Residence time[f] (years)
F	[15]	Thailand	Old regrowth	All species	—	1569 (—)	100 (—)	1^m	1.3
			Mature[g]	All species	—	128 (—)	80 (—)	1^m	0.1
			Transect from clearing into mature forest[f]	Six pioneer species	5	0–42 (5)*	0–7 (5)ns	7	0.0–24.6*
				Castanopsis	5	0 (5)	104–1674 (5)*	7	0.0
				Other species	5	2–9 (5)ns	24–70 (5)*	7	0.04–0.16ns
F	[43]	Papua New Guinea	Young re-growth (2 yr)	Total	5	6207 (5)	2120/337^k (5)	22/12^k	5.4/18.4^k
				Trees	5	525 (5)	163/147^k (5)	22/12^k	5.9/3.6^k
				Forbs	5	5426 (5)	1794/112^k (5)	22/12^k	5.6/48.4^k
				Graminoids	5	244 (5)	88/28^k (5)	22/12^k	5.1/8.6^k
				Climbers	5	12 (5)	74/49^k (5)	22/12^k	0.3/0.2^k
			Old regrowth (~55 yr)	Trees	5	630 (20)	395 (5)	22	2.9

[a]Six methods (A–F) were used to compare size of the soil seed bank with new input of seeds into the soil: A, paired in situ plots of sterilized and nonsterilized forest soil; B, paired in situ plots of sterilized and nonsterilized soil; C, paired in situ plots covered with fine mesh, seeds trapped on top were placed in cage or thrown away; D, paired soil samples, covered or not covered with fine mesh; E, seed traps of sterilized forest soil and an independent estimate of seed bank; F, seed trap containers and an independent estimate of seed bank. See text for discussion.

[b]References as in the Appendix. Vegetation types as in Tables 2 and 3.

[c]Depth of soil sample from surface.

[d]Mean seed density is used to compare the size of the soil seed bank and seed rain during the specified time interval. If the number of seeds in the seed bank exceeds the number input from seed rain during a 12-month period, then seeds are accumulating in the soil. Depending on the method used, the soil seed bank (Bank) and seed rain (Rain) were compared directly or compared to a combined measure of seed rain and seed bank (Bank + Rain). For methods A, B, and C, measures of the seed bank and seed rain were significantly different (*, $p < 0.05$) or not significantly (ns) different; for method F, significance of regression on distance along transect is given; N, number of samples.

[e]The period during which seed rain was measured.

176

fResidence time (seed density in soil/annual seed rain) is an estimate of the average time seeds reside in the soil seed bank (see text for discussion of assumptions). For methods A, B, E, and F, annual seed rain was calculated by dividing mean seed rain by the specified time interval. For methods A and B, annual seed rain was subtracted from Bank + rain to determine soil seed bank density. In methods E and F, where Bank > Rain + bank, annual seed rain was assumed to be very low and residence times long.

gIncludes hammocks, pinelands, and freshwater glades.

hThe data were contradictory (see *Cecropia ficifolia*; pp. 420 and 422, in Ref. 26).

iZero used instead of negative turnover time, when Bank < Rain.

jSeed rain was estimated from 21 1-month-long sampling periods over 3 yr; mean seed rain per year is given. The first number is mean seed rain per year or residence time for the entire 3-yr period. The second number is mean seed rain or residence time for the last (third) year only.

kSeed rain was measured in 11 2-month intervals. The first number is total seed rain or residence time for the entire 22-month interval; the second number is seed rain or residence time for the last 12 months only. Seed bank size was measured at the end of the 22-month period.

lTransect 25–175 m from clearing; range of densities along transect are given; regressions of density and residence time with distance from clearing were significant (*) or not (ns), my calculations; there was one replicate at each location.

mAn unspecified number of 1-month periods during the year were used; monthly seed rain averaged over the year was given.

n*In situ* measurement; depth indeterminate.

oDepth of soil placed in trays in forest clearings.

pSoil sterilized *in situ*.

qSubtropical.

residence time of seeds in the soil (soil seed bank density ÷ annual seed rain) is very long. Average residence times are long (>2 yr) in most mature forests and some young regrowth/farm sites (Table 5). Residence times of pioneer species (>10 yr) are much longer than those of primary species (<1.5 yr) in Venezuelan and Thai forests (Table 5, Refs. 15 and 26). For pioneer species in the Thai forest, size of the seed bank and average residence time increase with distance from a clearing filled with pioneer species, while the seed rain shows no change (Table 5). Presumably, higher light levels close to the clearing triggered germination and depleted the seed bank, but these differences might reflect temporal changes in seed rain. In contrast, residence times were short in subtropical vegetation in Florida and weedy grasses in slash-and-burn farm sites (Table 5, Refs. 17 and 21).

Simple calculations of residence time are very sensitive to sampling methods and departures from several assumptions (e.g., single determinations of seed bank and short-term estimates of seed rain accurately reflect year-long patterns, sites are nonsuccessional such that the seed bank and seed rain are in equilibrium, and the seed bank and seed rain are spatially and temporally homogeneous). In Costa Rican young regrowth, where seed rain was nearly constant over 3 yr (Table 5, Ref. 38), variation in estimated residence time (1.0–2.1 yr) reflects differences in depth of soil sampled; in forest, where seed rain decreased substantially over 3 yr, variation in estimated residence times (1.3–13.6 yr) reflects differences in both depth and seed rain (Table 5). Two years after forest clearing in Papua New Guinea, residence times of herbs and trees appear similar if seed rain from the 2-yr period is used to calculate residence time (Table 3, Ref. 43); however, because seed rain of herbs decreased and that of trees increased during succession, residence times of herbs are 13 times greater than those of trees if seed rain from the previous year only is considered (Table 3, Ref. 43). If primary species are included in the seed rain, residence times of seeds in the soil (primarily pioneers) will be greatly underestimated (e.g., residence time for all species in a Venezuelan forest is 3.6 yr but that of secondary species is >14.8 yr; Table 5, Ref. 26). In some studies (Table 5), it was not clear whether seed rain was measured before or after the soil samples were collected or at the same location. Although not without problems, these estimates of residence times strongly suggest that persistent seed banks occur under mature tropical forest.

V. Changes in Tropical Soil Seed Banks during Succession

A variety of disturbances initiate succession in the tropics. Falling branches or trees, clumps of trees felled by hurricanes, or forest patches lost during landslides or during wildfires create gaps of increasing size and severity but with decreasing frequency (Whitmore, 1978; Garwood *et al.*, 1979; R. B. Foster, 1980; Brokaw, 1985a; Leighton and Wirawan, 1986). Shifting agriculture, selective or clearcut logging, road building, and plantations are generally more severe disturbances than are natural gaps of the same size; their increasing frequency is fast changing the tropical landscape (Gómez-Pompa and Vázquez-Yanes, 1974).

A. Role of Persistent Seed Banks in Regeneration

Short-lived pioneer species appear rapidly after disturbance, often in <2 wk (Whitmore, 1983). Some argue that seeds, particularly of animal-dispersed pioneers, cannot be dispersed fast enough and in sufficient numbers to account for this dense rapid growth (e.g., Budowski, 1961). Others argue that, because gaps are often dominated by a single species, species dispersing seeds at the time of gap creation dominate regeneration (see Whitmore, 1978, 1983). The argument cannot be resolved at this level. Pioneers species can produce prodigious numbers of seeds over short periods, but dominance could be caused by a single species that dominates the persistent seed bank. If the soil seed bank is relatively unimportant, then the timing of seed dispersal will in large part determine the species composition of regeneration initiated at different times of the year (Hartshorn, 1978, 1980).

1. Origin of Seeds during Regeneration

The origin of seeds can be assessed by comparing seed germination in sterilized forest soil with that in unsterilized forest soil; seeds originate from seed rain only in the former and from both seed rain and the seed bank in the latter. Seeds sown in steam-sterilized soil germinated normally (Williams-Linera and Ewel, 1984; Putz and Appanah, 1987), but growth within 2 months was affected in eight of nine species. Different growth rates could cause differential mortality after many months, as could differences in initial seed density or the absence of mycorrhizae or pathogens in sterile soil. Thus, the power of this method depends on early and frequent censuses of germination.

In lowland Malaysian dipterocarp and Costa Rican elfin forests, ≈85% of regeneration came from the seed bank, but seed rain was important for several species (Table 5, Refs. 39 and 40). Similarly, in a slash-and-burn farm site, 95% of the woody seedlings orginated from the seed bank, but grasses, dispersed into the site after clearing (Table 5, Ref. 21). Unfortunately, these estimates are based on late and infrequent censuses (4–12 months).

The origin of seed during regeneration can be also assessed by excluding seed rain with fine mesh. However, mesh may reduce light intensity, increase soil moisture, and alter behaviors of dispersers, and may not exclude all seeds [e.g., of 45 small-seeded Panamanian species, including Melastomataceae (13 species), *Piper* (11), *Ficus* (12), and *Cecropia* (1), 78% of the species passed through 1.5-mm^2 mesh, 58% passed through ≈0.9-mm^2 mesh, and 16–18% passed through two different ≈0.1-mm^2 meshes; no *Piper* or *Ficus* but about half of the Melastomataceae passed through the two finer meshes (N. C. Garwood, unpublished data)].

In 75-yr-old Costa Rican forest, seed rain was unimportant in a small clearing without stems and litter during a 2-month period (Table 5, Ref. 38). (Controls for secondary effects of mesh were used, although dispersers might have avoided the bare clearing and cages.) Seed rain was also apparently unimportant in a French Guianan forest (Table 5, Ref. 20), but effects of mesh covers (≈1 mm^2) were not tested and seed rain was not independently measured. Melastomataceae, comprising ≈20% of all seedlings, appeared later than common larger seeded species (Prévost, 1981); they represented >50% of the contaminants in a later study (de Foresta and Prévost, 1986).

In forests of Venezuela, Costa Rica, Papua New Guinea, and Thailand, where there are fewer pioneer seeds in the current annual seed rain than in the upper 4–5 cm of the soil (Table 5, Refs. 15, 26, 38 and 43), seed rain should also account for only a small portion of regeneration if a disturbance occurred. This is also true for *Muntingia calabura* in a Costa Rican dry forest (Fleming *et al.*, 1985, and unpublished data of D. Thomas, cited therein). Conversely, seed rain should be more important in young Costa Rican regrowth and subtropical Florida vegetation because the number of seeds in the annual seed rain is equal to or greater than that in the soil (Table 5, Refs. 17 and 38). Similarly, after removal of shade trees from a Mexican coffee plantation, the slowly developing weed community originates from seed rain because these species are not in the seed soil bank (Goldberg and Kigel, 1986).

In all studies where origin of seed was determined directly, the seed bank accounted for most regeneration following disturbance. Seed rain appears to be important for some pioneer and weedy species, for primary species, and in subtropical vegetation. A variety of methodological problems, however, must be solved before broad generalizations can be made.

2. Population Models Evaluating Role of Soil Seed Banks

Demographic studies of tropical plants often use late secondary or primary species that have rapid germination (Sarukhán, 1980) and transient seed banks. Seed bank size has been measured in population studies of only a few pioneers (Aubréville, 1947; Alexandre, 1978; Holthuijzen and

Boerboom, 1982) and a dominant species in the Andean páramo (Guari-guata and Azocar, 1988). Two recent demographic studies of tropical species use models to assess the changing importance of seed banks in regeneration (Horvitz and Schemske, 1986; K. G. Murray, 1986, 1988).

Horvitz and Schemske (1986) combine models of the dynamics of successional forest patches with Lefkovitch stage-projection models of population growth to estimate mean population fitness of an understory herb. They evaluate three demographic life histories, ranging from high-ly shade intolerant to relatively shade tolerant, two disturbance regimes, and the effects of a seed bank on demographic response following tree-falls. The presence of dormant seeds in the soil is beneficial for all demographic models, but especially for relatively shade-intolerant pioneers. Population fitness increases as the sensitivity of the mechanism controlling facultative dormancy increases. The rate of forest disturbance is least important. Although their study species, *Calathea ovandensis*, best fits the relatively shade-tolerant demographic life history in which a seed bank was least important, seeds of this species persist in the soil of the shaded understory up to 3 yr (C. C. Horvitz and D. W. Schemske, personal communication).

K. G. Murray (1986, 1988) models the seed bank dynamics of three gap species from a Costa Rican montane forest. Using two models of plant fitness, he evaluates the relative importance of seed shadow, phenology, germination requirements, and gap size. Potential lifetime reproductive output (analogous to net reproductive rate, R_o) increases linearly as length of dormancy increases, as do differences among seed shadows generated by different birds. Relative fitness (analogous to intrinsic rate of natural increase, r) also increases initially as length of dormancy increases, but rapidly reaches an asymptote (within 2–5 yr). Differences among seed shadows are small. Although neither model is an unbiased estimator of fitness in iteroparous species, Murray (1988) concludes that short-term dormancy will increase fitness and magnify differences among disperser seed shadows, but extended dormancy will increase fitness little.

B. Mature Tropical Forests versus Secondary Regrowth and Farms

The soil seed banks of mature tropical vegetation, secondary regrowth, and disturbed/agricultural land are characterized in Tables 2, 3, 6, and 7. Because of enormous variation in the methods used (Section VIII, Appendix), rigorous comparisons among studies cannot be made and the conclusions should be considered very general. No studies have determined what sample size or frequency is best for different tropical vegetation types or successional stages. The typical study (based on the median values of the 43 studies summarized in the Appendix) consists of six samples collected at the same time in two habitats, each 950 cm² and 5 cm deep. This sampling intensity is inadequate to document intrasite variability. Moreover, species richness is often underestimated because seedlings are grouped into generic or familial taxa and the percentage of unidentified species or individuals is not given.

1. Seed Density

The most striking differences in seed bank density and composition come from comparisons of mature tropical forests with young secondary regrowth and farms. A Mann–Whitney U-test ($p < 0.001$) shows that seed density from mature forests (range = 25–3350 seeds m^{-2}, median = 384, $N = 41$; Table 2) is significantly lower than that of regrowth/farms (range = 48–18,900 seeds m^{-2}, median = 1650, $N = 41$; Table 3). (This comparison excludes studies of single species, uses the shallower of two soil depths, and uses the midpoint of ranges.)

2. Number of Species

The numbers of species found in forests (4–79; $N = 28$) and regrowth/ farms (8–67; $N = 28$) are similar (Tables 2 and 3), but the comparison is misleading. In forest sites, the number of species significantly increases as total area sampled increases ($p < 0.001$, excluding three outliers; $p > 0.33$, including outliers). In regrowth/farm sites, the number of species does not increase as total area sampled increases ($p > 0.39$, with or without two outliers). This suggests that seed banks are more spatially heterogeneous in forest than in regrowth/farm sites. (This comparison uses the midpoint of ranges.)

The comparison is further confounded because the range of total area sampled (excluding outliers) in forests (640–30,000 cm²) is greater than that in regrowth/farms (171–12,500 cm²). The number of species does not increase with total area sampled in forests if areas greater than 12,500 cm² are excluded ($p = 0.12$). For total area sampled of ≤12,500 cm², 67% of the regrowth/farm sites ($N = 26$) had >20 species, but only 29% of the mature forest samples ($N = 22$) had >20 species. This suggests that regrowth/farm seed banks are locally more diverse than forest seed banks when comparable areas are sampled. Proper species–area curves for mature forest and regrowth from the same region are needed to verify these generalizations.

3. Species Composition

Primary species represent only a small proportion of all individuals in the seed banks of tropical forests (0–16%, median = 3%) and regrowth/ farm sites (0–5%; Table 6). They represent a higher proportion of species (6–21%, median = 12%; Table 7) than individuals (<1–16%, median =

Table 6
Life-forms present in tropical soil seed banks and abundance of four most common species

Location	Ref. no.[a]	Vegetation type[a]	Edge[a]	N[b]	Primary species[c]	Percentage of individuals								
						Life-form[d]					Most common species[e]			
						Tree	Shrub	Herb (F/G)	Liana, vine	Uniden-tified				
Tropical forests														
Puerto Rico[j]	[6]	Forest + clearing	—	640[r]	—	48	19	3	1	30	42 T	19 S	3 T	3 Hf
Mexico[t]	[7]	High ever-green	C	250	Low	18	2	69/4	—	8*	28 Hf	28 Hf	8 T	7 T
	[7]	High ever-green	C	240	Low	49	2	42/0	—	7*	45 T	26 H	13 Hf	7 K
Costa Rica	[40]	Elfin forest	—	203	6	69	18	2	—	11	54 T	12 S	9 T	6 S
Panama	[25]	Tropical moist	Fx	635[p]	Low	88	11	<1	—	—[p]	54 T	18 T	5 S	4 T
Venezuela	[26]	Caatinga	—	243	2 T	78[m]		0	—	22	55 T	—	—	—
	[26]	Tierra firma	—	92	1 T	66[m]		1	—	31	70 T	—	—	—
	[21]	Tierra firma	—	512	3 T	—		—	—	29	64 T2	—	—	5 T
French Guiana	[20]	Lowland	—	302[i]	Low	93		4/1	2	0	52 T2	21 F	6 T	5 T
Nigeria[j]	[3]	Lowland (Usonigbe)	—	34	0	53	18	15/0	6	9	41 T	12 S	12 Hf	9 T
	[3]	Lowland (Omo)	—	72	6	75	12	6/4	3	0	61 T	12 S	4 T	4 H
Ghana[j]	[18]	SE outlier	B	107	11	21	6	18/43	13	0	30 Hg	14 Hf	13 Hg	10 C
	[18]	Southern marginal	B	384	3	21	54	4/1	20	0	34 S	15 T	12 S	5 S
	[18]	Dry semi-deciduous	B	696	<1	33	56	3/1	5	0	53 S	19 T	5 T	4 T

(continued)

Table 6 (*Continued*)

Location	Vegetation type[a]	Ref. no.[a]	Edge[a]	N[b]	Primary species[c]	Life-form[d] Tree	Shrub	Herb (F/G)	Liana, vine	Unidentified	Most common species[e]			
	Moist semi-deciduous	[18]	E	623[h]	<1	19	70	6/0	4	0	64 S	15 T	3 C	3 T
	Wet evergreen	[18]	F	163	0	91	6	2/0	0	0	71 T	3 T	3 T	3 S
	Upland evergreen	[18]	F	45	0	84	7	5/2	2	0	44 T	11 T	7 T	7 S
Malaysia	Lowland dipterocarp	[14]	E	164	6	—	—	—	—	40	—	—	—	—
	Lowland dipterocarp	[39]	Fx	74	—	24	34	20	4	18	16	14	7	7
				273[s]	—	23	—	—	—	—	16 Hf	8 T	8	7
	Lowland dipterocarp	[9]	—	1181[h]	16 TS	79	—	4	16	0	22 T	13 T	12 T	9 C
Papua New Guinea	Araucaria forest	[31]	—	199[h]	<1	57	—	22/2	1	19	—	—	—	—
	Lowland forest	[43]	—	199	—	74	—	10/10	7	—	—	—	—	—
Australia	Evergreen (four sites)	[24]	—	8322	6	47		43	4	<1	16 T	7 T	4 S	4 T
	Semi-deciduous	[29]	—	823	Low	62	9	17/1	8	3	32 T	10 T	9 T	6 C
	Evergreen	[28]	—	296	10	80[k]		10[s]	—	—	—	—	—	—
	Range[f]				0–16	18–91	2–70	0–73	0–20	0–40*	16–71	3–28	3–13	3–10
	Median[f]				3	49	12	6	4	2	45	14	6	5

Young secondary regrowth or agricultural lands (age in years)

Location	Ref	Type (age)		Area											
Mexico	[7]	Regrowth (<1)	C	1211	—	2	3	90/<1	2	6*	64 Hf	21 Hf	5 K	2 Hf
	[7]	Regrowth (7)	C	1100	—	<1	3	61/<1	<1	35*	35 K	26 Hf	25 Hf	6 Hf
Belize	[11]	Farms & pasture	—	1479	—	2	28[l]	26/21	—	23[o]	27 S	8 F	5 Hg	4 Hf
	[12]	Corn field	—	490	—	0	1	67/26	—	6	19 Hg	19 H	19 H	12 H
	[12]	Pasture	—	650	0	<1	10	42/46	—	2	26 H	15 Hg	8 Hg	8 Hg
Costa Rica	[32]	Forest (70)	D	146	—	33	18	5/1	18	24	27 T2	15 S	14 C	10 F
	[38]	Forest (75)	—	119	—	15	34	29/2	7	13[g]	18 S	15 T2	11 Hf	9 Hf
	[38]	Regrowth (3.3)	—	450	—	6	41	25/11	12	5[g]	12 S	9 S	7 Hf	6 S
Venezuela	[38]	Regrowth (11)	—	648	0 T	5	24	47/5	13	6[g]	22 Hf	9 Hf	7 Hf	6 S
	[26]	Pasture (6)	—	500	—		1[m]	54/45	—	0	—	—	—	—
	[26]	Cow droppings	—	6560	0 T	0	0	2/98	—	0	—	—	—	—
	[26]	Farmland (3)	A	350	0 T	6[n]		15/53	—	26	—	—	—	—
	[21]	Burned slash (3 months)	A	126	3 T			—	—	22	61 T2	—	—	—
Senegal	[4]	Farmland	—	114,326	—	—	—	1/93	—	6*	36 H	33 H	10 H	9 H
Ghana	[36]	Regrowth (~20)	D	~2575	—	—	—	>31	—	—	81 Hf	—	—	—
Nigeria	[27]	Farmland	—	11,648	2	—	—	75/25	—	>1	43 H	15 H	15 H	5 H
Malaysia	[14]	Regrowth (4)	BC	95	2	—	—	—	<1	5	—	—	—	—
Papua New Guinea	[30]	Clear-cut (0 months)[u]	—	5125	—	93	—	6/1	<1	0	—	—	—	—
	[43]	Clear-cut (0 months)[u]	<1	994	93		—	6/1	<1	—	—	—	—	—
	[43]	Regrowth (0.5–2)[u]	—	4810–11,539	—	2–8	—	74–82/4–22	<1–5	—	—	—	—	—
	[43]	Regrowth, 2 sites (1.5–2.5)	—	2381–3912	—	2	—	73–92/5–23	1–3	—	—	—	—	—

(continued)

185

Table 6 (*Continued*)

Location	Ref. no.[a]	Vegetation type[a]	Edge[a]	N[b]	Primary species[c]	Percentage of individuals					Most common species[e]
						Life-form[d]					
						Tree	Shrub	Herb (F/G)	Liana, vine	Unidentified	
	[43]	Regrowth, 8 sites (3.5–11)	—	315–1263	—	25–75	—	12–61/1–14	<1–6	—	— — — — —
Australia	[43]	Regrowth (55)	B	3745	~5	81	—	11/6	2	—	— — — — —
	[28]	Forest, 3 sites (~30)	—	167–720	~5	49–87[k]	—	8–48[q]	—	—	— — — — —
—	[28]	Four disturbed sites	—	122–1558	0–5	0–20[k]	—	92–98[q]	—	—	— — — — —
				Range[f]	0–5	0–93	1–41	6–100	<1–18	0–26	
				Median[f]	—	5	18	75	3	6	

[a] References as in the Appendix; headings and terms as in Table 2.

[b] The actual number of seedlings germinated from soil samples (not number m^{-2}).

[c] As classified by the authors cited, but usually species whose seedlings can establish in the shaded understory. When restricted to one to two life-forms, these are indicated (see codes in footnote e).

[d] As classified by the authors or by myself from general sources. Herbs are divided into forbs (F) and graminoids (G). Values between columns represent percentage of individuals in the combined life-forms of adjacent columns. Percentage of individuals in each category is calculated from total number of seedlings (not number of identified seedlings) to avoid biases against species in less well known life-forms. Unidentified seedlings include unknowns from the references as well as some identified species (usually herbs and shrubs) which I did not classify to life-form; I marked the latter with an asterisk when my unidentified portion was ≥5%.

[e] Species are ranked from most to least abundant (left to right); the fifth most common species was always <10% of the total. Life-forms included trees (T), shrubs (S), herbs (H), climbers (C), and either herbs or shrubs (K). Where known, herbs were separated into forbs (Hf) and graminoids (Hg). Multiple-species taxa included two species in the same genus (usually *Cecropia*) and life-form (2), and species in the same family with mixed life-forms (F). Percentages, based on total sample (N), were from references cited (modified if necessary) or were calculated from species lists therein.

[f] For common species, combined taxa were excluded.

[g] Includes epiphytes (≤1%).

[h] Ferns excluded.

^{i}Composition based on the 50% of seedlings which survived 2 months after soil seed densities were determined.

^{j}Sun-germinated samples only.

^{k}Pioneer and late secondary species; not included in range and median.

^{l}Although in Ref. 8 it is stated that all are herbs except *Cecropia*, I consider *Lindenia crustacea*, *Miconia*/*Clidemia*, and *Hamelia patens* to be shrubs.

^{m}Only trees are given in the table, but shrub species are discussed in the text.

^{n}Data are contradictory; $N = 211$ in text and $N = 199$ in an appendix listing the species.

^{o}Unidentified monocots, probably graminoides, comprise 19% of unknowns.

^{p}Percentages are of identified species; unidentified species comprise <10% of total.

^{q}Weedy species (including some sclerophyllous woody species not usually found in forest gaps). Not included in range and median.

^{r}Total estimated from graph.

^{s}Based on methods summarized in Table 5, not in the Appendix.

^{t}Salmerón (1984; cited in Vázquez-Yanes and Orozco-Segovia, 1987) reports more seeds of pioneer trees at the same site and suggests that the methods used in Ref. 7 underestimated tree abundance.

^{u}The same site was sampled five times in 2 yr; the discrepancy in N (based on mean density) between Refs. 30 and 43 for the just clear-cut site is not discussed in Ref. 43.

Table 7
Abundance of life-forms in tropical soil seed bank floras

Location	Ref. no.[a]	Vegetation[b]	N[c]	Primary species [trees]	Life-form[d] Trees	Shrubs	Climbers	Herbs (F/G)	Unidentified
Mature forests									
Mexico	[7]	Evergreen (2)	33*	—	27	9	—	36/12	15
Panama	[25]	Semideciduous (1)	48	—	23	17	—	6	54
French Guiana	[20]	Lowland	15[l]	—		60	20	13/7	—
Nigeria[e]	[3]	Lowland (2)	39*	[10]	38	10	18	21/5	8
Ghana[e]	[18]	Various (6)	99*	12 [6]	24	12	26	19/9	0
Malaysia (Sabah)	[9]	Lowland (1)	29*	21		59	31	10/0	0
Papua New Guinea	[31]	Upland (1)	18	— [6]	44	—	11	44	—
Australia	[24]	Evergreen (4)	131*	18 [8][f]	39	—	—	—	61[i]
Australia	[29]	Evergreen (1)	58*	—	41	9	19	24	7
Regrowth/farmland									
Mexico	[7]	Regrowth (2 months)	14*	—	15	11	4	41/7	22[k]
Mexico	[7]	Regrowth (5 yr)	21	—	5	24	10	48/5	10[k]
Costa Rica	[38]	Regrowth (3.3 yr)	40[g]	—	5	32	10	32/18	0[g]
Costa Rica	[38]	Regrowth (11 yr)	47	—	6	30	13	30/17	4
Costa Rica	[38]	Forest (75 yr)	21[g]	—	5	38	19	29/5	0[g]
Senegal	[4]	Farmland	45*	—	2	18[i]	4	29/22	24[k]
Nigeria	[27]	Farmland	25*	—	0	0	—	60/36	4
Papua New Guinea	[30]	Clear-cut logged (0 months)	43[h]	—	60	—	5	21/7	7

[a]References as in the Appendix.

[b]Numbers in parentheses denote number of sites (for mature forests) or age (for regrowth/farmland).

*c*Sample size (N) is number of species (or taxa). It is often unclear whether number of species given in a particular reference is total number of species or number of identified species. Studies which included complete lists of all species, including unknowns, or clearly specified the number of unknown but distinguishable species are marked with an asterisk. To reduce further biases against poorly known life-forms, I excluded studies if >20% of seedlings were not identified to species or life-form (see Table 6) or if seedlings were frequently lumped into generic or familial taxa.

*d*See Table 6.

*e*Percentage was based on the combined sun- and shade-germinated samples because they could not be separated or the number of species was small.

*f*Another 9% are epiphytic trees.

*g*N and percentage unidentified may be underestimated because 5–12% of seedlings were not identified to life-form (see Table 2, Ref. 37).

*h*N was given as 40 in table and 43 in text; I assumed three species were not identified to life-form.

*i*These were either shrubs or herbs.

*j*Species were identified but data were summarized as secondary and weedy groups, not life-forms (except trees).

*k*Primarily identified species that I could not classify to life-form.

*l*About one-third were multispecies taxa.

6%; Table 6) in five forests where both are documented. Although pioneer species dominate the seed bank, the portion residing in a persistent seed bank is generally unknown. In perhaps an extreme example, an introduced weed, *Eupatorium odoratum*, with maximum longevity of only 2 yr, formed ≈25% of the seed bank in a forest in Ghana and ≈81% of that in regrowth where it was fruiting (Epp, 1987).

Mature tropical forests and regrowth/farm sites differ greatly in the proportion of species (Table 7) and individuals (Table 6) of different lifeforms present. Forest seed banks are dominated by trees (18–91%, median = 49%; Table 6). Low tree abundance in Ghana and Mexico, but not Malaysia, is associated with proximity to large stretches of secondary regrowth (Table 6). Regrowth/farm seed banks are dominated by herbs (6–100%, median = 75%; Table 6). Older regrowth (30–75 yr) has more trees and shrubs than young regrowth/farms (Tables 6 and 7).

Forest soil seed banks are dominated by a single species (16–71%, median = 45%; Table 6), often one species of tree. Thus, single-species dominance of the seed bank may account for the single-species dominance often reported in regeneration (Enright, 1985). In regrowth/farmland, single-species dominance is less pronounced (12–81%, median = 28%; Table 6), with a second species about half as abundant as the first (median = 15%); the dominant species are more frequently herbs and shrubs than trees. In all forest and regrowth/farm sites except one, the most common species is in the most common lifeform (Table 6), and usually accounts for >50% of the individuals in that life-form. Thus, the two or three most common taxa (Table 6) account for most of the variation in abundance of life-forms.

Dominance might be a regional phenomenon caused by ubiquitous dispersal, a local phenomenon caused by local dispersal and sampling in small areas, or a seasonal phenomenon caused by seasonal dispersal and infrequent sampling. In French Guiana, de Foresta and Prévost (1986) found both local dominance of pioneer species near adults and widespread occurrence of the common dominants throughout the site (e.g., *Cecropia* species and Melastomataceae). Seed bank floras from widely separated forest sites in the same region are more similar to each other than to the flora of the forest from which they were sampled, and are more similar than the forest floras are to each other (Table 8; Hopkins and Graham, 1984a; de Foresta *et al.*, 1984). Species present in the seed bank but absent from the forest are most frequently pioneer and weedy species, but sometimes are species from lower altitude forests (Lawton and Putz, 1988) or are unknown in the region (Hall and Swaine, 1980). This suggests a ubiquitous, homogeneous pioneer flora. In contrast, composition of seed bank floras of late secondary forest (Hopkins and Graham, 1984a), young regrowth (de Foresta *et al.*, 1984; Young *et al.*, 1987), and farms (Kellman, 1974; M. K. Marks, 1983a,b) reflects that of the standing vegetation. Some pioneer and weedy species (e.g., *Eucalyp-*

Table 8

Similarity among seed bank floras, among vegetation floras, and between seed bank and vegetation floras

Location	Ref. no.[a]	Vegetation	Jaccard's coefficient[b]: range (median)			Distance between sites (km)	Number of sites
			Among seed banks	Among floras	Between flora and seed bank		
Ghana	[18]	Mature forest	2–31 (13)	0–16 (3)	0–9 (2)	40–255	6
Mexico	[7]	Mature forest	30	20	2–3	<2	2
		Regrowth	33	15	0–20	<2	2
		Forest vs. regrowth	17–37	7–12	—	<2	4
Australia	[24]	Mature forest	40–43	—	—	8–27	4
Costa Rica	[19]	Cut and burned regrowth[c]	25–31	—	—	0	3

[a]References as in the Appendix.

[b]Jaccard's coefficient of community similarity (JC): JC = $(C / N1 + N2 - C) \times 100$, where C is the number of species common to two sites and $N1$ and $N2$ are the number of species at each site.

[c]Floras of regrowth sites before cutting, after cutting, and after burning were compared.

tus, Macaranga, and *Imperata*), however, are absent from the regrowth they dominate (Hopkins and Graham, 1984a).

Dominance of species and life-forms may change during the year. For example, in a Mexican forest site (Guevara and Gómez-Pompa, 1972), the tree *Robinsonella mirandae* was the most abundant species, comprising 45% of the 240 seedling germinated during the year-long study. It was dominant in 5 of 8 months sampled (47–92%, N = 13–40), but three different herb and shrub species were dominant in other months (39–78%, N = 18–47). Hence, we should be wary of characterizing a soil seed bank by one species or one life-form if it is sampled only once during the year. Of 43 studies, 79% were sampled only once (Section VIII, Appendix).

Dominance in the seed bank might also arise from variation in seed size. Seed size and number of seeds produced are presumably inversely correlated. Assuming equal abundance of adults, small-seeded species should have higher seed densities in the soil than larger seeded species, even if the latter have persistent seeds (S. A. Foster, 1986; Putz and Appanah, 1987). Because pioneer and late secondary tree species have smaller seeds than primary trees (Foster and Janson, 1985), it is not surprising that they dominate the seed bank. On the other hand, rarity of larger seeded species in the seed bank is not evidence that they lack long-term viability or that their seed bank is unimportant in regeneration.

C. Changes during Succession

I present a model of change in seed banks during succession (Fig. 3) to summarize the descriptive studies of seed banks discussed below. Later, I use this model to explain variation in seed banks among tropical communities and to predict patterns in little-sampled communities. A general model must include both density and species composition. I use the ratio of pioneer to weedy species, a crude estimate of species composition, to focus on seed bank change from mature forest (dominated by short-lived pioneers) to regrowth/farms (dominated by weeds). For actual data sets, clustering or ordination techniques would be appropriate (see Hopkins and Graham, 1984a). A more specific model of temporal change in seed bank density on farmland is presented by Young *et al.* (1987).

1. Treefall Gaps and Unburned Artificial Clearings

There are no studies that follow change in the seed bank of treefall gaps or small clearings as regeneration proceeds. Consequently, such changes must be inferred from studies of sites of different ages. Seed bank densities decrease rapidly immediately after a gap occurs through germination and death (Hopkins and Graham, 1984b; Fig. 3: forest to gap).

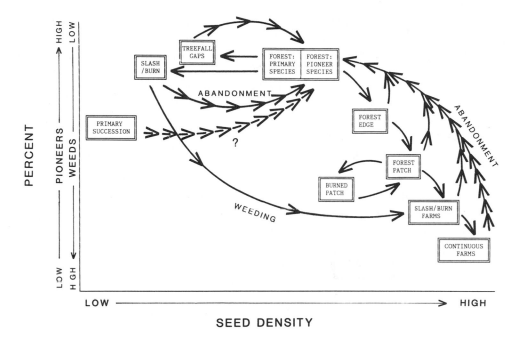

Figure 3.   Changes in tropical soil seed bank density and species composition during succession. Species composition is represented as reciprocal changes in pioneer or weeds species, because primary species are rare. Pioneers or weeds are given as percentage of all seeds in soil. Seed density is total number of seeds of all species in the soil. The number of arrows along successional paths is inversely related to rate of change in seed bank; where there are few arrows, change is rapid. All possible paths are not shown.

Within a gap, density decreases most in disturbed soil and least under thick litter (see Section III,D.).

The rate at which the seed bank returns to predisturbance size and composition (Fig. 3: gap to forest) is influenced by degree of isolation, size, severity of disturbance, and the regeneration strategy of the colonizing species. In Puerto Rico and Malaysia, seed densities were lower in small clearings and treefalls (1–5 yr old) than in forest (Tables 2 and 3, Refs. 6 and 39); but, also at the same Malaysian site, seed densities in a larger clearing (4 yr old) were similar to those in the forest edge and the forest (Tables 4 and 5, Refs. 14 and 39). In French Guiana, the seed bank of a small isolated clearing (4 yr old) was smaller than that in mature forest, but compositions were similar (Tables 2 and 3, Ref. 33; de Foresta and Prévost, 1986). In larger clearings in the same area, the seed bank was larger and was dominated by shrubs and trees reproducing in older sites (6–9 yr old) or by pioneer trees from adjacent forest in younger sites

(2 yr old) where local reproduction was just beginning (Tables 2 and 3, Ref. 33). Weedy species were locally common in the largest site only where burning or soil compaction occurred (de Foresta and Prévost, 1986).

In gaps filled by short-lived pioneer species, seed bank density may remain high because seed rain is high and/or seeds are persistent, until individuals die 10–50 yr later. Conversely, in gaps filled by longer lived species, density may remain low for >100 yr (Fig. 3, pioneer versus primary). Soil seed density in two old gaps (>25 yr) filled by *Cecropia* and *Miconia*, respectively, was significantly higher than in mature forest soil 30 m away (Table 4, Ref. 33; *t*-test, $p < 0.05$, my calculations); seed banks were enriched in *Cecropia* and *Miconia*, respectively, which were ubiquitous elements of the forest seed bank (de Foresta and Prévost, 1986). Seed density in a third gap filled by *Laetia* and *Bellucia* was not significantly different from forest soil; *Laetia* was rare in the soil under the adult and throughout the forest.

Seeds dispersed to and accumulating in the seed bank in young gap centers may contribute little to regeneration or individual fitness because the next treefall will not occur soon (Schupp *et al.*, in press). Rather, because new treefalls are more likely to occur next to old treefalls than elsewhere (Hubbell and Foster, 1986a; Lawton and Putz, 1988), seeds deposited in gap edges (where many frugivores process seeds collected in older gaps) should contribute more to regeneration and individual fitness (Schupp *et al.*, in press). This reinforces Murray's conclusion (Murray, 1988) that long-term dormancy confers little increase in plant fitness.

Changes in the seed bank following extensive clearing were studied in Papua New Guinea, where >36,000 ha were cleared over 11 yr, by monitoring recently cleared areas for 2 yr and inferring later changes from well-dated older stands (Saulei and Swaine, in press). Pioneer tree seed density in the soil decreased rapidly in the first year, as occurs in treefalls; density recovered to predisturbance levels within ≈3 yr of disturbance and remained relatively stable for the next ≈8 yr. In contrast, weedy herb seed density in the soil rapidly increased within 6 months and reached a maximum in <2 yr, accounting for >90% of the seed bank. Although the seed rain of weedy herbs reached a maximum within 6 months of disturbance and dropped to low values within 2 yr, they comprised ≈40–50% of the seed bank 6–11 yr after disturbance.

2. Burned Agricultural Clearings

In slash-and-burn agriculture, the forest is typically cut and burned during the dry season, crops are planted as the rains begin, and fields are actively cultivated and weeded for several years then abandoned (e.g., Uhl *et al.*, 1981). During burning, temperatures often exceed 200°C at the soil surface and 100°C 1 cm below the surface (Uhl *et al.*, 1981; Ewel *et al.*, 1981). After the burn, soil surface temperatures are higher on bare soil (Hopkins and Graham, 1984a) and near charcoal (Uhl *et al.*, 1981) than under unburned slash. High temperatures lower the soil seed density and favor heat-resistant or heat-stimulated species (Fig. 3, forest to slash/burn; Tables 2 and 3, Refs. 19 and 21).

Seed bank density and dominance by weedy species rapidly increases during farming (Tables 3, 6, and 7; Fig. 3, slash/burn to farm). Woody pioneer species germinate rather synchronously from seeds surviving in the soil, but most are removed in the first weeding. In contrast, weedy species originate from the seed rain over long periods; many germinate after the first weeding and reproduce profusely before the second (Uhl *et al.*, 1981, 1982a; Kellman, 1970a). At the time of abandonment, seed density and the number of woody colonists are higher under remaining shade trees and slash than in bare areas (Table 3, Ref. 22), perhaps reflecting more dispersal from perches in the trees and/or lower germination in the shade.

If a burned site is abandoned without being weeded or farmed, it should more rapidly return to the original density and composition (Fig. 3, slash/burn to forest). Three years after burning but without farming, soil seed density, depth profiles, and life form and species composition of a burned site were similar to that of the young Costa Rican regrowth from which it had been created (Young *et al.*, 1987). Recovery may be faster on logged sites, which are burned but not weeded, than on farms.

Seed bank densities and dominance by weeds increases with continuous farming (Fig. 3, farm to continuous farm). Both pastures and farms maintain large seed banks dominated by weedy herbs, especially grasses (Tables 3 and 6, Refs. 11, 12, and 27). Density was not correlated with age (4–20 yr) in over 50 pasture and farm sites (Kellman, 1974b). Because continually farmed sites are usually larger than slash-and-burn farms, seed sources of forest pioneer species will be farther away, resulting in slower buildup of woody pioneers in the seed bank.

After abandonment, the rate at which seed banks revert to pre-disturbance densities and composition (Fig. 3, farm to forest) depends on the duration of agricultural activity, severity of initial disturbance, frequency of later disturbances, and distance to forest pioneer seed sources, as does recovery of the vegetation (Purata, 1986). In Australian secondary regrowth (>20 yr old) and mature forest, seed bank size and species richness are greatest in forest sites not burned for >5–10 yr and lowest in sites burned annually or biennially (Tables 2, 3, and 6, Ref. 28). Pioneer and late secondary species dominate less disturbed sites, while weedy species dominate repeatedly burned and disturbed sites. Sites with the most similar disturbance histories have the most similar seed banks, as determined by a clustering analysis using both species presence and abundance (Hopkins and Graham, 1984a). During logging, road building, and mining, vegetation recovery is delayed because bull-dozers remove large quantities of topsoil, removing the seeds as well, and compact the soil in their path, preventing seeds from germinating (Uhl et al., 1982b; Swaine and Hall, 1983; de Foresta et al., 1984).

The seed density and proportion of weedy species should increase in forest seed banks as distance to farms and weedy regrowth decreases, edges become more common, farms become more abundant, and size of forest patches dwindles (Fig. 3, forest to forest edge to forest patch). Janzen (1983b) suggests that the influence of large areas of young secondary forest stretches >5 km into apparently mature forest, given the flight distances of many bats and birds and the large home ranges of some terrestrial mammals. Within a mature French Guianan forest, seed densities 50 m from the edge of a large patch of regrowth (25 ha, 6 yr old) are nearly twice those 500 m to 6 km away (Table 2, Ref. 33). (t-tests for samples with unequal variances indicate that only the 50- to 500-m pair of adjacent samples are significantly different; $p < 0.05$.) Seed density in primary forest in Costa Rica significantly decreases with distance (10–500 m) from the edge of secondary forest (Table 2, Ref. 34); 10 m from the forest/regrowth edge, density is four times higher in the secondary forest than it is in primary forest (Tables 2 and 3, Ref. 34). In Malaysia, densities differed little among several samples from a small clearing, forest edge, and forest 100–500 m from the edge (Tables 2 and 3, Ref. 14). In Papua New Guinea, two-thirds of the pioneer tree seeds in the seed rain and seed bank of a small patch of old regrowth forest (<1 ha, ~55 yr) were from species abundant in the surrounding young regrowth (<2 yr) but absent from the patch (Saulei and Swaine, in press). However, two-thirds of the species represented in both the seed rain and seed bank were growing within the patch. Weedy herbs, however, comprised only a small portion of the seed rain and seed bank of the patch, although they dominated the seed rain and seed bank of the surrounding young vegetation. In contrast, pioneer seed density in a Thai lower montane forest significantly increased with distance (25–175 m) from a pioneer-filled clearing (Table 5, Ref. 15).

3. Primary Seres

A seed bank should be absent at the start of primary succession, but many substrates of "primary" succession, such as beaches and landslides, are reprocessed soils that may contain seeds. If present, the seed bank must be small because early regeneration is sparse (Fig. 3, primary succession). On earthquake-induced landslides, plant density is very low and pioneers comprise >90% of the individuals (Garwood et al., 1979; Garwood, 1985). On Amazonian floodplains, weedy beach species, pioneers, and late secondary species that will dominate the forest for >200 yr germinate together at low densities (six seedlings per square meter) on beach levees (R. B. Foster *et al.*, 1986). Beach species are diverse (59% of the flora) but not dominant (9% of the individuals).

4. Regional Variation

Regional differences in fertility, drainage, disturbance regimes, seasonality, or other factors should cause predictable changes in density and/or composition and shifts in the relative position of the dynamic system. I discuss these below, although available data (Tables 2, 3, 6, and 7) are too variable to test specific predictions.

In communities with low productivity, such as nutrient-poor or elfin forests, total seed production (Janzen, 1974b) and the size of the seed bank should also be low. In Southeast Asian heath forests (Riswan, 1982, cited in Whitmore, 1983) and many upper montane and elfin forests (e.g., Byer and Weaver, 1977; Sugden *et al.*, 1985), most regeneration is from sprouting rather than from seed, and pioneers are rare. Facultative dormancy was not found in 36 species from infertile Amazonian campinas (Macedo, 1977). Analogous to mechanisms increasing leaf life-span in nutrient-poor sites (see Marín and Medina, 1981; Chabot and Hicks, 1982), changes in structure, dormancy, or toxicity of seeds might increase longevity. Conversely, large seeds with extensive reserves but little dormancy might predominate.

Increased soil saturation, caused by poor drainage, higher rainfall, or river flooding, might cause a decrease in seed density or a change to tolerant species. In tropical freshwater swamp forests, the potential importance of nurse logs (Hartshorn, 1978), reports of vivipary (Leite and Rankin, 1981), and the prevalence of large buoyant seeds suggest that regeneration from a seed bank is unimportant, although prolonged seed dormancy also occurs (Coutinho and Struffaldi, 1971; Hartshorn, 1978). Some savanna annuals exploit seasonally waterlogged sites by germinating after excess water drains off and reproducing before the soils dry out (Sarmiento and Monasterio, 1983).

Treefall regimes vary greatly among forests (Brokaw, 1985a); where gaps are large and frequent, pioneers should be common and seed bank densities should be high. Compared to a Panamanian forest, Malaysian trees are more narrow crowned, more often die standing, and create thinner treefalls which support fewer pioneers, presumably lowering overall pioneer seed production and accounting for the low seed bank density and rareness of pioneer trees (Putz and Appanah, 1987). Whether similar dynamics explain the low densities in other forests in Asia, Africa, and the Neotropics (Table 2) is unknown. If seeds of Malaysian pioneers are larger than those in Panama (Putz and Appanah, 1987), this could further explain lower seed densities. Random samples of pioneer species and accurate measures of seed size, not fruit size, are needed to test the hypothesis.

Average seed density, within-year variance in density, and the abundance of species with seasonal-transient seed banks should increase as the length of the dry season increases. Annual herbs that rely on a seasonal-transient seed bank are often common in savanna (Sarmiento and Monasterio, 1983) but are absent from rain forest; some perennial savanna species also have seasonal-transient seed banks (Sarmiento and Monasterio, 1983). Species with seasonal dormancy might be preadapted for long-term persistence in the soil. As frequency of fire increases, the proportion of fire-tolerant or fire-stimulated species should also increase.

D. Effect of Gap Size

It is widely believed that gap size is a major factor controlling the relative contribution of the seed bank to regeneration (Schulz, 1960a,b; Baur, 1968; Hartshorn, 1978, 1980; Whitmore, 1978; Denslow, 1980; Bazzaz and Pickett, 1980; Alexandre, 1982; Bazzaz, 1984; Brokaw, 1985a; but see Herwitz, 1981; Rollet, 1983a,b), although few data exist from a range of gap sizes in the same forest. Light intensity, temperature, and drought stress at the soil surface (but not at deeper levels) increase with size of gaps (Bazzaz and Pickett, 1980; Denslow, 1980, 1987), but overall heterogeneity should be large in all but the smallest gaps (Ricklefs, 1977; Hartshorn, 1978; Denslow, 1980, 1987; Orians, 1982; Bazzaz, 1984). In Costa Rican elfin forest, two correlated measures of gap size, area disturbed and gap aperture (arc between horizons), are also correlated (to different degrees) with size of fallen tree/branch, light intensity, and area of nurse logs and bare mineral soil (Lawton and Putz, 1988).

The predominant regeneration pathway will change as gap size increases (Fig. 4A). The importance of canopy ingrowth decreases rapidly with gap size (Fig. 4A). Sprouting is of considerable importance in a variety of cut, unburned forests (Byer and Weaver, 1977; Kartawinata *et al.*, 1980; Uhl *et al.*, 1982b; Riswan, 1982, cited in Whitmore, 1983; Sugden *et al.*, 1985; Saulei, 1984), but the role of resprouted, broken trees decreases with gap size in treefalls in Panama (Fig. 4A; Putz and Brokaw, in review). Pioneer species become more abundant as gap size increases in many forests (Aiyar, 1932a,b; Kramer, 1933; Schulz, 1960a,b; Brokaw, 1985a,b, 1987; Lawton and Putz, 1988). Thus, regeneration from the seed rain or the seed bank of pioneers predominates in larger gaps (Fig. 4A) and regeneration of primary species from the seedling bank or advance regeneration predominates in smaller gaps. Bazzaz (1984) presents a conceptually similar model.

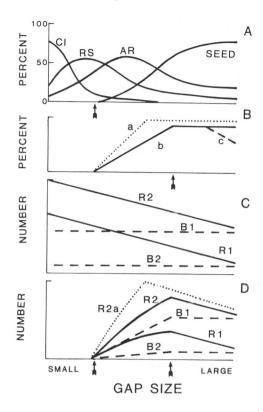

Figure 4. Relationship of gap size to changing roles of different regeneration pathways in lowland tropical forest. (A) Percentage contribution to regeneration from different pathways: canopy ingrowth (CI), root and shoot sprouts (RS), seedling bank and advance regeneration (AR), and seed bank and seed rain combined (SEED). The arrow indicates the minimum gap size for establishment of pioneers from seed. (B) Percentage of available pioneer seeds from the seed bank or seed rain that will germinate, successfully establish, and contribute to regeneration. The arrow indicates the gap size at which the maximum contribution to regeneration first occurs. The maximum contribution and gap size at which this occurs may differ among taxa (a versus b). If mortality is greater in larger, more environmentally stressful gaps, the percentage of seeds contributing to regeneration will decrease (c). (C) Number of seeds available for regeneration at the time of gap creation. The number arriving in the seed rain (R1 or R2) will decrease as gap size increases because distance from forest edge and seed sources increases. The number of seeds in the seed bank (B1 or B2) will be independent of size of the gap just created. (D) Number of seeds contributing to regeneration from seed rain and the seed bank. Number contributing at each gap size is the product of number available (Fig. 4C) and percentage contributing (Fig. 4B: b for R1, R2, B1, B2; a for R2a). In B–D, percentage and number increase in arbitary units along the axis.

The generalization that the relative contribution of the seed bank to regeneration is greater than that from seed rain (see Section V., A.) should be reexamined, given the expected changes in regeneration from seed with gap size (Fig. 4A) and successional changes in the seed bank (Fig. 3). The percentage germination and successful establishment of pioneer seeds from either the seed bank or seed rain should increase with gap size (Hopkins and Graham, 1984b; Lawton and Putz, 1988; Murray, 1988) from a very low percentage at the minimum critical gap size of pioneers (Brokaw, 1985a) to a maximum value in larger gaps (Fig. 4B). The size of the seed bank should be independent of gap size at the time of gap creation (Fig. 4C, B1 or B2), whereas, seed rain should decrease with increasing gap size as distance from pioneer adults in the forest increases (Fig. 4C, R1 or R2). The number of seeds that will contribute to regeneration (Fig. 4D) is a multiplicative function of the number of seeds available for regeneration (Fig. 4C) and the percentage that can establish (Fig. 4B). This simple model suggests that the contribution from seed rain is highest at intermediate gap sizes (Fig. 4D, R1 or R2). The contribution from the seed bank reaches its maximum at the same gap size but stabilizes at this level (Fig. 4D, B1 or B2). If recently dispersed seeds establish relatively better than seeds in the soil (Fig. 4B, a versus b), because they respond more rapidly to germination cues than buried seeds or because they are larger, then their relative contribution would increase (Fig. 4D, R2 versus R2a).

In gaps formed in previously undisturbed forest, the size of the seed bank is much greater than incoming seed rain (Fig. 4C, B1 >> R1); consequently, pioneer regeneration from the seed bank will be greater than that from seed rain over all gap sizes (Fig. 4D, B1 >> R1). The percentage contribution of the seed bank to regeneration increases with gap size because the absolute contribution from seed rain decreases (Fig. 4D, R1).

After gap formation, the size of the soil seed bank is drastically reduced (Fig. 4C, B2 << B1). The seed rain into the gap does not increase (Fig. 4C, R1) until local reproduction occurs within the gap perhaps 2–4 years later [unless wind-dispersed seeds preferentially disperse into new gaps (see Augspurger and Franson, 1988; Schupp et al., in press)]. Hence, if a second disturbance occurs before local reproduction begins, seed rain will contribute relatively more to regeneration than the seed bank over all gap sizes (Fig. 4D, R1 > B2). Continued disturbance further reduces the seed bank and increases the role of seed rain. As pioneers in gaps mature, local seed rain increases and replenishes the seed bank within the gap and along its edges.

The model (Fig. 4A) best describes the early stages of gap dynamics in lowland forest following treefalls of increasing size. Because short-lived pioneers die out in larger gaps and suppressed late secondary and primary species gain dominance (Baur, 1968), the apparent contributions of each pathway change through time. In other forests, one pathway may dominate regeneration in all gap sizes while others are unimportant (e.g., sprouting predominates in heath and elfin forests). Particular disturbances may eliminate one pathway and enhance another. For example, fire destroys seedlings and sprouts, kills many seeds in the soil, stimulates others, but has no effect on seed rain. The degree to which one pathway can replace another depends on whether these curves are generated by the physiological tolerances of species, their competitive abilities, or predators and pathogens. If replacement does not occur, recovery will be slow. The relative position of each pathway along the gap size axis may differ among forests because the minimum gap size for pioneer establishment varies among forests (Brokaw, 1985a)

VI. Spatial and Temporal Heterogeneity

A. Intrasite and Intersite Variation

Much of the small- and large-scale spatial heterogeneity of soil seed banks within tropical vegetation (not associated with succession) may be related to patterns of seed dispersal, abundance of predators, local edaphic conditions (e.g., fertility or drainage), or other factors that have been little studied. Intrasite variation in seed density is considerable as measured by high coefficients of variation (8–88%, median = 40%; Tables 2 and 3). Seeds are clumped in woody species from forest soil (Uhl et al., 1981) and weeds from fields and pastures (Miège and Tchoumé, 1963; Kellman, 1978).

The seed shadows of tropical species are not well known, but are probably leptokurtic (Augspurger and Hogan, 1983; Howe, 1986; Augspurger and Franson, 1987; but see Murray, 1988). Behavioral, morphological, and physiological differences among bird, bat, monkey, elephant, rodent, fish, ant, and other dispersers (Howe, 1986) should cause very different seed shadows and clumping at different scales. These can be substantially altered through secondary dispersal by ants (Roberts and Heithaus, 1986), dung beetles (Estrada and Coates-Estrada, 1986), or heavy rainfall or runoff (Miège and Tchoumé, 1963; Hopkins and Graham, 1984b; Fleming et al., 1985). Within the limits of the seed shadow, I expect the degree of clumping to increase with distance from parent and wind-dispersed species to be less clumped than animal-dispersed species. The abundance of dispersal types has been little studied, but animal-dispersed species are common and wind-dispersed species are rare in the seed banks of upland evergreen forest in Ghana

(Swaine and Hall, 1983) and in young regrowth and forest in Costa Rica (Young *et al.*, 1987). This might be caused by predominance of animal-dispersed species in pioneer seed rain (as is the case in Costa Rica) or greater longevity of animal-dispersed seeds.

There is considerable spatial heterogeneity, largely unexplained, in seed banks among forest sites in the same region. Two mature Mexican forest sites <1 km apart had similar seed densities but different dominant life-forms and species (Tables 2 and 6, Ref. 7). Densities in two Venezuelan tierra firme forests several kilometers apart varied fourfold, but those in two caatinga forests were similar (Table 2, Refs. 21 and 26). Size and species richness of the seed bank differed significantly among four Australian forests 8–27 km apart (Table 2, Ref. 24), possibly reflecting differences in soil texture (Hopkins and Graham, 1983). In Ghana, variation in density from six forests 40–255 km apart (Tables 2, 6, and 8, Ref. 18) may reflect site-specific accumulation, the germinability of seeds in different soil types (Hall and Swaine, 1980), or the distance from regrowth and farms (Tables 2 and 6, Ref. 18).

B. Variation among Continents

Seed density in forest soil seed banks (Table 2) is highest in the Neotropics (60–4700 seeds m^{-2}), intermediate in Africa (45–900 seeds m^{-2}) and Australia–New Guinea (398–1120 seeds m^{-2}), and lowest in Southeast Asia (2–243 seeds m^{-2}). Low densities in Malaysia were attributed to site-specific gap dynamics (Putz and Appanah, 1987). Differences should be interpreted with caution, given the variability in methods. In addition, the neotropical minimum and penultimate maximum values come from the same French Guianan site (Table 2, Refs. 20 and 33), the maximum value comes from a forest site 10 m from regrowth, and nearly half the range of variation among all sites is displayed by the seasonal variation at one Mexican site (Table 2, Ref. 7). We need careful studies well-replicated in time and space carried out across the tropics using the same methods to document the existence of regional differences.

C. Temporal Variation

The size and species composition of the community seed bank should reflect seasonal changes in abundance of transient and seasonal-transient seed banks. These may explain the considerable within-year variation in size and species composition of a relatively aseasonal Mexican forest and the *Cecropia* seed bank in Surinam (Table 2, Refs. 7 and 23). In a Costa Rican forest, seed densities in June and October were not significantly different but species composition changed (Young *et al.*,

1987). Within-year variation was lacking in the *Trema* seed bank in the Ivory Coast (Alexandre, 1978).

Evidence for between-year variation in size or species composition of the seed bank is scanty. A. P. Smith (1987, and personal communication), in a well-replicated study (Appendix), found that seed density of understory herbs in a Panamanian forest varied little between years (0.3–0.9 seeds m^{-2}). In contrast, Cheke *et al.* (1979) report greater than fivefold differences in seed bank density between years in a Thai montane forest (Table 2), but sample sizes were small.

VII. Summary

Abundant seeds of pioneer species are present in the soil of mature tropical forests. Seed densities are much greater in soil of farms and secondary regrowth, where seeds of weedy species predominate. Many tropical vegetation types have not been sampled. The presence of seeds in a single soil sample is insufficient to establish the type of seed bank present, how long seeds remain viable, or role of the seed bank in regeneration.

The seed bank generally contributes more to regeneration than the seed rain, with pioneer species persisting in the soil for >2 yr. Due to the paucity of studies, this generalization may not hold for all types of tropical vegetation or in all successional stages. Thorough community studies are needed to understand intra- and intersite spatial and temporal heterogeneity of seed banks and their role in regeneration. The intensity of sampling needed to understand heterogeneity will be time consuming and labor intensive. Success will depend upon well-designed sampling programs and facility with seedling identification.

Population studies are desperately needed to determine the relative importance of the seed bank versus seed rain in regeneration for even one species and how this changes with forest type, gap size, and/or age. A combined observational and experimental approach would be needed.

The fate of seeds in the soil is still largely a mystery, including the rate and mechanisms of burial, the depth from which they will respond to gap creation, role of predators and pathogens, and the factors causing changes in seed physiology and longevity during burial. There are some very simple but extremely important experiments waiting to be done in this area. However, factors causing changes in seed physiology and longevity in the soil will require more complex, sophisticated approaches.

VIII. Appendix

Because the methods summarized in this Appendix vary greatly, Tables 2–8 in the preceding sections are extensively footnoted. Some data were reanalyzed to improve comparisons (e.g., ferns and contaminants were deleted). Hence, some values in my tables differ from those in previous shorter summaries (Kellman, 1974b; Ewel *et al.*, 1981; Uhl and Clark, 1983; Whitmore, 1983; Putz and Appanah, 1987) or from those in the original articles. Seed bank measurements were often

Table A1
Summary of methods used in tropical soil seed bank studies

			Collection of soil samples[p]							
			Soil samples			Replication[e]				
Reference[a]	Date	Country	Area[b] (cm²)	Depth[c] (cm)	Litter[d]	Sites	Types	Samples	Frequency	Design[f]
1. Symington	1933	Malaysia	40,469	3	Y	1	2	2[dd]	1	U
2. Aubréville	1947	Ivory Coast	40,000	5	—	1	1	1	1	U
3. Keay	1960	Nigeria	1871–3871[u]	6	—	2	1	1	1	—
4. Miège and Tchoumé	1963	Senegal	10,000	—[r]	Y	1	1	6	3	R, P
5. Blum	1968	Panama	—[q]	—[q]	—	1	3	10	1	H
6. Bell	1970	Puerto Rico	2500[v]	12	Y	1	3	1–3	1	PS
7. Guevara and Gómez-Pompa	1972	Mexico	40	12	—	1	2 × 2	16	8	RS
8. Lepe Gómez and Jiménez Avila	1972	Mexico	—[q]	60	—	1	2	4	1	P
9. Liew	1973	Malaysia	40,459	15	—	1	1	5	1	R
10. Vázquez-Yanes	1974, 1976a	Mexico	—[q]	3	—	1	3	2	1	—
11. Kellman	1974b	Belize	29	4	—	1	2	37–41	1	R
12. Kellman	1978	Belize	25	10*	Y	1	2	20	1	TS
13. Alexandre	1978	Ivory Coast	—	5	—	2	—	—	>1	—
14. Ashton	1978	Malaysia	10,000	—	Y	1	3	2–3	1	U, SD
15. Cheke *et al.*	1979	Thailand	2500–7500[w]	5	—	5	1	1	1	P
			10,000	5*	—	1	1	6	1	TD
16. Ewel and Conde	1979	Florida, USA	2500	3	—	1	7	3	1	TS
17. Ewel *et al.*	1979	Florida, USA	—	—	—	1	10	—	>1	U
18. Hall and Swaine	1980	Ghana	5000[u]	4	Y	6	1	2	1	PS

a small part of larger studies. Several studies in this Appendix are not summarized in Tables 2–8 because the data were incompatible with the chosen formats. Not included in this Appendix are (1) a few references that mention unpublished seed bank studies but give no quantitative data (Webb *et al.*, 1972; Fleming *et al.*, 1985; Goldberg and Kigel, 1986), (2) two relevant theses that I have not seen (Salmerón, 1984; Guevara, 1986), and (3) studies that contain no new or quantitative data (e.g., Juliano, 1940; Richards, 1952; Aikman, 1955; Schulz, 1960a,b; Budowski, 1961; Kellman, 1970a; Liew and Wong, 1973) that have been cited as if they do.

Methods[g]	Germination techniques[p]						Data summary[n]
	Trial length[h] (wk)	Census frequency[i] (wk)	Soil depth[j] (cm)	Cover over sample[k]	Type of control[l]	Seedlings removed[m]	
G	26	—	—	ON	N	—	D
E	na	na	na	na	na	na	TS
G*	52	13–26	3	ON	N	R	TS
G[r]	13–35	3	—[r]	ON	N	—	TR
G	—	—	—	—	N	—	F
G*	26	2	7–10	OM	SD	—	TG
G	6	6	—	EC	VN	—	MS
G, E	9	—	—	I	N		TR
G	22	4	16	EG	N	—	TR
G	4	1	—	I	N	—	TR
E + G	48	1	<1	I	N	I	M
E + G	≥26	1–2	<1	I	N	I, M	TR
E	na	na	na	na	na	na	MM
G	9	3–4	—	OM	N	—	TR
G, E	35	—	—	EM	SS	T	TS
G	39	—	—	EM	SS	T	TR
G	52	—	—	EC	N	T	M
G	—	—	—	EG	—	I	M
G*	14	1	3	ON	SF	R	TS

(continued)

Table A1 *(Continued)*

			Collection of soil samples[p]							
			Soil samples			Replication[e]				
Reference[a]	Date	Country	Area[b] (cm²)	Depth[c] (cm)	Litter[d]	Sites	Types	Samples	Frequency	Design[f]
19. Ewel et al.	1981	Costa Rica	114[x]	4	Y	1	3[t]	10	1	R
20. Prévost	1981	French Guiana	2500[y]	15	—	1	1	20[y]	1	U
21. Uhl et al.	1981	Venezuela	400	5	—	1	1	17–20	1	P, U
22. Uhl et al.	1982a	Venezuela	400	5	—	1	2	13–15	1	U
			452[z]	5	—	1	3	1	1	U
23. Holthuijzen and Boerboom	1982	Surinam	625	2*	Y	1	1	5	6	PS
24. Hopkins and Graham	1983	Australia	2500	5*	—	1	4	12	1	RS
25. Putz	1983	Panama	855	10*	N	1	1	10	1	RF
26. Uhl and Clark	1983	Venezuela	400	5	—	1	4	10–30	1	T, U
27. Marks, M. K.	1983a, b	Nigeria	4000[aa]	4	—	1	1	1	13	US
28. Hopkins and Graham	1984a	Australia	1250	5	—	1	8	4	1	RS
29. Hopkins and Graham	1984b	Australia	2500[bb]	3	—	1	1	8	1	—
30. Saulei	1984	Papua New Guinea	2500	5	—	1	1	5	1	U
31. Enright	1985	Papua New Guinea	400	8*	Y	1	1	4	1	PS
32. Young	1985	Costa Rica	110[cc]	20*	—	1	1	6[cc]	1	PS
33. de Foresta and Prévost	1986	French Guiana	625	2	N	1	6	4–16	1	RS, D
34. Soderstrom	1986	Costa Rica	100	5	N	1	2	—	1	TS
35. Smith	1987*	Panama	312[u]	2	Y	1	10	12	1*	R
36. Epp	1987	Ghana	100	5	—	1	5	25	1	R
37. Sabiiti and Wein	1987	Uganda	10,000	4*	—	1	2	10	1	R
38. Young et al.	1987	Costa Rica	57–314	4*	Y	1	3	3–12[s]	1–2	V
39. Putz and Appanah	1987	Malaysia	707	10*	—	1	2	4–8	1	HF
40. Lawton and Putz	1988	Costa Rica	1000	10	—	1	1	5	1	U
41. Guariguata and Azocar	1988	Venezuela	900	5	—	1	1	60	2	RT

	Germination techniques[p]						
Methods[g]	Trial length[h] (wk)	Census frequency[i] (wk)	Soil depth[j] (cm)	Cover over sample[k]	Type of control[l]	Seedlings removed[m]	Data summary[n]
G	11	4	1	EG	GN	—	M
G	22	—	15	OH	N	—	TT
G	>9	<1	—	ON	SO, V	M	M
G	>9	<1	—	ON	SO	M	MV
G	>9	<1	—	ON	SO	M	TS
G	9	<1	T	EC	DN	—	MR
G	14	2	2	EC	SH	I	MV
G	26	—	3	EC	GN	I	MV
G	>9	1	—	ON	SO	M	MV
G	4	1	4	EG	N	—	MS
G	31	1	2	EC	SH	I	MV
G	14	1	1	EC	U	—	MV
G	26	1	—	O	DS	I	M
G	10	1	1	EC	N[ee]	—	MV
G	17	2	1–2	EM	SS	I	MV
G	13	6	1	OM	GN	—	MV
G	6	6	<4	E	GH	—	MV
G*	52	—	—	EG	SS	—	M
G	7	<1	<1	O	N	MR	MV
E	na	na	na	na	na	na	MV
G	10–19	1	1	EG	DM	M	MV
G	22	—	2	OC	SS	I	MV
G	>9	—	3	OM	SM	I	MV
E	na	na	na	na	na	na	M

(*continued*)

Table A1 (*Continued*)

			Collection of soil samples[p]							
			Soil samples			Replication[e]				
Reference[a]	Date	Country	Area[b] (cm²)	Depth[c] (cm)	Litter[d]	Sites	Types	Samples	Frequency	Design[f]
42. G. Williams-Linera	Unpublished	Panama	900	5	N	1	1	10	1	T
43. Saulei and Swaine	in press	Papua New Guinea	2500	5	—	1	13	2–20	2–3*	UP
	Range[o]		25–40,469	2–60	—	1–6	1–10	1–120	1–13	
	Median[o]		950	5	—	1	2	6	1	

[a]References are listed in order of publication. *, Additional data by personal communication.

[b]Surface area of sample germinated in high light conditions.

[c]Depth of soil sample from surface; studies which analyze multiple depths are indicated by asterisk (*).

[d]Litter/humus layer included (Y) or excluded (N) from sample.

[e]Number of replicates at each level. Sites are widely separated locations (>50 km), usually with different regional vegetation types; types are vegetation types, habitats, successional stages, or treatments at one site; samples are soil samples; frequency is number of sampling periods during year, with an asterisk denoting replication in 2 yr. Additional replication within a level is indicated as A × B.

[f]Arrangement of replicates in study area. First letter indicates that replicates were placed along a transect (T), at random (R), within a randomized block design (B), haphazardly (H), in a nonrandom regular or clearly described pattern (P), in various (V) or unspecified (U) arrangements. The second letter indicates that samples were in transects or quadrats of specified size (S), different distances from clearings or young forest (D), or ≥20 m from gaps or forest edge (F).

[g]Methods used to estimate number of seeds in soil sample were germination (G), seed extraction (E), or extraction followed by germination (E + G). Samples were germinated under high light conditions; *, additional samples were germinated in shade.

[h]Length of time during which germination was observed.

[i]Length of interval between censuses.

[j]Depth to which soil was spread; T, thin layer.

[k]First letter indicates whether samples were placed in an open clearing (O), a screened or walled exclosure (E), or an incubator (I). Second letter indicates whether samples or exclosures were not covered (N) or that the covering was mesh (M), mesh over some samples only (H), transparent solid roof (C), unspecified type (mesh or solid, transparent or shaded) over greenhouse or shadehouse (G).

[l]First letter indicates that no control was used (N) or that soil (S), vermiculite (V), sand (G), soil or compost over sand (D), or an unspecified substance (U) was used. Second letter indicates that control medium was not sterilized (N); not sterilized but put in the shade (D); or sterilized with unspecified substance (U), methyl bromide (M), heat (H), oven-dry heat (O), flames (F), or steam (S).

[m]Seedlings were removed after identification (I); some or all were removed after an unspecified time (R); seedlings were transplanted when large enough (T); all were presumably removed before soil was mixed to bring ungerminated seeds to surface (M).

[n]A qualitative description (D); mean (M); mean and maximum (MM); mean and range (MR); seasonal mean (MS); mean and SD, SE, or CI (MV); totals of replicates, but mean and variance can be calculated (TR); total for site (TS); total for treatment (TT); total from site, but mean and variance can be estimated from graphs of replicates (TG); and frequency (F).

[o]Ranges in individual studies were averaged before calculating the overall range and median.

[p]Unavailable information (—) and nonapplicable categories (na) are distinguished.

Methods[g]	Germination techniques[p]				Type of control[l]	Seedlings removed[m]	Data summary[n]
	Trial length[h] (wk)	Census frequency[i] (wk)	Soil depth[j] (cm)	Cover over sample[k]			
G	13	1	<5	EM	SS	R	MV
G	26	—	—	E	SU	I,M	TT

[q]Area indeterminate: Ref. 5 filled 9-oz cups from a shovel-full of soil; Ref. 8 used 500-g samples; Ref. 10 pooled soil from four areas then divided the 20-kg sample into three treatments.

[r]Soil watered in place during dry season; depth indeterminate.

[s]Some replicates paired within blocks.

[t]Sequential samples of precut, postcut, and postburn forest soil; treatment effects might be compounded by seasonal effects.

[u]Original sample twice as large, but half separated and germinated in shade.

[v]Size of flat used for germination; actual area of soil sampled not specified.

[w]Original sample was 10,000 cm^2, but 2500 cm^2 was removed and analyzed by the extraction method.

[x]The sum of two equal-sized cores pooled together.

[y]Five 1-m^2 samples were collected; each was divided into 4 parts, two of which were covered with fine mesh.

[z]Sixteen 28-cm^2 subsamples pooled for each treatment.

[aa]Ten 400-cm^2 samples were mixed then placed into five trays for germination.

[bb]Eight greenhouse-germinated subsamples from a portion ($\frac{1}{6}$) of a 12-m^2 soil sample.

[cc]Five subsamples of 22 cm^2 were pooled for each of the six replicates.

[dd]Half of the replicates were covered with 0.5-in. wire netting.

[ee]New Guinea samples were germinated in a New Zealand greenhouse.

Acknowledgments

I thank T. M. Aide, N. Brokaw, D. De Steven, N. Devoe, K. Hogan, E. G. Leigh, F. E. Putz, A. S. Rand, E. W. Schupp, and S. J. Wright for constructive comments on the manuscript; S. Appanah, B. Brown, J. Ewel, M. Keller, C. Horvitz, R. Lawton, J. Molofsky, K. G. Murray, A. Orozco Segovia, L. Ponce de León, F. Putz, S. M. Saulei, D. Schemske, A. Smith, R. Stallard, M. D. Swaine, C. Vázquez-Yanes, T. C. Whitmore, G. Williams-Linera, and K. Young for use of manuscripts in press or unpublished data, valuable discussions, and/or help with references; J. Sevenster and C. Haverkate for Dutch translations; and the Smithsonian Institution Libraries staff at the Smithsonian Tropical Research Institute for obtaining countless references. This was supported in part by National Science Foundation Grant BSR-8517395.

CHAPTER 10

Impacts of Seed Banks on Grassland Community Structure and Population Dynamics

Kevin J. Rice

Department of Agronomy and Range Science
University of California
Davis, California

I. Introduction

At first glance, the structure of most grassland communities appears dominated by a single type of plant morphology. The more careful observer comes to know the diversity of structure and life history that exists within the grasses. In addition, closer inspection reveals a large number of dicot species that are quite diverse in morphology, phenology, and ecology. Very few people, however, come to appreciate the complexity hidden in the soil beneath a grassland community.

The seed banks in grasslands have generated a long-standing interest in basic and applied ecology. The economic necessities of understanding the potential of a seed bank to alter grassland composition and productivity have motivated researchers to compare the composition of the surface vegetation to seed reserves hidden in the soil. In this review, several questions motivated my examination of differences and similarities between above- and belowground composition. For example, if differences exist between the composition of the seed bank and the vegetation, what factors affect the magnitude of this difference? Are rates of disturbance important? Do dominant species within the vegetation play a large part in causing differences between above- and belowground flora? Are differences in seed bank composition among grasslands determined by the relative abundance of certain groups of species with long-lived seed banks?

I also consider the implications of seed banks for the population dynamics of grassland species. Survival in belowground seed populations is poorly understood. What might be the potential impact of granivory, microenvironment, and genetic variation on seed bank decay patterns? How might germination cueing in response to environmental heterogeneity affect the spatial distribution of seed banks? How much do we know about the effects of seed aging on the relative vigor of seedlings emerging from differently aged seed cohorts?

Finally, some modeling results on the population dynamics of annual plants with a seed bank are presented to illustrate demographic implications of seed bank age structure. Specifically, how might qualitative variation in seed carry-over and reproductive patterns among seed bank cohorts affect the dynamics of age structure in a seed bank?

II. Seed Banks and Community Structure in Grasslands

A. Aboveground versus Belowground Botanical Composition

Species composition in the grassland seed bank flora and the aboveground plant community is strikingly dissimilar (Chippendale and

Milton, 1934; Champness and Morris, 1948; Douglas, 1965; Major and Pyott, 1966; Johnston *et al.*, 1969; Roberts, 1972b; Hayashi and Numata, 1975). The implications of this difference in species composition depend on the research objectives and orientation of a particular study. Discrepancies between above- and belowground composition can make the description of a grassland flora for a particular locale extremely difficult (Major and Pyott, 1966), because a complete description of the plant community should include the seed bank. The relative accuracy of seed bank estimates depends on a number of factors, such as the number of samples, the density and distribution patterns of the species within the vegetation, and the season during which samples are taken (Major and Pyott, 1966). For example, the correspondence between the above- and belowground flora in a sagebrush grassland depended on the season of sample collection, whether the sample was collected under shrub canopy, whether the sample site had been recently burned, and, if recently burned, the relative intensity of the fire (Hassan and West, 1986). In applied ecological studies, differences between above- and belowground populations of a desirable forage species can indicate potential management problems. Douglas (1965) found that an important pasture grass, *Lolium perenne*, was not present in the seed bank despite its vegetative dominance in the standing vegetation. He suggested that any attempts to use aboveground composition to assess the success of sward renewal by cultivation or herbicide application would be grossly inaccurate because the seed bank was dominated by indigenous grasses and weeds.

In British perennial grasslands, Brenchley (1918) noted that only in permanent grassland was there a good correlation between seed in the soil and aboveground vegetation. Differences in composition between the seed bank and the standing vegetation were caused by arable weed species and generally decreased with age of the pasture or grassland. In moderately aged pastures (60 yr old), seed banks of arable weeds, although still present, are smaller than in pastures recently (10 yr old) released from cultivation (Brenchley, 1918; Chippendale and Milton, 1934; Douglas, 1965)

The degree of divergence between the species composition of the seed bank and the vegetation may vary among grassland types. This may depend on the size of the seed bank formed by the dominant taxa. Canopy dominants, such as *Lolium perenne* or *Dactylis glomerata*, often are poorly represented within the seed bank, while other indigenous grasses, such as *Agrostis* spp. and *Poa* spp., form substantial seed reserves (Chippendale and Milton, 1934; Champness and Morris, 1948; Douglas, 1965; Jalloq, 1975). In Japan, the vegetation of meadows dominated by *Miscanthus sinensis* or *Arundinella hirta* is poorly correlated with seed bank composition, while in meadows dominated by *Zoysia japonica* there is a general correspondence (Hayashi and Numata, 1975). In Great Britain, upland pastures dominated by *Agrostis tenuis* exhibit closer correlation between seed bank composition and surface vegetation than do lowland sites (Champness and Morris, 1948).

B. Seed Bank Longevity in Annuals versus Perennials

In grasslands, annuals are more likely than perennials to produce a seed bank. In a bunchgrass community in California, Major and Pyott (1966) found that the relative contribution of annuals and perennials to the seed bank was substantially different. Seed banks for the perennial grasses (primarily *Stipa* spp.) were absent (Fig. 1). In contrast, annual grasses had well-developed seed banks despite their relatively minor contribution (< 10%) to total cover. A survey of a variety of grazed lands in Great Britain indicated that species with a large contribution to surface vegetation cover and a small contribution to the seed bank were almost exclusively perennials (Champness and Morris, 1948). A similar survey of grasslands and meadows in Japan suggests that annuals and biennials consistently formed more extensive seed banks than did perennials (Hayashi and Numata, 1975).

Obviously, the sharpness of this distinction between perennials and annuals may depend on the relative allocation of the perennials to seed production. As noted above, dominant grasses, such as *Lolium perenne*, that reproduce primarily by vegetative means are almost completely absent from the seed bank (Chippendale and Milton, 1934; Champness and Morris, 1948; Douglas, 1965; Jalloq, 1975). In contrast, well-developed seed reserves are found for perennial grasses that reproduce pri-

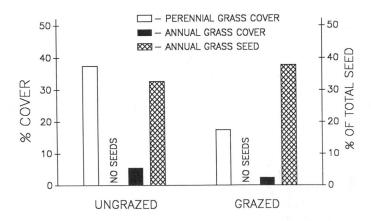

Figure 1. Differences between vegetative cover and seed bank composition for perennial and annual grass species in a California perennial bunchgrass (*Stipa* spp.) community (data from Major and Pyott, 1966).

marily by seed, such as *Zoysia japonica*, *Poa trivialis*, *Poa pratensis*, and *Agrostis* spp. (Chippendale and Milton, 1934; Champness and Morris, 1948; Hayashi and Numata, 1975).

C. Seed Bank Longevity in Grasses versus Forbs

As a group, forbs (i.e., any herbaceous dicot that is not a legume) seem more likely to form seed banks than do grass species. In a broad survey of seed banks, Roberts (1981) examined the relative contributions of different plant groups to the seed banks of several grassland communities. Averages taken across 13 different communities indicated that grasses contributed approximately 30% of the total viable seeds, while forbs dominated, contributing 50%. In a description of the persistent seed bank of a Californian annual grassland, Young *et al.* (1981) estimated rates of seed carry-over from one year to the next for annual grasses, annual forbs, and annual legumes (Fig. 2). Within the annual flora, carry-over rates were low, but data over the 5 yr of the study indicated that annual forbs produced a more persistent seed bank than did annual grasses. The increased carry-over (percentage) for all three types of annuals in the final year of the study suggests that seed bank persistence may vary over time. Among colonizing species that invade molehills in British perennial grasslands (Fig. 3), the weedy forbs have more extensive seed banks than do the weedy grasses or rushes and sedges (Jalloq, 1975). Champness and Morris (1948) also commented on the dominance of seed banks by forbs that often were rare in the surface vegetation. In contrast, Chippendale and Milton (1934) found that in certain types of

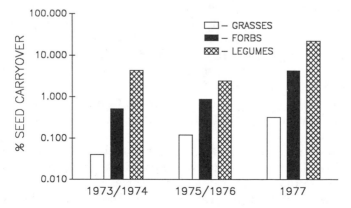

Figure 2. Seed carry-over for different annual species groups within a California annual grassland (note log scale). Numbers of germinable seed from soil samples taken after autumn germination were used to estimate potential rates of seed carry-over (data from Young *et al.*, 1981).

Figure 3. Comparison of the vegetative cover of sown and indigenous grassland species in undisturbed sward and the seed bank composition of these species in soil disturbances (molehills) (data from Jalloq, 1975).

grasslands the germinable seed banks of many forbs were low relative to seed pools of the dominant grasses (*Poa annua*, *Poa trivialis*, *Holcus lanatus*, and *Agrostis* spp.). The reliance of these particular grass species on reproduction by seed may explain this reversal in numerical dominance.

D. Legumes in Seed Banks

Seed coat impermeability to water (hard seed) is a form of seed dormancy (see Baskin and Baskin, Chapter 4, this volume) widespread in legumes (Rolston, 1978) and is a major factor in promoting the formation of legume seed banks (Quinlivan, 1968). In California annual grasslands, carry-over rates in annual legumes (Fig. 2) are an order of magnitude greater than in the other dicots and two orders of magnitude greater than in annual grasses (Young *et al.*, 1981). Legume genera, such as *Lotus*, *Trifolium*, and *Medicago*, had consistently higher seed populations than would be expected from their occurrence in British grassland vegetation, and no legume species were present among those that had a disproportionately low contribution to the seed bank (Champness and Morris, 1948).

Seed longevity may explain, in part, the abundance of legumes in grassland seed banks. Hull (1973) compared the germination rates for a number of North American rangeland species that had been stored in unheated sheds for 14 to 41 yr. Seeds of legumes retained their viability longer and had higher rates of germination than did most of the grass and forb species tested, regardless of storage time. A study by Lewis (1973) on the effects of soil type and burial depth on the viability of seeds enclosed in mesh bags indicated that only a few weed species had higher longevity than legumes. At least some seeds of most legumes persisted the length of the experiment (20 yr). Longevity of legume seed was slightly higher in loam soil than in acid peat and also increased with burial depth.

Differences among legumes in seed bank longevity can often be related to variation in seed dormancy. High persistence of *Lotononis bainesii* and *Trifolium repens* in subtropical pastures corresponded with high percentages of hard seed, 97% and 70%, respectively (Jones and Evans, 1977). In contrast, *Desmodium intortum* had a low percentage of hard seed, 9%, and was absent from heavily grazed areas where replenishment of seed reserves by seed production was severely reduced. Differences in seed bank longevity caused by differences in hard seed percentages may also exist among cultivars of a single legume species. The potential management implications of such differences are illustrated by an Australian study of *Trifolium subterraneum* cultivars (Beale, 1974). The Yarloop cultivar of *Trifolium subterraneum* was the primary cultivar in seed mixtures from 1957 to 1963 because it was a superior forage. After the Yarloop cultivar was found to induce infertility in sheep and to be susceptible to clover scorch (*Kabatiella caulivora*), many pastures were ripped up and replanted to other cultivars. Although many of these replanted pastures showed successful establishment of the new cultivars in the year of resowing, a reversion to the Yarloop cultivar often occurred in the following year. Beale (1974) demonstrated that this reversion was caused by regeneration from persistent seed reserves in the soil and that these well-developed seed banks would make conversion to the new cultivars extremely difficult.

E. Weedy Species in Seed Banks

Weedy or fugitive species, those that establish principally after disturbance, are among the largest component of grassland seed banks. As noted above, a large portion of the resident seed bank in a grassland may be arable weeds persistent from previous periods of cultivation (Brenchley, 1918; Chippendale and Milton, 1934; Champness and Morris, 1948; Douglas, 1965). In addition to arable weeds, native weedy species in grasslands also are characterized by large seed banks. For example, badger mounds in native tallgrass prairie create gaps in the vegetation that are colonized frequently by *Oenothera biennis* (Platt, 1975). Seeds of this species can remain viable in the soil for 80–100 yr (Kivilaan and Bandurski, 1973). Another prairie species, *Sporobolus cryptandrus*, can create extensive seed reserves in the soil (up to 20,000 seeds m^{-2}) following invasion after disturbance (Lippert and Hopkins, 1950). A series of defoliation experiments conducted in permanent grassland in

Great Britain demonstrated that seedling establishment of the grassland daisy, *Bellis perennis*, is enhanced significantly if the resident vegetation is disturbed by clipping or herbicide treatments (Foster, 1964, cited in Harper, 1977). Large and persistent seed bank populations of *Bellis perennis* (Chippendale and Milton, 1934; Champness and Morris, 1948), *Oenothera biennis*, and *Sporobolus cryptandrus* illustrate a temporal disperser colonization strategy. However, studies on the population dynamics of three *Ranunculus* species indicate that there is not always a consistent correlation between seed bank persistence and the ability to colonize soil disturbances or vegetation gaps (Sarukhán, 1974; Sarukhán and Gadgil, 1974). Demographic and mathematical analyses suggest that *Ranunculus repens* inhabits relatively stable environments with little variation in population growth rates in space and time, in contrast to *Ranunculus bulbosa*. However, seed bank longevity for the fugitive species, *Ranunculus bulbosa*, is much shorter than for *Ranunculus repens*, the species found growing in less disturbed habitats.

Seeds of fugitive species are able to use various environmental cues, such as temperature and light quality, to detect gaps in the vegetation (Wesson and Wareing, 1967; Grime and Jarvis, 1975; Thompson *et al.*, 1977; Fenner, 1978; Silvertown, 1980a; Rice, 1985). Because of the sensitivity and specificity of germination cueing behavior, many of the traditional techniques for estimating seed reserves may be inaccurate. An example is provided by an annual forb, *Erodium botrys* (Rice, 1985, 1986). Germination tests of soil cores collected in early autumn before field germination indicated that large numbers of germinable seeds were present (Fig. 4). Plant number in the spring roughly correlates with

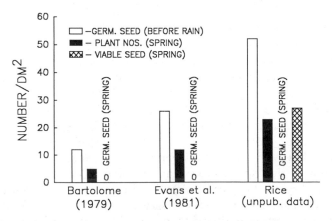

Figure 4.   Seasonal variation in numbers of germinable seed of *Erodium botrys*. The presence of viable but not germinable seed in spring soil samples provides evidence for a persistent *Erodium* seed bank.

these estimates of germinable seed. In contrast to the autumn results, soil cores collected in the spring lacked germinable seed (Fig. 4), suggesting that there is little carry-over from one year to the next. However, when soil from cores collected in the spring was sieved, a large number of viable *Erodium botrys* seeds were recovered (Fig. 4) that germinated only when mechanically scarified (K. Rice, unpublished data). *Erodium* seed dormancy is caused by a seed coat impermeability to water and is broken by wide diurnal temperature fluctuations that exist only during the rainless summer months (Rice, 1985). Thus, the pool of germinable seeds increases through the summer until the cooler temperatures that begin with the onset of winter rains. Temperature fluctuations within a greenhouse during germination testing of soil cores are unlikely to break *Erodium* seed dormancy.

A summary of the composition of the seed banks of several grassland communities illustrates wide variation in both seed density and the number of species present in the seed pool (Table 1). Although much of this variation may be caused by the particular sampling protocol of a study (e.g., time of year sample was taken), some general trends are apparent. Although dominant in the vegetation, grasses are relatively underrepresented in the seed bank. On the other hand, dicots and grasslike plants (when present at all) are major contributors to seed pools in grasslands. Seed bank species diversity, as measured by number of species, also appears higher for dicots than in the grasses.

III. Seed Banks and Population Dynamics in Grasslands

A. Patterns of Seed Bank Depletion

Patterns of seed mortality in collections kept under air-dry storage (Harrington, 1972) indicate that the loss of viability during shelf storage follows a negative cumulative normal distribution. A depletion curve with this functional form suggests the importance of seed senescence where the probability of mortality depends on seed age. Roberts (1972b) suggests that such curves may result from the gradual accumulation of random cell deaths to some critical threshold after which the seed becomes inviable. In contrast, several studies on the depletion patterns of weed seeds in the soil (Roberts and Dawkins, 1967; Roberts and Feast, 1973a; Warnes and Andersen, 1984) suggest that the rate of seed bank depletion is constant, resulting in a negative exponential distribution for seed loss. This type of depletion pattern indicates that the probability of seed loss is independent of seed age. A constant depletion rate suggests that age-independent losses to germination, granivory, and disease are more important than losses caused by senescence.

Table 1
Contributions of different taxa to seed bank density (germinable seed)
and species diversity in several types of grassland communities

Location	Grassland type	Grasses	Rushes and sedges	Legumes	Other dicots	Total
			Number of species per sample[a]			
Kansas	Short-grass prairie	6	0	1	9	16
	Mixed grass prairie	10	0	1	10	21
	Mid-grass prairie	1	0	0	11	12
Missouri	Tall-grass prairie	7	2	0	15	24
Saskatchewan	Mid-grass prairie	1	6	1	7	15
Alberta	Fescue prairie	4+	0	0	4+	16+
	Mixed-grass prairie	4+	1+	0	2+	11+
California	*Stipa* grassland					
	(ungrazed)	5	0	0	5	10
	(grazed)	6	0	3	8	17
California	Annual grassland					
	Before autumn germination	7	0	4+	5+	16+
	After autumn germination	3	0	3+	1+	7+
Wales	Acid grassland	9	2	0	7	18
Wales	*Molinia* grassland	6	6+	2	5	19+
	Nardus grasslands	4	5+	0	6	15+
	Calluna grassland	1+	6+	0	4	11+
	Marsh grassland	9+	10+	3	29	51+
Bohemia	Peat meadows (mean of 8 sites)	10+	2+	7	43	62+
Tasmania	*Poa gunnii* grassland	2	3	0	8	13
Japan	*Zoysia* grassland	2	4	1	9	19

[a]Species numbers followed by a plus sign represent minimum estimates.

For many weedy grassland species the survivorship of seeds in storage is often a very poor predictor of potential longevity in the field. Seeds of *Ranunculus repens* and *Chenopodium album* were completely inviable after 20 yr of storage under granary conditions, but after 20 yr in mineral soil, viability estimates were 51 and 32%, respectively (Lewis, 1973). Conversely, viability remained high in granary collections of *Geranium dissectum* and *Geranium molle*, but longevity in buried populations of these species was less than 4 yr. Collections of *Bromus tectorum* stored for 10 yr showed virtually no loss of viability (Hull, 1973); however demographic studies in steppe communities indicated very little carryover from one year to the next (Mack and Pyke, 1983).

Grasses	Rushes and sedges	Legumes	Other dicots	Unknown	Total	Reference
	Number of germinable seeds per m^2					
377	0	4	380	0	761	Lippert and Hopkins
122	0	11	273	0	406	(1950)
7	0	0	280	0	287	
492	3026	0	1488	1362	6368	Rabinowitz (1981)
2697	0	0	5533	0	8230	Major and Pyott
4743	0	326	7208	0	12,227	(1966)
18,050	0	6350	3000	0	27,400	Young et al. (1981)
700	0	900	100	0	1700	
207	445	64	675	0	1391	Archibold (1981)
1303	0	0	1752	0	3055	Johnston et al. (1969)
566	0	0	474	0	1040	
5570	85	0	3458	72	9185	King (1976)
484	1184	32	9612	0	11,312	Chippendale and
430	9020	0	5737	0	15,187	Milton (1934)
43	1421	0	8396	0	9860	
3143	20,387	2260	5554	0	31,344	
1102	408	172	1585	0	3267	Mika (1978)
351	781	0	1072	0	2204	Howard (1974)
9770	4460	10	9160	30	23,430	Hayashi and Numata (1975)

B. Effects of Germination Cueing on Seed Bank Depletion

Although depletion rates in weed seed populations may remain constant for a particular environment, seed bank depletion rates are sensitive to the frequency of soil disturbance (Roberts and Feast, 1973a). This suggests that the ability of many grassland species to germinate in response to soil disturbance or gap formation may result in a coupling between rates of seed bank depletion and spatial variation in germination cues such as temperature and light.

In certain grassland species, germination depends on variation in light quantity and quality. Wesson and Wareing (1967) contributed much of the earlier information on light-promoted germination and the induction of a light requirement by seed aging. Later studies by Grime and Jarvis (1975) support the generalization that many grasses do not require light for germination. In contrast, germination of many forb species is significantly reduced in the dark or under light filtered by plant canopy. This response is especially pronounced in species that colonize bare ground and emphasizes the importance of light quality as a germination cue in plants that require vegetation gaps for successful establishment. Silvertown (1980a) found that 17 of the 25 grassland dicot species he tested had significantly lower germination under light filtered by a leaf canopy than in the dark.

In a survey of 112 native herbaceous species, Thompson *et al.* (1977) found that germination of grassland species with persistent seed banks was often sensitive to temperature fluctuations in darkness. They speculated that such sensitivity allows buried seeds to detect canopy gaps. Rice (1985) reported that variation in diurnal temperature range for three types of microsites in annual grassland (under grass litter, bare soil, and buried under gopher mounds) significantly affected rates of dormancy release in *Erodium botrys*. Microsite variation in germination cues might have a strong influence on seed bank depletion by affecting the loss rate due to germination. Using data from Rice (1985) and assuming a constant depletion rate for each type of microsite, one would predict much more persistent *Erodium* seed banks under gopher mounds than in bare soil areas (Fig. 5). Thus, a large amount of spatial heterogeneity in physical parameters, such as temperature and light, may create a parallel amount of spatial variation in the longevity of a species' seed bank.

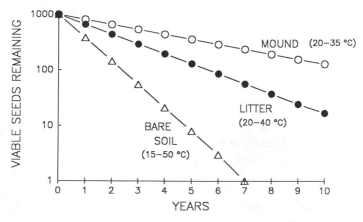

Figure 5.   Predicted effects of varying microsite temperature regimes on seed bank longevity in *Erodium botrys* within California annual grasslands (note log scale) (data from Rice, 1985).

C. Impacts of Granivory on Seed Banks

Beginning with the work of Chippendale and Milton (1934), there has been much speculation on the importance of granivory in the spatial and temporal dynamics of grassland seed banks. Unfortunately, the general difficulty in identifying sources of seed mortality has forced much of the discussion to remain highly speculative. A notable exception is provided by a study of small mammal grazing on the seed banks of *Avena fatua* and *Avena barbata* in California annual grasslands (Marshall and Jain, 1970). Rates of loss to predation were highly variable. At one site there were no detectable losses, but at another site 65% of the seed reserve was consumed. Variation was such that at one site the degree of predation was independent of species while at another rates of seed loss were higher for *Avena barbata* than for *Avena fatua*.

The potential effects of earthworm foraging on grassland seed banks was explored by examining the selectivity of earthworms (*Lumbricus terrestris*) and the viability of seeds recovered from earthworm casts (McRill and Sagar, 1973). Ingestion rates and the percentage of ingested seeds recovered from casts varied widely among plant species. For example, earthworms ingested 60% of the *Poa annua* seed offered but only 3% of *Lolium perenne*. In addition, the recovery rate of seeds from earthworm casts was lower for *Poa annua* (28%) than for *Lolium perenne* (67%). Even among species preferred by the earthworms, large differences occurred in the survival of ingested seeds. *Agrostis tenuis* was the second most preferred species (50% of offered seeds ingested), and no seeds of this species were ever recovered from worm casts. Additional germination trials in this study suggested that earthworms also may deplete seed banks by increasing germination rates. Germination rates of *Poa trivialis*, *Bellis perennis*, and *Trifolium repens* increased significantly after passage through the earthworm gut.

Taken together, these studies suggest the potential importance of vertebrate and invertebrate consumers on the longevity of seed banks and the spatial distribution of seed reserves. There is a critical shortage of field studies that examine selective seed predation, differences among microhabitats in probability of predation, and seasonal or annual variation in predation pressure.

D. Genetic Variation in Seed Bank Depletion Rates

Few studies have focused on the genetic basis of variation in seed bank longevity as it relates to seed dormancy. Interpopulational variation for seed dormancy within a species is often substantial. For several annual grassland species, Jain (1982) detected significant intraspecific differences in seed dormancy among populations located along a rainfall gradient in California. In addition, heritability estimates suggested significant within-population variability for *Trifolium hirtum* seed dormancy. Sexsmith (1967) reported significant differences in degree of seed dormancy both among and within varieties of *Avena fatua*, and phenotypic expression of these genetic differences was affected by the physical environment during seed maturation. Seed dormancy variation in *Taeniatherum caput-medusae* results from an interaction of environmental conditions during seed maturation and genetic differences between populations (Nelson and Wilson, 1969). Varietal differences in seed dormancy also is well documented for several annual grassland legumes (Cameron, 1967; Quinlivan, 1966, 1968). Quinlivan (1968) found that for *Trifolium subterraneum*, an interaction of variety and temperature determined the rate at which the seed coat became permeable. Data from germination trials conducted under one daily temperature fluctuation regime (15–40°C) illustrate how varietal variation in seed dormancy could affect seed bank depletion rates (Fig. 6). Assuming that losses from the seed bank result only from germination of permeable seeds and that the rate of such losses remains constant, there are large potential differences in seed bank longevity among the four cultivars (Fig. 6). In addition, genotype and environment interactions also are suggested because changes in diurnal temperature regimes had differential effects on loss of dormancy. Though limited in number, these studies suggest that patterns of seed bank depletion in many grassland species may result from a complex interplay of genetics and spatial variation in the seed microenvironment.

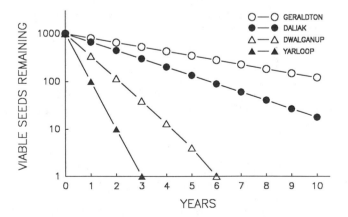

Figure 6.   Predicted effects of varietal variation in seed dormancy on *Trifolium subterraneum* seed bank longevity (note log scale) (data from Quinlivan, 1968).

E. Variation in Vigor among Seed Bank Cohorts

Only a handful of studies provide any data on the effects of seed age on postgermination plant growth, development, and reproduction. How germination rate or the growth rate of emergent seedlings may vary with respect to seed age are questions that are not considered. Seeds of grasses, legumes, and weedy species that survived 20 yr of burial in undisturbed soil produced plants that were apparently normal in development and reproduction (Lewis, 1973). Hull (1973), however, found that, in addition to differences between species, speed of germination within each species is affected significantly by seed age. Seedlings emerged more rapidly from younger seeds than from older ones. Villiers (1973) speculated that such delays in emergence may be caused by the necessity, upon hydration, for repair of membrane damage accumulated during dormancy before further cell elongation and development can occur. Given that even small delays in emergence can result in a significant competitive disadvantage in dense populations (Black and Wilkinson, 1963), the demographic consequences of delayed emergence in older seed bank cohorts deserve further study.

IV. Annuals with Seed Banks: Modeling the Implications of Age Structure

Carry-over of seed in an annual plant population between years has highly significant demographic consequences. Seed carry-over in persistent seed banks results in overlapping generations and creates age structure within the seed bank cohorts. The population dynamics of annuals with a seed bank were modeled by Schmidt and Lawlor (1983) using a matrix projection technique developed by Leslie (1945). They studied the sensitivity of the finite rate of increase (dominant positive eigenvalue of the matrix) to variation in life history parameters. The finite rate of increase represents the rate of exponential growth when the population reaches a stable age distribution. They found the finite rate of increase to be more sensitive to plant survival and fecundity than to germination fraction. Their analysis also demonstrates the formal equivalence of a seed bank annual matrix to the matrix form developed by Leslie (1945) for an iteroparous organism. By focusing on the response of the finite rate of increase, the analysis of Schmidt and Lawlor (1983) is based upon what is known as the limiting behavior of the population (i.e., the population growth rate attained in the limit as time increases and a stable age distribution is realized).

Another approach to the dynamics of age-structured populations is to examine the transient dynamics of the population that occur before the population reaches a stable age distribution. This form of analysis is especially appropriate for colonizing plant (Caswell and Werner, 1978) or insect (Taylor, 1979) species because changes in environmental conditions often prevent such populations from ever reaching a stable age distribution. Using the number of generations required to reach a stable age distribution as an index, one can examine how changes in life history parameters influence transient population dynamics (Taylor, 1979). Increased time required to reach a stable age distribution reflects a tendency within a population for more prolonged oscillations in age structure before reaching a stable age distribution (Lefkovitch, 1971).

To examine how changes in seed bank life history parameters might influence oscillations in seed age structure, I used a form of the matrix model developed by Schmidt and Lawlor (1983) that contained five different seed age classes. Probabilities of germination, seedling survival, and reproductive output for a particular seed age class were varied. This allowed examination of the effects of qualitative changes in reproductive patterns on transient behavior in seed age structure. Seed carry-over patterns were also varied to explore their impact on transient behavior. Using the number of generations required to reach a stable age distribution as an index of oscillatory behavior, the interaction of seed carry-over and delays in reproduction on transient dynamics of seed bank age structure can be studied. Four types of reproductive patterns were examined (Fig. 7): (1) no delay (constant)—no delay in constant reproduction (i.e., true annual pattern) such that newborn seeds (age class 0) survive after germination to reproduce at rates comparable to other seed age classes, and rates of reproduction are constant among all seed age

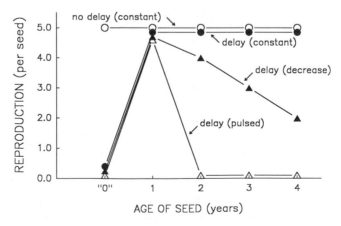

Figure 7. Reproductive patterns for seed age cohorts used in the matrix model of the transient dynamics of age structure. The 0 age class represents newborn seed. (See text for details.)

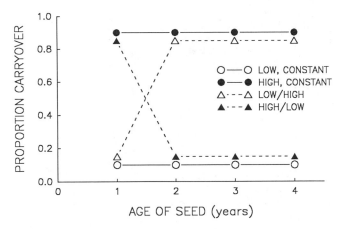

Figure 8. Patterns of seed carry-over used in the matrix model of the transient dynamics of age structure. (See text for details.)

classes; (2) delay (constant)—1-yr delay where seedling survival and reproduction in newborn seed is relatively lower than in other age classes, but reproduction is constant beginning with 1-yr-old seed; (3) delay (pulsed)—1-yr delay that is similar to (2) but where reproduction is concentrated in the 1-yr-old age class; and (4) delay (decrease)—1-yr delay that is also similar to (2) but where reproduction after the 1-yr-old age class gradually decreases. There were also four types of carry-over patterns examined (Fig. 8): (1) low and constant rates of seed carry-over; (2) high and constant rates of seed carry-over; (3) low rate of carry-over in the first year followed by high rates in succeeding years; and (4) a high rate of carry-over in the first year followed by low rates in older age classes. These particular patterns of reproduction and seed carry-over were chosen more for their qualitative differences in configuration than for any direct application to a particular species. Although the interaction of carry-over and reproductive patterns during convergence to a stable age distribution are complex (Fig. 9), certain general patterns emerge. Delays in reproduction, caused by either reduced reproduction in newborn seed or by high carry-over rates to later age classes, dramatically increase the time required to attain a stable age distribution. Given some delay in reproduction, oscillatory dynamics also are promoted by concentrating reproduction on a single seed age class. This effect is demonstrated by increased time to a stable age distribution in the pulsed reproductive pattern and in the high/low carry-over pattern where most germination occurs in the 1-yr-old seed age class. These results are similar to the those reported by Taylor (1979) in models of insect populations. He found that rates of convergence to a stable age distribution

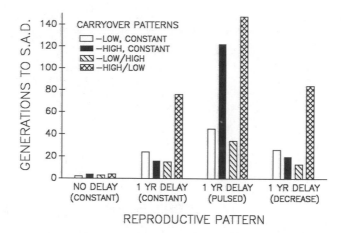

Figure 9. Interactive effects of reproductive patterns (Fig. 7) and carry-over (Fig. 8) on convergence to a stable age distribution (S.A.D.) in the matrix model of seed bank age structure.

were very sensitive to changes in the age of first reproduction and to changes in the variance of the age-specific birth rate. The importance of the pattern of the age-specific birth rate on the dynamics of these models further emphasizes the need for better information on the growth and reproductive vigor of different seed age cohorts within seed banks.

Delays in reproduction that result from the formation of seed banks effectively increase the generation time of an annual species. The qualitative relationship between generation time (T), net reproductive rate (R_0), and the finite rate of increase for the seed bank matrix model is shown in Fig. 10. The rapid decline in the finite rate of increase with increasing generation time would suggest little selective advantage for the formation of a seed bank. However, selection for delayed reproduction may depend on whether a population is increasing or decreasing (Mertz, 1971). From the equation in Fig. 10, it can be seen that in increasing populations ($\ln R_0 > 0$), any increase in generation time (T) will decrease the finite rate of increase. However, in a declining population ($\ln R_0 < 0$), an increase in T will slow the rate of population decline. Thus in populations that experience long periods of gradual increase in population size separated by brief, rapid periods of decline, one might expect selection to act against the formation of seed banks. In contrast, one might anticipate selection for a seed bank in populations (e.g., weedy species) that experience rapid bursts of population growth followed by long periods of population decline.

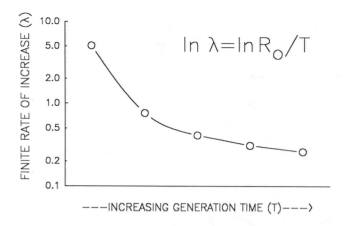

$$\ln \lambda = \ln R_O / T$$

———INCREASING GENERATION TIME (T)———⟩

Figure 10.   Effects of increasing generation time on the finite rate of increase in an annual species with a seed bank (data from the matrix model of seed bank age structure, Fig. 9.)

V. Summary

In many grassland communities there is a distinct difference between seed reserves and the surface vegetation. Dissimilarities between above- and belowground populations are caused, in part, by differences among taxa in the occurrence and persistence of seed banks. Seed banks appear to be (1) more developed in annuals than in perennials; (2) more extensive in forbs than in grasses; (3) widespread in leguminous species; and (4) common in weedy or fugitive species that colonize disturbances or gaps in the vegetation. These generalizations encompass designations that are by no means independent; the trend that forbs have more persistent seed banks than grasses may be related, in large part, to the weediness of many grassland forbs.

The demography of seed banks of most grassland species is still poorly understood. The work on arable weeds indicates that seed bank decay rates are constant within a particular environment; this possibility needs to be tested for typical grassland species before realistic models of seed bank dynamics can be created. How seed banks are affected by the interaction of genetic variation in dormancy and spatial variation in the microenvironment is poorly known even for important pasture species. The scant information available on the selectivity and intensity of granivory suggest that seed predation pressure is highly variable in space and time. There is a striking lack of information on the effects of seed aging on the relative growth, reproduction, and competitive ability of differently aged seed cohorts. This information would be valuable especially because the behaviors of models of seed bank dynamics in annual plants are affected significantly by potential variation in reproductive output among seed bank cohorts. Finally, selection for a seed bank strategy in a population may depend on whether the population usually is increasing or decreasing in size. Seed banks might be expected to arise in weedy species that experience bursts of population growth in response to openings in the canopy by disturbance. After each isolated episode of rapid growth, the populations of the weedy species decline as sward vegetation slowly recovers from disturbance. In this situation, delays in reproduction caused by a seed bank might be selectively advantageous because an increase in generation time would slow rates of population decline.

Acknowledgment

This work was supported in part by a grant from the Alfred P. Sloan Foundation.

CHAPTER 11

Seed Banks in California Chaparral and Other Mediterranean Climate Shrublands

V. Thomas Parker

Department of Biology
San Francisco State University
San Francisco, California

Victoria R. Kelly

Institute of Ecosystem Studies
The New York Botanical Garden
Mary Flagler Cary Arboretum
Millbrook, New York

I. Introduction

Mediterranean summer-dry climates are dominated by sclerophyllous evergreen shrub communities. This vegetation type is found in California (chaparral), in central Chile (mattoral), around the Mediterranean Sea (e.g., macchia, maquis, and garrique), on the south and west coasts of South Africa (fynbos), and in southern and southwestern Australia (e.g., mallee–heath). Although these ecosystems are diverse and unique in the plant and animal communities that compose them and in other respects (Cody and Mooney, 1978), climatic stress, nutrient limitations, and recurrent fires each limit the number and type of species that can survive. This is especially true at the critical stage of seedling establishment. Recent reviews of various aspects of evolution, fire, physiology, and nutrient relationships can be found in di Castri and Mooney (1973), Mooney and Conrad (1977), Cody and Mooney (1978), Miller (1981), Conrad and Oechel (1982), and Kruger et al. (1983).

These diverse vegetations share similarities at the seedling establishment stage. Two principal establishment environments exist temporally within chaparral. One, undisturbed chaparral, often in older stands, has several stresses, including low levels of resources, growth-inhibiting compounds, and high levels of predation (McPherson and Muller, 1969; Christensen and Muller, 1975a). This undisturbed chaparral includes the small gaps that may occur following death of individuals. The second environment, disturbed by fire, has resources such as light, water, and often nutrients that are temporarily more available (Christensen and Muller, 1975b). In the former, seedling establishment is spatially and temporally rare, while in the latter, en masse germination and establishment can occur for a number of species.

The seed bank dynamics among the species of these vegetations varies depending on the type or extent to which these seedling establishment environments are used. While woody and herbaceous species have been variously classified in relation to how they survive fire (Naveh, 1973, 1974; Hanes, 1977; Keeley and Zedler, 1978; Gill, 1981a; Christensen, 1985), from the point of view of seed bank formation, species may be divided into three broad groups: those with transient seed banks, those that form persistent seed banks, and those that produce both a transient and persistent seed bank due to polymorphic seed production (Table 1).

Our goal in this chapter is to review and synthesize the literature on chaparral seed banks, and to make comparisons where possible with seed banks in other Mediterranean shrub communities. Specifically, we consider seed bank formation and accumulation based on seed production, dispersal into the seed banks, and seed predation. We compare

Table 1
Classification of woody and herbaceous chaparral
species according to their seed bank dynamics

Species with exclusively transient seed banks
  Woody, obligate sprouters (e.g., *Quercus dumosa*); seedling establishment
    usually in older stands
  Geophytes and perennial grasses; seedling establishment usually in postfire
    period
  Annual species; generally introduced species
Species with essentially persistent seed banks
  Woody, obligate seeders (e.g., many *Arctostaphylos* and *Ceanothus* spp.); fire-
    stimulated germination and postfire seedling establishment
  Most facultative, woody sprouters (e.g., some *Arctostaphylos* and *Ceanothus*
    spp.); fire-stimulated germination and postfire seedling establishment
  Herbaceous, postfire species (e.g., *Emmenanthe penduliflora*); mostly annuals
    with fire-stimulated germination and postfire seedling establishment
Species with both transient and persistent seed banks
  Other species (e.g., the faculative sprouter, *Adenostoma fasciculatum*); both a
    yearly transient fraction and a persistent fraction stimulated by fire with
    postfire seedling establishment

---

germination cues and seedling establishment patterns of species with
and without persistent seed banks. Finally, we consider seed bank patterns at the ecosystem level in relation to dominance and biogeographic
patterns within chaparral.

II. Species Seed Bank Dynamics

Studies of chaparral have principally examined seed banks of individual
species in the soil. First we consider seed bank size. Then we examine
factors influencing seed banks, including seed production, dispersal,
predation, longevity, dormancy, and patterns of germination and
establishment.

A. Seed Bank Size

Seed bank size varies among and within species (Table 2) due to several
compounding factors: when and where seed banks were collected
(Davey, 1982), current year seed production, predator activity (Davey,
1982; Kelly and Parker, 1989), seed size (Kelly and Parker, 1989), and
differences in sampling techniques (Smith, 1942; e.g., Keeley, 1977a,
compared with Davey, 1982, or Kelly and Parker, 1989). For example,

Table 2

A summary of seed bank sizes of woody dominants in California chaparral and other Mediterranean-type ecosystems

Species	Seed bank size ($\bar{x} \pm SE$)	Viability (%)	Reference
California			
Adenostoma fasciculatum	5385–320,000 seeds m^{-2}	4	Stone and Juhren (1953)
	2000–21,000 m^{-2}	100	Zammit and Zedler (1988)
	450–3745	100	V.T. Parker (unpublished)
Arctostaphylos glauca (1972)	346 ± 138	54	Keeley (1977a)
Arctostaphylos glauca (1982)	298 ± 85	54	Keeley (1987a)
Arctostaphylos glandulosa (S. Calif., 1972)	4116 ± 982	7	Keeley (1977a)
Arctostaphylos glandulosa (S. Calif., 1982)	3038 ± 731	7	Keeley (1987a)
Arctostaphylos glandulosa (N. Calif.)	8422 ± 1575	3	Kelly and Parker (1989)
Arctostaphylos canescens	4500 ± 822	10	Kelly and Parker (1989)
Arctostaphylos pechoensis	4490 ± 658	6	Kelly and Parker (1989)
Arctostaphylos crustacea	978 ± 144	13	Kelly and Parker (1989)
Arctostaphylos viscida	28,177 ± 4743	9	Kelly and Parker (1989)
Arctostaphylos mewukka	6463 ± 1237	4	Kelly and Parker (1989)
Ceanothus greggii	369	71	Keeley (1977a)
Ceanothus leucodermis	87	96	Keeley (1977a)
Australia			
Eucalyptus incressata	800 ± 45[a]		Wellington and Noble (1985)
Eucalyptus populnea	3200		Hodgkinson et al. (1980)
Eucalpytus marginata	767 ± 178		Vlahos and Bell (1986)
Acacia suaveolens	12.4 ± 2.3[b]	100	Auld (1986b)
Banksia attenuata	133 seeds per plant[a]	42	Cowling et al. (1987)
Banksia leptophylla	2133[a]	63	Cowling et al. (1987)
Banksia menziesii	67[a]	3	Cowling et al. (1987)

Banksia prionotes	400[a]	31	Cowling et al. (1987)
Banksia ericifolia	1368[a]		Carpenter and Recher (1979)
Banksia spinulosa	20[a]		Carpenter and Recher (1979)
Banksia aspleniifolia	15[a]		Carpenter and Recher (1979)
Banksia ornata	783 ± 459 seeds per plant[a]	100	Gill and McMahon (1986)
Banksia ornata	1002 ± 585 seeds m^{-2}[a,d]	100	Gill and McMahon (1986)
South Africa			
Acacia longifolia[c]	2901 ± 415 seeds m^{-2}[a]	99	Pieterse and Cairns (1986)
Protea aurea	61.0 seeds per plant[a]	99	Bond (1985)
Protea eximia	45.4	100	Bond (1985)
Protea lorifolia	177.8[a]	100	Bond (1985)
Protea punctata	165.3[a]	96	Bond (1985)
Protea repens	68.2[a]	48	Bond (1985)
Leucodendron album	1199.2[a]	12	Bond (1985)
Leucodendron conicum	78.4[a]	51	Bond (1985)
Leucodendron eucalyptifolium	407.3[a]	68	Bond (1985)
Leucodendron rubrum	166.1[a]	79	Bond (1985)
Leucodendron uliginosum	800.3[a]	28	Bond (1985)

[a]These values represent canopy-stored seed pools.

[b]Average of seven sampling sites of soil samples 0–5 cm in depth across 3 yr of sampling; does not include seeds from ant nests.

[c]This species is a nonnative weed in South Africa.

[d]Average of six sampling sites, of six different ages (6–50+ yr).

Arctostaphylos species in California have soil seed banks ranging from 298 to 28,177 seeds m^{-2}. In a comparison of six species, the smallest seeded species had the largest seed bank (*Arctostaphylos viscida*, 13.3 mg seed^{-1}, 28,177 seeds m^{-2}), whereas with those producing larger seeds (26.2–52.9 mg seed^{-1}) had considerably smaller seed banks (978–6463 seeds m^{-2}) (Kelly and Parker, 1989). This seed bank/seed size relationship may not hold for other genera, e.g., *Ceanothus*. Seed bank size for the same species may differ between California sites [e.g., *Arctostaphylos glandulosa* in Marin County, 7840–9825 seeds m^{-2} (Kelly and Parker, 1989), versus *Arctostaphylos glandulosa* in San Diego County, 3038–4116 seeds m^{-2} (Keeley, 1977a); *Adenostoma fasciculatum* in Lake County, 13,000 seeds m^{-2} (Stone and Juhren, 1953), versus *Adenostoma fasciculatum* in San Diego County, 2000–21,000 m^{-2}, (Zammit and Zedler, 1988), versus *Adenostoma fasciculatum* in Contra Costa County, 450–3745 seeds m^{-2}, (V. T. Parker unpublished)] but may be similar at the same site over time (e.g., *Arctostaphylos glandulosa*, Keeley, 1987a) (Table 2).

Soil seed banks of dominant species in Australian and South African communities are usually small except for *Acacia*, *Cassinia*, *Bedfordia*, *Pultenea*, and some others (Carroll and Ashton, 1965; Barbour and Lange, 1966; Howard and Ashton, 1967; Purdie, 1977; Bond, 1980; Wellington and Noble, 1985; Auld, 1986b; Pieterse and Cairns, 1986). A number of other species, however, also rely on soil seed banks (Bond, 1980; Purdie, 1977). For those cases where total seed banks are reported, the range of densities is as great as for other plant communities (Mott and Groves, 1981). In one survey of Australian sclerophyllous communities, woody species were usually a small component of the soil seed bank, 5.4% in mallee, but in some forests understory shrubs could comprise >58% of the soil seed bank (Carroll and Ashton, 1965). Total seed banks sizes in these communities ranged from 667 to 35,855 seeds m^{-2} and were principally herbaceous species. Dominant species missing from the soil seed bank were in the Proteaceae or the genus *Eucalyptus* and were serotinous species with aboveground seed banks (Carroll and Ashton, 1965; Howard and Ashton, 1967).

B. Serotiny and Aboveground Seed Banks

Seed retention in serotinous cones or in infructescences stored in plant canopies is a characteristic of some species in chaparral and other Mediterranean shrub communities. Serotiny is most widespread in South African and Australian taxa, especially in the Proteaceae (e.g., *Protea*, *Leucospermum*, *Leucodendron*, and *Banksia*), the Myrticaceae (*Eucalyptus*), and Cupressaceae (*Widdringtonia* and *Callitris*). Serotiny is less prominent in California and the Mediterranean area but is found there in the Pinaceae (*Pinus*) and Cupressaceae (*Cupressus*). The number of seeds retained in cones varies widely among and within species and with geographic location (Bond, 1980, 1985; Gill and McMahon, 1986; Cowling and Lamont, 1985b; Cowling *et al.*, 1987). For these plants, however, a seed bank of considerable size (15 to 2133 seeds per plant) can be stored in the canopy (Table 2). Fire may be required for the opening of cones or capsules (Bradstock and Myerscough, 1981). Timing and quantity of seed release are correlated with fire temperature, and are enhanced by repeated wetting and drying of burnt cones and cool postfire temperatures (Bradstock and Myerscough, 1981; Cowling and Lamont, 1985a). In Proteaceae of South Africa, however, where serotiny is relatively weak, seeds are seldom stored for more than 5–6 yr, and fires are not essential for seed release (Bond, 1985). Total stored seed declines as population density declines in older stands (Bond, 1980). For Australian *Banksia* species, Zammit and Westoby (1987a,b) found seed release to be higher in a sprouting species than in an obligate seeding species in the absence of fires, but, at a different site, Cowling *et al.* (1987) found that the degree of serotiny did not correlate with sprouting and nonsprouting habits.

Variation in serotiny illustrates the role of fire in the selection for dormancy and long-term seed storage. In *Pinus coulteri*, for example, the tendency toward serotiny was present in chaparral and other fire-prone habitats, but not in habitats where the frequency of canopy fires was low (Borchert, 1985). For serotiny to be selected, large, stand-killing fires, relatively short, fire-free intervals, and fire sizes large enough to prevent significant dispersal from unburned areas are necessary (McMaster and Zedler, 1981).

C. Flower, Fruit, and Seed Production

Flower and fruit production may vary significantly within species between years (Keeley, 1977a, 1987a; Kelly and Parker, 1989), within species between sites (Lamont, 1985), and between species within genera (Cowling *et al.*, 1987). Flower and subsequent fruit production for species that produce dormant flower buds may be determined largely by rainfall during the previous year when the buds were set (Keeley, 1977a). Reproductive output among species that do not set dormant flower buds may not be related to annual precipitation (Cowling *et al.*, 1987).

Comparisons of flower, fruit, and seed production of sprouting and nonsprouting congeneric species reveal no consistent pattern (Table 3; Keeley, 1977a, 1987a; Lamont, 1985; Cowling *et al.*, 1987; Kelly and Parker, 1989). Australian *Banksia* species that produced large seeds showed lower follicle set (Cowling *et al.*, 1987). A nonsprouting *Banksia* species had higher fruit set than did a sprouting species (51 versus 28%), yet the

Table 3
Summary of flower and seed production in plants of California chaparral and other Mediterranean-type ecosystems

Species	Production	Reference
California	**Flowers m^{-2}**	
Arctostaphylos canescens	497–1344	Kelly and Parker (1989)
Arctostaphylos glandulosa	408–1164	Kelly and Parker (1989)
Australia	**Florets plant^{-1}**	
Banksia leptophylla	4446	Cowling *et al.* (1987)
Banksia menziesii	1540	Cowling *et al.* (1987)
Banksia prionotes	1904	Cowling *et al.* (1987)
California	**Seeds m^{-2}**	
Ceanothus greggi	0–8340	Keeley (1977a, 1987a)
Ceanothus leucodermis	5–1285	Keeley (1977a, 1987a)
Arctostaphylos glauca	0–~2000	Keeley (1977a, 1987a)
Arctostaphylos glandulosa	0–~7000	Keeley (1977a, 1987a)
Arctostaphylos canescens	591–992	Kelly and Parker (1989)
Arctostaphylos glandulosa	392–709	Kelly and Parker (1989)
Australia	**Seeds plant^{-1}**	
Acacia suaveolens	12.3[a]	Auld and Myerscough (1986)
Acacia elongata	14.9[b]	Auld (1986b)
Acacia linifolia	216.9[b]	Auld (1986b)
Acacia longifolia	190.4[b]	Auld (1986b)
Acacia myrtifolia	58.6[b]	Auld (1986b)
Acacia terminalis	101.1[b]	Auld (1986b)
Acacia ulicifolia	51.3[b]	Auld (1986b)
Mediterranean		
Cistus salvifolius	20,141	Troumbis and Trabaud (1986)
Cistus villosus	10,488	Troumbis and Trabaud (1986)
South Africa		
Leucospermum cordifolium	708–808,586	Lamont (1985)
Leucospermum cuneiforme	752–1854	Lamont (1985)
Leucospermum erubescens	341–1102	Lamont (1985)

[a]Mean number for eight populations times 2 yr per population (minus 1 yr for one population).

[b]Average of three populations.

number of seeds per inflorescence containing seeds was not significantly different (Zammit and Westoby, 1987a). Fruit set of California non-sprouting and sprouting *Arctostaphylos* species showed no consistent pattern (Kelly and Parker, 1989). Keeley (1987a) followed fruit production over a 10-yr period for two *Arctostaphylos* species and found great variability from year to year, ranging from near zero to ~1100 m^{-2}.

D. Seed and Fruit Dispersal

Several types of fruit dispersal occur in California chaparral. Fruit and seeds of many species do not have an obvious mechanism for dispersal; most, including the dominants (e.g., *Arctostaphylos* spp. and *Adenostoma* spp.), have predominantly dry fruit at maturity that usually land beneath or close to the parent plant. *Ceanothus* is unique because seeds are explosively dispersed 1–3 m from the mature capsules. Typically, these dominants form persistent seed banks.

The potential for long-range dispersal also occurs, e.g., the wind-dispersed achenes of *Cercocarpus* spp. Fleshy fruit are found in other species and, along with acorns of *Quercus*, are dispersed by vertebrates. Few direct studies have examined the dynamics between dispersers and fruit and seed distribution. Generally, these exhibit transient seed banks (except *Rhus ovata*; Stone and Juhren, 1951).

Horizontal movement of seeds after release from parent plants is commonly a result of animal activity or wind. Seeds may be cached by ants (Berg, 1966; Bullock, 1981) or vertebrates. In Australia and South Africa an unusually great proportion of seeds are dispersed by ants (Berg, 1975; Rice and Westoby, 1981). Myrmecochory may result in clumped seed banks, sometimes buried too deep for germination stimulation by fire (Auld, 1986a,b). Many proteaceous species produce seeds with wings, plumes, or other pubescence that aid in wind dispersal. Among *Leucodendron* species of South Africa, seeds of serotinous species contain various structures for wind dispersal, whereas nonserotinous species more commonly are myrmecochorous (Slingsby and Bond, 1985). Dispersal distances for myrmecochorous species are usually shorter than those for wind dispersed species. For example, maximum ant dispersal distance of *Leucospermum conocarpodendron* seeds was 9.84 m, as compared with wind dispersal distances of 26.3 m for *Protea laurifolia*, a nonmyrmecochorous species (Manders, 1986).

Vertical movement of seeds depends largely on size of seed, soil, and litter particle, but may also be related to other factors such as rain or cracks in the soil due to heat or frost. In one study, significant vertical movement (2–4 cm) of *Arctostaphylos* seeds occurred within 12 months, but did not change thereafter (Kelly, 1986). Although vertical movement into the soil may enable seeds to escape surface-foraging predators and survive fires, long-term ecological consequences are not known.

E. Seed Longevity and Dormancy Characteristics

Seeds of many herbaceous chaparral species must remain dormant in the soil for extended periods, at least the length of a fire cycle (e.g., 50–100 yr, *Emmenanthe penduliflora*; Sweeney, 1956). Seeds of woody chaparral species have been found in the duff of conifer forests (Quick, 1956), and most likely were deposited when chaparral dominated those areas. This indicates that seeds remain viable and dormant for 40 to >200 yr (Gratkowski, 1962; Zavitkovski and Newton, 1968; Quick, 1956, 1975). Quick and Quick (1961) found high rates of germination for *Ceanothus* stored in a laboratory >24 yr. In Australia, seeds of *Acacia suaveolens* buried in the soil were estimated to have a half-life of 10.7 yr, with 3.9% remaining viable after 50 yr (Auld, 1986b).

Few studies (e.g., Hilbert, 1988) have attempted to estimate how dormancy influences seed bank dynamics. Woody species continue to produce seed and add to the seed bank between fires. Only in the case of species whose populations die out prior to a fire—as with postfire annuals, some *Ceanothus* species, or montane chaparral species—does seed longevity become critical to their persistence. In contrast, the potential seed bank size is influenced directly by differences in longevity and in the rate at which viability is lost.

The importance of longevity and death rates can be assessed in a simple model. Seed bank sizes are compared among populations with different longevity when demographic characteristics, including changes in population size, rate of density-dependent thinning, and seed production, are the same. The size of the seed bank is determined by current production added to past production, where past production is reduced annually by a year-specific death rate, or

$$S_t = \left(\sum_{j=1}^{t-1} Y_j V_{jt} \right) + (N_t P_t Y_t)$$

where Y_j = seeds added to seed bank in year j, V_{jt} = fraction of seed produced in year j that remains viable in year t, N_t = number of plants in year t, P_t = fraction of population reproducing in year t, and Y_t = seeds produced per plant in year t. This model illustrates that seed bank size can be modified significantly by changes in longevity (Fig. 1) and death rates (Fig. 2). The pattern of death (Fig. 2) determines the average seed longevity for a population, and thus the potential seed bank size.

Among chaparral annuals, seed production is limited to the first few years after fire. Persistence in the seed bank and the response to the next fire will depend upon the longevity and death rates. Differences in the death rate over a fire cycle can compensate for differences in seed production (Fig. 3). In this hypothetical example, species B has a greater potential population size after time X even though rates of seed production in the initial postfire years were less than that of species A. Variation in fire frequency would influence the relative proportion of these two species, and, if the fire interval was too long, would result in the

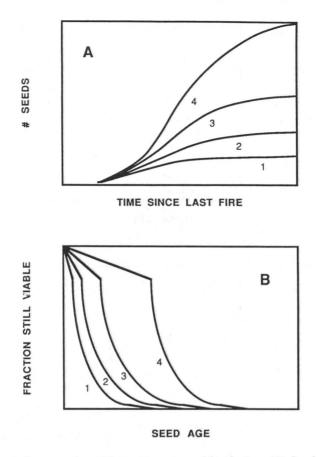

Figure 1.   Influence of seed longevity on seed bank size. (A) Seed bank
sizes resulting from different seed longevities. (B) Four patterns of longevity for a single year's cohort through time. The numbers 1–4 match seed
bank sizes to seed longevity patterns.

loss of one or both species. Bond (1980), for example, reported large
declines in dicots in older stands in South African cape fynbos, shifting
the soil seed bank composition toward monocots.

The number of propagules and energy allocated to each are important measures of reproductive effort; this is enhanced by lack of or low
rate of germination from the seed bank, usually the factor causing the
greatest loss of seed from a buried population (Karssen, 1982). The
genetic implications of differential mortality have not been explored,
although potentially significant genetic changes in populations may result from variation in fire frequency.

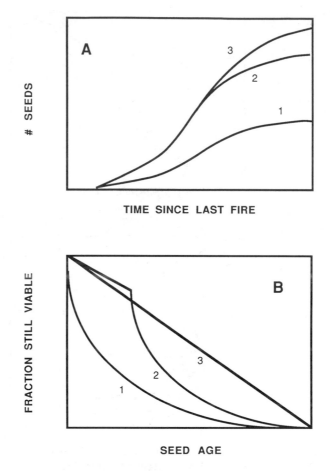

Figure 2. Influence of seed death rate on seed bank size. (A) Seed bank sizes resulting from different seed death rates. (B) Three different death rates for a single year's cohort. The numbers 1–3 match seed banks sizes to seed death rate patterns.

F. Seed Predation

Both predispersal and postdispersal seed predation may have a significant impact on the seed bank. Reported loss of seeds before dispersal is due to predation by insect larvae, and may be substantial. For example, loss of *Yucca whipplei* seeds to *Yucca* moth larvae (*Tegeticula maculata*) averaged 24.5% (Keeley *et al.*, 1986), and loss of *Acacia suaveolens* seeds to a weevil (*Melanterious corosus*) ranged from from 9.6 to 60.6%, with an additional loss of 2.1–21.2% to other insect seed grazers (Auld and Myerscough, 1986). Likewise, weevils were the major predispersal preda-

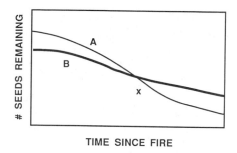

Figure 3. Hypothetical changes in number of seeds stored in soil for two herbaceous species differing in rates of viability loss. Seed production only occurs in the first several years after fire.

tors of seven *Acacia* species (Auld, 1986a). Seed predation varied among and within species, with seed loss related to seed production and predator satiation.

Serotinous species also suffer considerable predation. In South Africa, 84% of the seed produced by *Protea repens* were damaged by lepidopteran and coleopteran larvae (Coetzee and Giliomee, 1987). Insect predation prevented one-third of the inflorescenses of Australian *Banksia* from setting seed and caused damage to or loss of up to 31% of the seeds present on plants (Scott, 1982; Zamitt and Westoby, 1987b). Seed and inflorescence predators included larvae of lepidoptera and Curculionidae, with species in the former being more numerous. Insecticide treatment produced a 40% increase in the number of seeds per inflorescence. Predation was lowest for the strongly serotinous species (Cowling *et al.*, 1987). It increased with cone age (Zammit and Hood, 1986; Zammit and Westoby, 1987a; Cowling *et al.*, 1987), but not with stand age (Gill and McMahon, 1986; Cowling *et al.*, 1987; Zammit and Westoby, 1987a).

Short-term studies reveal that the number of seeds produced by some dominant chaparral plants can exceed the number of seeds in seed banks by as much as 99% (Keeley, 1977a, 1987a). Rates of predation of seeds in soil are as high as 60% within 6 months for *Arctostaphylos* species (Keeley and Hays, 1976; Kelly and Parker, 1989), nearly 100% in 6 months for *Ceanothus crassifolius* (Davey, 1982), and 100% within 3 days for South African and Australian species (Bond and Breytenbach, 1985; Wellington and Noble, 1985). In contrast, Stone and Juhren (1953) found up to 323,000 seeds m^{-2} in the soil beneath *Adenostoma fasciculatum*, suggesting lower levels of seed predation. Because seeds of many *Arctostaphylos* species are large and have a very thick endocarp, probable seed predators are vertebrates instead of ants. In comparison, important consumers of *Ceanothus* seeds are harvester ants (Davey, 1982) as well as birds (Davey, 1982) and rodents (Smith, 1942).

Predators removed a higher percentage of *Arctostaphylos glauca* seeds, which were larger than *Arctostaphylos glandulosa* seeds, from cleared areas (Keeley and Hays, 1976). When Kelly and Parker (1989) mixed seeds with soil and litter, a higher proportion of the smaller seeds of *Arctostaphylos canescens* disappeared. Overall, however, predation of the larger seeded *Arctostaphylos glandulosa* was greater than *Arctostaphylos canescens*.

Ants are the most important consumers of seeds of many Australian species. For *Eucalyptus incrassata*, during nonfire years there was complete removal of seeds by harvester ants (Wellington and Noble, 1985). However, because *Eucalyptus incrassata* is serotinous, seeds are released *en masse* after fires, apparently satiating predators and allowed seedling recruitment. Other predators may be important. These include lygaed bugs (Anderson and Ashton, 1985) and vertebrates, the latter preying on larger fruits with high nitrogen and phosphorous levels (Abbott and Van Heurck, 1985).

Ants, which serve as dispersers for myrmecochorous species, may reduce seed predation and be critical for the maintenance of effective seed banks (Bond and Slingsby, 1984). Ant removal of elaiosomes from *Mimetes pauciflorus* and *Leucospermum glabrum* made seeds difficult to find, thereby reducing vertebrate predation (Bond and Breytenbach, 1985). Removal of elaiosomes from *Leucospermum conocarpodendron* had no effect on germination in the laboratory, but reduced seedling recruitment in the field (Slingsby and Bond, 1985). This species is obligately dependent on seed burial by ants to reduce predation.

G. Germination and Seedling Establishment Patterns

Germination cues reflect the fact that germination and seedling establishment occur either in undisturbed chaparral or after fire. The three seed bank syndromes (Table 1) constitute different combinations of transient and persistent seed fractions.

1. Patterns within Undisturbed Vegetation

In undisturbed vegetation, transient seed banks predominate among woody species, although some species with persistent seed banks may respond to small gap formation (e.g., *Salvia mellifera*, Keeley, 1986b; *Cistus villosus* in Greek maquis, Troumbis and Larigauderie, 1987). Seed dispersal occurs from late spring to winter with the germination period from late fall to early spring corresponding with the rainy period. Some species with transient seed banks appear to lack any form of dormancy and disperse seed just prior to germination (e.g., *Heteromeles arbutifolia* or some *Quercus* species). Within-year dormancy for other species with transient seed banks may be based on stratification or afterripening requirements. Such germination behavior will fail to produce persistent seed banks, thought by some to be ideal for this fire-prone vegetation. Selection for dormancy is probably not occurring among species that have a high fire survival rate and resprout after fire.

Many woody species may be constrained from developing persistent seed banks by a number of factors. Accumulating persistent seed banks may require changes in both dispersal characteristics and seed size. Fruit characteristics of species with transient seed banks, such as animal-dispersed fleshy fruit of *Heteromeles* or *Rhamnus* or wind-dispersed plumed achenes of *Cercocarpus*, contrast greatly with those of most chaparral genera with persistent seed banks, such as *Ceanothus* (Stebbins, 1974); modifications in such characteristics may not be readily possible. Small seed size, often related to persistent soil seed banks (Thompson, 1987), may not be a reasonable evolutionary expectation for species from many genera, e.g., *Quercus*, especially under conditions of high predation (Christensen and Muller, 1975b). Indeed, large seed size and mast fruiting counter predation (Janzen, 1971a; O'Dowd and Gill, 1984; Wellington and Noble, 1985).

Woody species such as *Heteromeles* with transient seed banks are often understory components in the more mesic parts of chaparral. Shade tolerance would be important in these habitats. Seedlings usually can be observed in the understory in the spring; *Heteromeles arbutifolia* germinated in one area at densities up to 45 m^{-2}, but none persisted through the summer (V. T. Parker, personal observation). Seedlings of a variety of species can be found, especially on north-facing slopes, although numbers can be low (0.043 seedlings m^{-2}; Patric and Hanes, 1964). Distribution, lack of seedling survival, height, and increases in density with stand age have led to the speculation that older stands are required for much regeneration of these species (Patric and Hanes, 1964; Keeley, 1986a) because gap frequency and size may increase.

Seedlings of *Quercus* generally establish more readily than do other woody chaparral species with transient seed banks. Seedlings of *Quercus dumosa* were found in densitites of 0.13 m^{-2} while those of four other obligate resprouters together totaled 0.10 m^{-2} (Keeley, 1986a). Densities of *Quercus wislizenii* var. *frutescens* within an *Arctostaphylos*-dominated stand were ≤1.2 m^{-2} (V. T. Parker, unpublished). The size class distribution of these seedlings suggests that individuals can reach the canopy with time; in fact, some may have reached the canopy since the last fire (Fig. 4). Thus, *Quercus* species can establish seedling banks within undisturbed vegetation, even with low nutrient status, high pre-

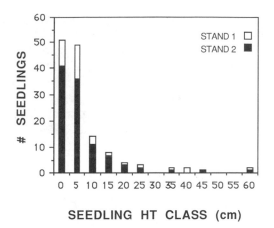

SEEDLING HT CLASS (cm)

Figure 4. Height class distribution for oak seedlings (*Quercus wislizenii* var. *frutescens*) found in two chaparral stands in northern California. Total numbers of seedlings are presented for each 5-cm height class (V. T. Parker, unpublished).

dation, low water availability, and low light availability for seedlings (Christensen and Muller, 1975b).

2. Postfire Patterns

Persistent seed bank species respond to environmental disturbances. Of the germination cues identified, almost all are environmental changes related to fire; these produce *en masse* seedling establishment following fires. Germination cues, operating alone or in combination, include heat (Wright, 1931; Quick, 1935; Quick and Quick, 1961; Sampson, 1944; Stone and Juhren, 1951, 1953; Sweeney, 1956; Hadley, 1961; Keeley, 1987b; Keeley and Keeley, 1987; Parker, 1987b), chemicals leached from charred wood (Wicklow, 1977; Jones and Schlesinger, 1980; Keeley, 1987b; Keeley and Keeley, 1987; Parker, 1987b), release from toxic compounds (Muller *et al.*, 1968; McPherson and Muller, 1969; Christensen and Muller, 1975a), increased light (McPherson and Muller, 1969; Christensen and Muller, 1975a; Keeley, 1987b; Keeley and Keeley, 1987), and stratification (Quick and Quick, 1961).

Seedling density following fire is characteristically variable (e.g., Sampson, 1944; Horton and Kraebel, 1955; Hanes, 1971; Keeley and Keeley, 1981; Bond *et al.*, 1984) because of localized differences in wind, fuel, temperature, and moisture. The great variability in seedling establishment is hardly surprising. In Australia, for example, the heat pulse into the soil can stimulate seeds of *Acacia suaveolens* only over a small range of soil depths directly related to intensity and duration of fire

(Auld, 1986b). Higher rates of seedling emergence are usually observed following autumn fires; during other seasons establishment is reduced (Orme and Leege, 1976) by predation (Bond *et al.*, 1984) or mortality in moist soil (Parker, 1987b).

Fire frequency can have a severe impact on seed bank renewal. If the interval is too long, seed longevity may be exceeded, while if too short, reproductive maturity may not be reached (Zedler *et al.*, 1983a; Fox and Fox, 1986). Fire frequency can have an impact on the relative dominance of sprouting and nonsprouting species (Keeley, 1986a; Hilbert, 1988).

Intensity, duration, season, and frequency are collectively termed the fire regime of a species (Gill, 1975; Gill and Groves, 1981). Small changes in any one of these factors can have a differential effect. Variability in vegetation composition may reflect historical differences in fire regime among sites (Malanson and O'Leary, 1985). Among co-occurring serotinous species in Australia, for example, a fire of a given intensity may favor the germination and establishment of one species over another; the differences in the timing and rate of follicle opening associated with a particular fire intensity (Bradstock and Myerscough, 1981) result in patchy distribution.

Fires may select species with certain characteristics. Seeds with thick coats or cuticles have a better chance of surviving fire and germinating. For germination to occur a heat pulse is required that modifies the seed coat in the area around the hilum (Stone and Juhren, 1951), making it permeable (see Baskin and Baskin, Chapter 4, this volume). Cracking of the seed coat occurs with heat in *Rhus ovata* (Stone and Juhren, 1951) and in *Ceanothus* spp. (Quick, 1935, 1959; Quick and Quick, 1961; Gratkowski, 1962). Many hard-seeded species, especially legumes, establish from soil seed banks after fires in Australia (Purdie, 1977). Among many thick-coated herbaceous species, germination occurs readily following mechanical scarification that replaces heat treatment (e.g., Sweeney, 1956; Ammirati, 1967). Heat also opens serotinous cones or other aboveground reproductive structures (Bradstock and Myerscough, 1981).

In many species a portion of the seed bank can germinate following exposure to compounds leached from incompletely burned wood (Wicklow, 1977; Jones and Schlesinger, 1980; J. E. Keeley and Nitzberg, 1984; S. C. Keeley and Pizzorno, 1986; Parker, 1987b). This response is widespread among chaparral species (Keeley *et al.*, 1985; Keeley, 1987b; Keeley and Keeley, 1987). In fact, a water-soluble oligosaccharide, a hemicellulose breakdown product, has been shown to stimulate germination (Keeley and Pizzorno, 1986). Desert populations of *Emmenanthe penduliflora* that rarely if ever experience fire also respond to charred wood, although not to the same degree as chaparral populations (Jones and Schlesinger, 1980). The full evolutionary significance of this germination cue must await an understanding of the physiological processes involved.

For species with persistent seed banks, germination cues seem rather specific. In nearly all, however, the potential exists for some germination and seedling establishment without fire (e.g., Sampson, 1944; Stone and Juhren, 1953). Some species have seeds that respond to several cues (e.g., *Salvia mellifera*, Keeley, 1986b). Germination in undisturbed chaparral may be due to changes in dormancy related to aging (e.g., Hadley, 1961). In Australian species with serotiny, fruits may spontaneously open as they age, resulting in seedling establishment between fires (Zammit and Westoby, 1987a).

Changes due to aging may be an important aspect of dormancy and longevity of seeds in chaparral seed banks, and may be related to the difficulty in germinating many herbaceous postfire species (e.g., *Dicentra ochraleuca*, Keeley *et al.*, 1985; V. T. Parker, unpublished data, and *Phacelia* spp., V. T. Parker, unpublished data). Seeds of *Arctostaphlylos canescens* extracted from soil germinated readily with addition of a charred wood extract, but freshly collected seed (<1 yr old) required addition of gibberellic acid to the charred wood extract to stimulate germination (Parker, 1987b). Seed from different populations of *Emmenanthe penduliflora* vary in germinability (Keeley *et al.*, 1985), indicating that there may be great between-population differences or polymorphism (Keeley, 1986b). *Adenostoma fasciculatum* produces a polymorphic seed population each year, some of which germinate readily in midwinter, and others remain dormant, contributing to a persistent seed bank (Stone and Juhren, 1953). For this species, establishment may occur the season following fire and also in unburned, undisturbed chaparral.

Studies using prescribed burning indicate that winter burns reduce the seedling density of species that rely on seed banks; when burns occur from late winter into the spring, germination may be delayed until the following year or may not occur at all (Bond *et al.*, 1984; Riggan *et al.*, 1986; Parker, 1986, 1987a, unpublished data). These effects are greatly influenced by soil moisture (Parker, 1987b). Herbaceous species generally fall into two groups, those that absorb moisture beyond 20% of their dry weight, and those that do not. The former group is quite sensitive to heat when imbibed; the maximum temperature and duration tolerated are both greatly reduced (Beadle, 1940; Sweeney, 1956; Parker, 1987b, unpublished data). Seeds of woody species behave in a similar manner, although mortality may not be as pronounced. For hard-seeded species of *Ceanothus*, the temperature of winter burns may be too low to stimulate germination (Tom White, USFS, personal communication).

Geophytes and perennial grasses exhibit another postfire pattern. Geophytes can be conspicuous following fire (Sampson, 1944), because flowering and reproduction are apparently stimulated (Stone, 1951; Howell, 1946, 1947, 1970; Stocking, 1962, 1966). Common California genera (e.g., *Brodiaea*, *Chlorogalum*, *Calochortus*, *Xerophyllum*, and *Zigadenus*) remain in the understory as a woody cover reestablishes, with production declining and plants often going dormant. Any reproduction in mature chaparral is limited to vegetative means. This pattern has parallels in Australia (e.g., *Xanthorrhea australis* and the orchid *Lyperanthus nigricans*; Gill and Ingwersen, 1976; Gill and Groves, 1981), in the Mediterranean region (e.g., *Bellevalia*, *Ornithogallum*, *Narcissus*, and *Ophrys*; Naveh, 1973, 1974), and in South Africa (e.g., "fire lilies" in the genera *Cyrtanthus* and *Bobartia*; Levyns, 1966; Martin, 1966; Kruger, 1977). In addition, perennial grasses are important in some areas of Australia and South Africa (Gill, 1981a; Kruger, 1983) and California (Parker, 1987a). Transient seed banks characterize this group (Fiedler, 1987; Keeley and Keeley, 1987). No germination occurs in the first postfire year due to lack of any reserve. Even with considerable seed production in the second or later postfire years (Stocking, 1962, 1966; Purdie, 1977; but see Kruger, 1977), vegetative reproduction predominates.

III. Seed Banks and Ecosystem Processes

Axelrod (1973) concluded that the major changes occurring in chaparral vegetation over the past 10,000 yr have included an expansion of the dominance of *Adenostoma fasciculatum*, radiation and increase in dominance of species of *Arctostaphylos* and *Ceanothus*, and a general decline in most other woody species. Persistent seed banks are common among species of these genera. Indeed, persistent seed banks are a principal ecological characteristic of most dominant woody species, successional shrubs, and postfire herbs. The importance of persistent seed banks in each of these ecological groups distinguishes Mediterranean vegetations (e.g., chaparral, mallee–heath, and fynbos) from other communities.

Persistent seed banks have played a key role in the fire response of chaparral vegetation. The cueing of germination to single large-scale disturbance events of the size of chaparral fires distinguishes the dynamics of these persistent seed banks. Mineral nutrients released by fire are quickly absorbed by postfire annuals and suffrutescents arising from seed banks (Sampson, 1944; Nilsen and Schlesinger, 1981; Rundel and Parsons, 1984) in a "nutrient dam" effect (Muller and Bormann, 1976). As a method of regeneration, persistent seed banks are fundamental to the maintenance and expansion of a large number of species (Sampson, 1944; Horton and Kraebel, 1955; Sweeney, 1956; Purdie and Slatyer, 1976; Keeley, 1977b, 1986a; Bond *et al.*, 1984).

A. Genetic and Adaptive Patterns

Models of nonchaparral annuals emphasize the conservative role of seed banks in population dynamics (Cohen, 1966; Brown and Venable, 1986). Populations are protected from local extinction in variable environments caused by reproductive failure in one poor year (Cohen, 1966; Brown and Venable, 1985). Persistent seed banks constitute an evolutionary memory that buffers against quick responses to environmental fluctuations (Epling *et al.*, 1960; Harper and White, 1974; Clegg *et al.*, 1978; Templeton and Levin, 1979; Brown and Venable, 1986). The longer the dormancy, the smaller will be the impact of a single year's drift in response to environmental extremes (Levin, 1978). In chaparral, on the other hand, postfire germination *en masse* results in no regenerative buffer to local extinction for annuals or obligate-seeding shrubs (e.g., *Ceanothus crassifolius*; Horton and Kraebel, 1955) or to strong environmental selection during seedling establishment. This population behavior better fits that of nonchaparral annuals lacking seed dormancy (e.g., *Clarkia rubicunda*, Bartholomew *et al.*, 1973).

Indeed, the lack of buffering provided by persistent seed banks is consistent with and contributes to the great diversity, endemism, and often narrow adaptation among obligate-seeding species of *Arctostaphylos* and *Ceanothus* in California (Nobs, 1963; Wells, 1969; Stebbins, 1974), *Banksia* in Australia, and *Protea* and *Leucodendron* in South Africa. In some cases, edaphic radiation required large adaptive shifts, e.g., *Arctostaphylos montana*, *Arctostaphylos bakeri*, *Ceanothus ferrisae*, and *Ceanothus jepsonii*, to serpentine soils and *Arctostaphylos myrtifolia* to leached, acidic, tropical-derived soils. Similarly, Wells (1962) found strong patterns of speciation in *Arctostaphylos* associated with the edaphic diversity of central California. Nobs (1963), however, demonstrated more subtle gradients of edaphic and climatic adaptive patterns in two series of *Ceanothus* species in a small region of northern California.

The greater potential variability associated with larger effective population size in woody species with persistent seed banks may have been important in the development of xeric adaptations. Seedling establishment is most prominent among woody obligate seeders (Keeley and Zedler, 1978). Seedlings of obligate seeders as a group may be much more resistant to water stress, be more efficient at water uptake and use, and grow and photosynthesize at higher rates (Parker, 1984). The distribution patterns of some obligate seeders in both California and Australia suggest tolerance of more xeric conditions (Specht, 1981; Keeley, 1986a).

Yet, as a group, woody species with persistent seeds, and facultative as well as obligate seeders, have a syndrome of characters that aid in seedling establishment in the immediate postfire years, distinguishing them from obligate sprouters. Along the northern California coast where the climate is moderated by fog, maritime chaparral on poor soils is dominated by both obligate and facultative seeders (Griffin, 1978). In these circumstances, species with persistent seed banks often increase in dominance relative to obligate sprouters for the regional chaparral.

B. Composition and Habitat Dominance

Herbaceous and successional woody or subshrub postfire species with persistent seed banks are most prominent in lower elevation chaparral stands. California chaparral can develop considerable annual cover following fire (Brandegee, 1891; Sampson, 1944; Howell, 1946, 1947; Horton and Kraebel, 1955; Raven, 1960; Stocking, 1962, 1966; Keeley *et al.*, 1981). Diversity of annuals and cover is lower near the coast in northern California than it is farther inland (Brandegee, 1891; Howell, 1946, 1947; Sweeney, 1967). Sweeney (1967) speculated that firetype annuals tend to

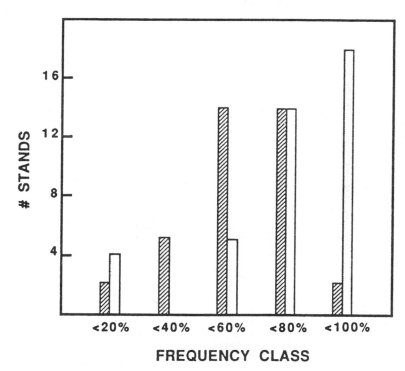

Figure 5. Species of chaparral with persistent seed banks as a proportion of total species at a site (hatched bars) or as a proportion of total cover (clear bars). Compiled from published and unpublished sources.

become infrequent or absent in coastal chaparral regions because of climate, moisture conditions, or reduced fire frequency.

Considerable variation in the herbaceous component of seed banks has been found even within a small region. Sweeney (1956) compared the herbaceous annual postfire response from 10 different stands and found that over a third of the species were found in only one stand, and over half occurred in 3 or less stands. Nonetheless, some annual species were quite frequent and almost 20% were found in at least 8 of the 10 sites.

The seed bank composition generally reflects that of the standing vegetation; approximately 66% of woody chaparral species produce persistent seed banks (estimated from Wells, 1969). Species producing persistent seed banks average ~50% of the woody species at any one site (Fig. 5). At the same time, the percentage of cover of these species proportionately increases to 60–100% (Fig. 5). Woody members of chaparral all have some means of surviving fire (e.g., Christensen, 1985), but the establishment of seedlings immediately after fire seems linked to processes that permit dominance of a site.

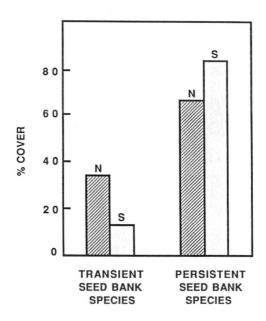

Figure 6. Percentage cover of chaparral species with transient or persistent seed banks by north-facing (N) (pole-facing) or south-facing (S) (equator-facing) slopes, summed for transects taken throughout California. Data from Critchfield (1971).

Within a stand of chaparral, woody species with transient seed banks tend to be patchy in their distribution (Bullock, 1978) and are more important on north-facing than south-facing slopes (Fig. 6, data from Critchfield, 1971; Patric and Hanes, 1964). North-facing slopes are more mesic, a factor critical for establishment of the transient seed bank species found there. Fire frequencies might also decline on these slopes, allowing more opportunities for seedling establishment between fires. The xeric characteristics of species with persistent seed banks is consistent with their domination of the more xeric south-facing slopes. Overall, species with persistent seed banks dominate both slope directions. While individual species can be exceptions, the nature of the seed bank appears well integrated into other life history aspects and is an important component of chaparral vegetation dynamics.

IV. Summary

Variation in seed bank dynamics is reflected in the composition, distribution, and dominance of chaparral species. These dynamics are associated with two very different seedling establishment environments, postfire and undisturbed chaparral. Three seed bank syndromes, transient, persistent, and transient + persistent, occur throughout Mediterranean shrub-dominated communities, with some important exceptions. Regional differences reflect the different resident floras, environmental conditions, including type and intensity of seed predation, and fire regimes. Shrub communities in Chile and the Mediterranean region contain relatively few species with persistent seed banks. California chaparral, demonstrates a larger reliance on persistent seed banks; postfire annuals are well developed only in this community. In contrast, serotiny is especially prominent among woody species in Australia and South Africa, perhaps a response of large-seeded species to extremely high rates of seed predation in otherwise resource-poor environments (Gill, 1981b; Kruger, 1983).

The diversity of regeneration syndromes, as shown by sprouters, obligate seeders, geophytes, and sapling banks, as well as by seed banks, points to the importance of disturbance. The particular seed germination cues, especially heat and leachate from charred wood, reflect the ecological and evolutionary importance of fire. Most dispersal appears local. Ant dispersal, resulting in burial, may be especially important to the maintenance of seed banks through decreased predation and impact of fire.

Woody dominant species have distribution patterns suggesting that persistent seed banks are the key to ecological success regardless of ability to resprout. The dominance of such species overall suggests the fundamental importance of fire and the postfire environment for seedling establishment. Their increase in importance in edaphic or climatic extremes underscores the role of persistent seed banks in evolutionary processes.

Most seed bank studies have focused on reproductive differences among congeneric woody sprouters and obligate seeders. Some studies report expected increases in reproductive effort among obligate seeders, but other studies fail to establish such patterns. The lack of consistency may lie in the fact that both sprouters and obligate seeders rely on persistent seed banks. The seedling establishment environment places strong selection on seedlings of the same life-form regardless of whether the adults resprout. Selection at this stage in the life history may mask other selective divergences.

Considerable data exist about germination cues and seedling establishment of chaparral species following fire. Nonetheless, several questions remain unanswered. Little is known about the size and dynamics of entire seed banks within chaparral or other Mediterranean-type ecosystems. Most studies have focused on the seed bank dynamics of individual dominant species. It appears that some chaparral species seed banks do not increase indefinitely (Stone and Juhren, 1951; Keeley, 1977a, 1987a), although others may (Stone and Juhren, 1953; Zammit and Zedler, 1988). Keeley (1977a) found that 99% of the seeds produced by dominant plants are not recruited as seedlings. However, no information exists on what proportion of seeds produced by an individual in a given year is retained in the seed bank and then recruited following fire. Data on longevity are primarily from laboratory studies, and are not available for most species. Knowledge of the seed banks of herbaceous species comes almost exclusively from postfire community analyses (e.g., Sweeney, 1956). Consequently, no information exists about between-fire demographics or depth distribution patterns.

Predation is a factor in the dynamics of chaparral seed banks. Size-selective postdispersal predation seems especially important. Predation within communities with different seed types, particularly those that lack dormancy, is largely unknown. As these species become a greater proportion of the vegetation, the questions are whether predation dynamics vary and whether species with dormant seeds have a greater chance of building up a seed bank quantitatively or spatially, as predicted by Janzen (1971a) for tropical species.

Similarly, the genetic composition of chaparral seed banks has not been examined. Genotypes may vary in longevity within a species and thus be differentially sensitive to fire frequency. Germination requirements are often variable among populations and may be genetically stable. Chaparral management practices may thus strongly affect the genetic diversity and natural genotypic combinations of species within communities. Fire suppression or ill-timed controlled burns may eliminate certain species or genotypes from chaparral communities.

Often management is practiced without regard to the sensitivity of seed banks. Frequency of prescribed fires must also be placed into the context of seed bank development (Fox and Fox, 1986). A number of years may be required before woody species have produced sufficient seeds to establish a seed bank that will regenerate the species. Not only is seed bank size important, but also its distribution within the soil column. Seeds close to the surface do not survive chaparral fires. From a conservation perspective, current practices are marginal in their success at maintaining species diversity and their ability to predict impact on the vegetation, especially for the rarer species.

CHAPTER 12

Seed Banks and Vegetation Processes in Deserts

Paul R. Kemp*

New Mexico Museum of Natural History
P. O. Box 7010
Albuquerque, New Mexico

I. Introduction

Seeds are a crucial and integral part of desert ecosystems. For annual (ephemeral) species, which may constitute 40% or more of the desert

---

*Present address: Systems Ecology Research Group, College of Sciences, San Diego State University, San Diego, California 92182.

Ecology of Soil Seed Banks Copyright © 1989 by Academic Press, Inc.

flora, seeds can be the most prevalent form of the species, and, during long droughts, the only form for several years. Seeds of most desert plant species represent the only means of dispersal and access to new regions. They constitute an important available food resource for some animal species (Brown *et al.*, 1979b). Finally, seeds are the source of variation for genetic differentiation and evolution in some of North America's most recent, and perhaps rapidly changing, ecosystems (Axelrod, 1983; VanDevender and Spaulding, 1983).

In view of the importance of seeds to the structure and function of desert plant communities, it is surprising that so little is known about desert seed banks and about their relationship to plant population dynamics. Our knowledge of desert vegetation is derived largely from studies of physiological and morphological adaptations of plants (Woodwell *et al.*, 1969; Mulroy and Rundel, 1977; Ehleringer, 1985; Caldwell, 1985) and studies of community structure and floristic composition (Shreve, 1942; Johnston, 1977; Yeaton and Cody, 1979; Wentworth, 1981). Much less is known about population dynamics and population processes, such as growth, recruitment, dispersal, and dormancy, that depend on the distribution and behavior of seeds (West *et al.*, 1979; Inouye *et al.*, 1980; Fonteyn and Mahall, 1981; Harper, 1977). Knowledge of seed banks and their relationship to the desert vegetation is a fundamental part of understanding the processes by which desert plants have become adapted to their harsh and uncertain environment. In this chapter, I discuss the size and variation of seed banks in desert soils and the importance of annual species, and offer some preliminary ideas about how seed banks are related to plant population dynamics and ultimately to community structure.

A. North American Desert Regions and Climate

Four different desert regions are recognized in North America (Fig. 1; Shreve, 1942; MacMahon, 1979; Brown and Lowe,1983). The Mojave, Sonora, and Chihuahua Deserts lie south of 37° north latitude and are called collectively warm or hot deserts, referring to their hot summers and relatively warm winters. Although nightly frosts are common in winter, particularly at higher elevations, mean monthly temperatures in winter are generally above 5°C. The Sonora Desert is both at a lower elevation and more maritime than are the Chihuahua and Mojave Deserts and, therefore, has the mildest winters. The hot deserts are rather different from one another in patterns of moisture distribution. The Mohave Desert receives a majority of its scant moisture in winter from Pacific air masses; average annual precipitation ranges from 50 to 200 mm (Turner, 1982b). The Chihuahua Desert receives 150 to 350 mm of precipitation annually, mostly in summer from air masses that originate

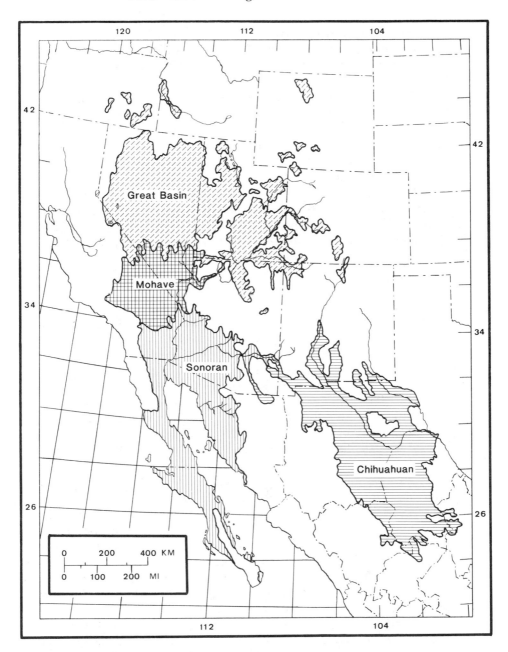

Figure 1.   Desert regions of North America as defined by plant communities (adapted from Kuchler, 1970, and Brown and Lowe, 1983). The boundary for the southern Chihuahuan Desert is approximate and based on climate (Schmidt, 1979).

in the Gulf of Mexico (Brown, 1982). The Sonora Desert, located between the Mojave and Chihuahua Deserts, has 50 to 400 mm of annual precipitation distributed in a distinctly bimodal pattern over much of its region (Turner and Brown, 1982).

The Great Basin Desert is referred to as a cold desert, relating particularly to its relatively cold winters. Mean monthly December and January temperatures are below 0°C, and the growing season is moderately short. The Great Basin is basically a rain shadow desert created by the Sierra Nevada Mountains to the west and, secondarily, the Rocky Mountains to the east. The total precipitation ranges from 125 to 350 mm, with a deficit of summer rainfall in the northern and central regions (Turner, 1982a; West, 1983a,b). The climate of the cold desert produces similar desert plant communities outside the Great Basin physiographic province, including much of the Colorado Plateau (Fig. 1), and I refer to all of these cold desert plant communities as the Great Basin Desert.

B. Floristics

The desert communities of North America are of recent origin, all having undergone their final assembly in the Holocene epoch (Wells, 1977; Axelrod, 1977, 1979, 1983; Turner, 1982a; VanDevender and Spaulding, 1983). Many desert species appeared in the Quaternary period, probably derived from drought-adapted ancestors of earlier communities (Axelrod, 1977, 1983). The hot deserts have a floristic similarity that reflects common ancestry from the Late Tertiary period dry tropical forests, oak woodlands, and thorn forests that previously occupied the region (Axelrod, 1979, 1983; MacMahon, 1979; Turner and Brown, 1982). The floras of the three hot desert regions have become differentiated from one another as a result of (1) geologic isolation in the Late Tertiary and Quaternary periods by the Sierra Madre Occidental and Mojave uplifts (Axelrod, 1983; MacMahon, 1979), (2) evolutionary and ecological responses to dissimilar climates (Shreve, 1942), and (3) differing impacts of adjacent floras and source regions (Johnston, 1977; McLaughlin, 1986).

Because of substantially different climate and historical biogeography, the Great Basin Desert is floristically distinct from the hot deserts (MacMahon, 1979; Axelrod, 1983; McLaughlin, 1986). By the Middle to Late Tertiary period, the Great Basin's mixed hardwood–conifer forests were isolated climatically from the warmer, dry tropical regions to the south (Axelrod, 1983). Drying associated with the Sierra Nevada uplift of the Late Tertiary period resulted in a displacement of the forests by arid-adapted shrubs and herbs of two floristic sources. The dominant perennial shrubs are related to species of temperate Asian deserts, whereas the subshrubs and herbs of the understory are related to the plants of the hot deserts (Axelrod, 1983).

C. Vegetation

There is both an ecological and a historical basis for differences in North American desert vegetation. In the Sonora Desert, bimodal moisture, mild winter temperatures, and a variety of floristic source regions foster a rich and complex vegetation with several synusia and many life-forms over much of its region (Shreve, 1951; Turner and Brown, 1982). There is great diversity of vegetation over the expanse of the Sonora Desert, and consequently it is often subdivided into several regions that have particular climates and vegetation (Turner and Brown, 1982; MacMahon and Wagner, 1985). In the low-elevation plains of the Lower Colorado and Gila Rivers (the Lower Colorado River Subdivision), *Larrea tridentata* and *Ambrosia dumosa* are dominant (Shreve, 1951; MacMahon and Wagner, 1985). Other important perennials are *Acacia* spp., *Prosopis* spp., *Encelia farinosa*, *Fouquieria splendens*, *Opuntia* spp., and *Hilaria rigida*. Along water courses small trees such as *Cercidium floridum*, *Olneya tesota*, and *Dalea spinosa* are abundant. Winter annuals are abundant in those years with sufficient rainfall (Shreve, 1951).

Bordering the Lower Colorado River Subdivision on the north and east is the Arizona Upland Subdivision (Shreve, 1951; Turner and Brown, 1982). Plant communities of this subdivision, which are the most complex of the Sonora Desert, are located on hills and bajadas (broad alluvial slopes) with coarse soils. The dominants are columnar cacti, such as *Carnegiea gigantea*, *Stenocereus thurberi*, and *Lophocereus schotti*, and small trees, such as *Cercidium microphyllum* and *Olneya tesota*. There are many shrubs, such as *Larrea tridentata*, *Ambrosia deltoidea*, *Simmondsia chinensis*, *Fouquieria splendens*, *Acacia* spp., and *Prosopis* spp., many cacti, and a great variety of both summer and winter annual species. For a description of the subdivisions of the Sonora Desert in Mexico, I refer the reader to Shreve (1951), MacMahon (1979), or Turner and Brown (1982). I do not consider the Mexican subdivisions because information about their seed banks is lacking.

The Chihuahua Desert is somewhat simpler than the Sonora Desert, mostly as a result of a less favorable winter growth period. Most communities have either *Larrea tridentata*, *Flourensia cernua*, *Prosopis glandulosa*, or *Acacia* spp. as dominants (Muller, 1947; Gardner, 1951; Brown, 1982; MacMahon and Wagner, 1985). Other important shrubs are *Parthenium incanum*, *Fouquieria splendins*, *Ephedra trifurca*, *Euphorbia antisyphyllitica*, and *Krameria* spp. Numerous rosettophyllous shrubs (*Yucca*, *Dasylirion*, *Agave*, and *Nolina*) are found at higher elevations (Shreve, 1942; Medellin-Leal, 1982). Cacti are diverse and locally abundant but are generally small, except in the southern portion of the desert where large cacti (*Opuntia* and *Myrtillocactus*) share dominance with *Prosopis* spp., *Acacia* spp., and large yuccas. Warm-season grasses (C_4 species of

Aristida, Bouteloua, Sporobolus, Muhlenbergia, Hilaria, and others) are important throughout much of the desert, reflecting the summer rainfall pattern and the late Quaternary change from a previous grassland–savanna system (Wells, 1977; MacMahon and Wagner, 1985). There also are numerous species of summer and winter annuals (Kemp, 1983).

Dry summers and winter frosts combine to limit the diversity of perennials in the Mojave Desert. The principal plant associations are dominated by low shrubs of *Larrea tridentata, Ambrosia dumosa, Atriplex spinosa,* or *Encelia farinosa* (Shreve, 1942; Vasek and Barbour, 1977; Mac-Mahon and Wagner, 1985). Other significant species include *Eriogonum fasciculatum, Lycium* spp., *Opuntia* spp., and the perennial grass *Hilaria rigida. Yucca* spp. (particulary *Yucca brevifolia*) are conspicuous, but are rarely dominant (Turner, 1982b). The driest basins may contain only one or two perennial species, notably *Atriplex polycarpa* or *Atriplex hymenelytra* (Hunt, 1966). There are numerous species of winter/spring annuals, many endemic to the Mojave Desert (Turner, 1982b).

The Great Basin Desert vegetation is characterized by low species diversity and few community types (Shreve, 1942; Turner, 1982a). Physiognomy is simple, with a single shrub species often the overwhelming dominant over broad areas (Turner, 1982a). Plant associations are usually dominated by one of the following low shrubs: *Artemisia tridentata, Atriplex confertifolia, Ceratoides lanata,* and/or *Artemisia spinescens* (Billings, 1949; MacMahon, 1979; Turner, 1982a; West, 1983a,b,c). Other important species include the shrubs *Coleogyne ramosissima, Purshia tridentata, Gutierrezia sarothrae, Ephedra* spp., *Chysothamnus* spp., and grasses (principally *Hilaria jamesii, Oryzopsis hymenoides, Bouteloua gracilis,* and *Sitanion hystrix*). The diversity of annual species is low in most areas (West and Ibrahim, 1968; Young and Evans, 1975), consisting mostly of spring-growing Eurasian introductions (e.g., *Bromus, Lepidium, Descurainia, Salsola, Kochia,* and *Lappula*).

II. Distribution and Dynamics of Desert Seed Banks

A. Patterns in Desert Seed Banks

Seeds in desert soils are distributed mostly near the surface, and it is these seeds that form the seed bank. From 80 to 90% of soil seeds are in the upper 2 cm of soil (Childs and Goodall, 1973; Reichman, 1975), and of those, most are in the litter or top few millimeters of soil (Dye, 1969; Young and Evans, 1975). The seeds of many desert annual species cannot germinate and emerge from below about 1 cm (Freas and Kemp, 1983; P. R. Kemp, unpublished data), and seeds of desert shrubs cannot emerge from below 4 cm (Williams *et al.*, 1974). Seeds are not foraged below 7 cm (Reichman, 1979). Thus, the small number of seeds that fall below 7 cm in the soil can be considered lost from the desert seed bank, as they would not be involved in germination, granivory, redistribution, or other processes that directly relate to seed bank dynamics.

Two unpublished data sets collected by Dr. Norman Scott of the National Ecology Research Center, U.S. Fish and Wildlife Service, Albuquerque, New Mexico (hereafter, N. Scott, unpublished data), illustrate well the patterns and dynamics of desert seed banks. Soil seed densities were investigated from 1976 through 1977 at a site in the Lower Colorado River Subdivision of the Sonora Desert within Cabeza Prieta National Wildlife Refuge (35 km southwest of Ajo, Pima County, Arizona; 330 m elevation) and at a site in the Great Basin Desert within Chaco Canyon National Historical Park (75 km south of Farmington, San Juan County, New Mexico; 1950 m elevation). At each site, from 12 to 20 soil samples (100 cm^2 area $\times$ 2 cm deep) were collected from random points approximately 10 m apart located within a 12.6 ha sample unit. There were four adjacent sample units, so that 48 to 80 replicates were collected per sample date.

The seed banks identified in these samples reveal a high degree of spatial heterogeneity. The frequency distribution of seeds per sample has a highly kurtotic pattern (Fig. 2), revealing that some samples have a great number of seeds, but most have only a few or none. Highly clumped distributions of seeds in soil are common for desert seed banks (Nelson and Chew, 1977; Reichman, 1979, 1984; Price and Reichman, 1987).

In addition to spatial variability, there is great seasonal and annual variability in the seed bank. Figure 3 reveals the degree of change in the seed bank that can occur from one year to the next. Soil seed densities in the Sonora Desert (Lower Colorado River Subdivision) were similar throughout 1976, averaging about 400 seeds m^{-2}. Densities increased nearly 20-fold to 7711 seeds m^{-2} in March, 1977, and then decreased to 715 seeds m^{-2} by October, 1977. This change is equivalent to a turnover time of approximately 0.5 yr, based on a maximum pool size of 7711 seeds m^{-2}, and indicates a lack of persistance or long-term stability of the seed bank as a whole. Other hot desert regions exhibit similar temporal variability in their seed banks. At a site in the Arizona Upland Subdivision of the Sonora Desert, Reichman (1984) found a 6-fold variation between 1974 and 1976 in soil seed densities of open areas and a nearly 10-fold variation in microhabitats that tended to accumulate seeds, such as depressions and wind shadows. Nelson and Chew (1977) reported a 10-fold annual change in the seed bank under shrubs and a 23-fold change in open areas at a Mojave Desert site. P. R. Kemp (unpublished data) found a 5-fold change in the seed bank, averaged over

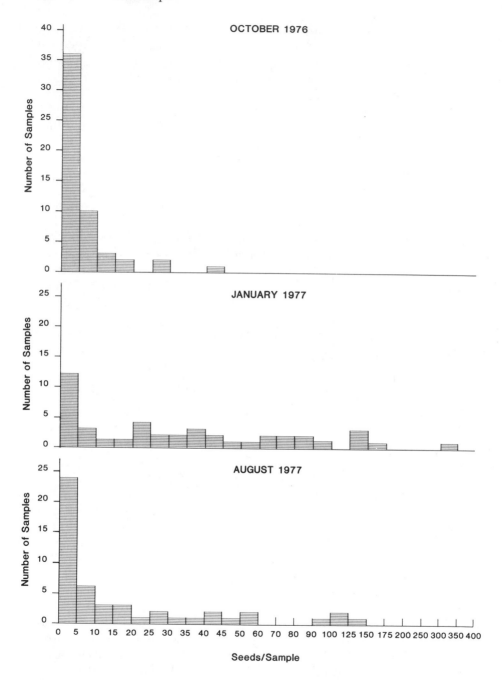

Figure 2.   Histograms of the distribution of samples containing specific numbers of seeds for three different sampling periods in the Lower Colorado Subdivision of the Sonora Desert near Ajo, Arizona (N. Scott, unpublished data). Note the change of scale along the abscissa.

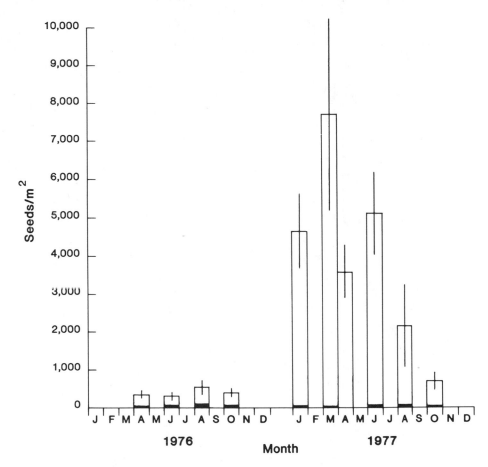

Figure 3.   Densities of seeds in random samples from the Lower Colorado Subdivision of the Sonora Desert near Ajo, Arizona, during 2 yr (N. Scott, unpublished data). Open portion of bar delineates annual species; solid portion of bar delineates perennial species. Vertical lines are ± 1 SE, $n = 48$–80.

different microhabitats, during a single year in the northern Chihuahua Desert.

For the most part, seed banks of the Great Basin Desert also undergo large temporal fluctuations. In a shrub–steppe community of northwestern New Mexico, the quantity of seeds in the soil increased nearly 4-fold from 2450 seeds m^{-2} in April, 1976, to 8431 seeds m^{-2} in May, 1976, followed by a decline to 2560 seeds m^{-2} in September, 1976 (Fig. 4). This change translates into a turnover time of about 0.75 yr, based on the maximum pool size of 8431 seeds m^{-2}. Young and Evans (1975) reported a large annual change in the quantity of germinable seeds in

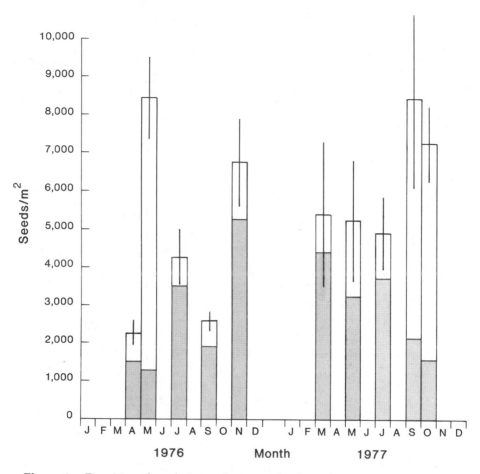

Figure 4. Densities of seeds in random samples from the Great Basin Desert, near Chaco, New Mexico during 2 yr (N. Scott, unpublished data). Open portion of bar delineates annual species; stippled portion of bar delineates perennial species. Vertical lines are ± 1 SE, $n = 48$–80.

the soil of the three principal annual species of an overgrazed sagebrush (*Artemisia tridentata*) community in northwestern Nevada. Germinable seed densities of *Bromus tectorum* showed the greatest variability, with 4000 seeds m^{-2} in November versus 400 seeds m^{-2} 1 month later. The only perennial with germinable reserve in the soil was *Chrysothamnus viscidiflorus*, with 10–370 seeds m^{-2} over the year. Hassan and West (1986) found a large annual change (5-fold or greater) in the seed bank of a shrub–steppe community in Utah. However, total seed densities were very low; peak density was 92 seeds m^{-2} in December, 1981. In contrast, Parmenter and MacMahon (1983) found no change in the total seed density between samples taken 1 yr apart (3590 versus 3940 seeds m^{-2}) in a southwestern Wyoming shrub–steppe community. However, single annual samples may not reveal the temporal dynamics of the seed bank. For example, in the New Mexico shrub–steppe community (Fig. 4), soil seed densities were similar in July, 1976 and July, 1977 (4275 versus 4917 seeds m^{-2}, respectively), but fluctuated throughout the year.

B. Processes Governing Seed Bank Dynamics

The spatial and temporal dynamics of desert seed banks can be understood, in principle, by considering the processes that govern additions to the seed bank, redistribution, and depletion of seeds from the seed bank (Harper, 1977). While it is relatively easy to identify the major processes that regulate seed banks, it is much more difficult to quantify them. Additionally, the role that different processes play in regulating seed banks can vary with the spatial or temporal scale that is considered. In the following text, I discuss the processes that affect seed banks at broad, intermediate, and local spatial scales.

In desert regions and major subdivisions within a desert, climate governs life-form distributions and primary productivity, which, in turn, affect seed banks. Much of the temporal variation in seed banks of hot deserts is associated with the variability in primary production of annual species (Nelson and Chew, 1977; N. Scott, unpublished data). In the Mojave Desert (southern Nevada), following 2 yr of low herbage production, soil seed densities in October, 1972, averaged 3578 seeds m^{-2} under shrub canopies (*Larrea tridentata* and *Lycium andersoni*) and 269 seeds m^{-2} in open areas (Nelson and Chew, 1977). In spite of these low seed densities abundant winter/spring rains (1972/1973) produced the largest herbage biomass in 10 yr, resulting in seed banks in October, 1973, of 37,259 seeds m^{-2} under shrubs and 6151 seeds m^{-2} in open areas. Total seed density was generally five times greater under shrubs than in open areas, reflecting the differences in densities and productivity of annual species in the two microhabitats (Nelson and Chew, 1977; Patten, 1978). Similarly, in the Lower Colorado River Subdivision of the Sonora Desert, seed banks were relatively small in 1976 (Fig. 3) following very low rainfall and sparse production of annuals in 1975 (N. Scott, unpublished data). Rainfall and production of annuals was also low in the first half of 1976, except for rather numerous *Plantago insularis* that apparently did not set seed (N. Scott, unpublished data). Abundant summer rainfall in 1976 resulted in relatively high densities of several annual species and augmentation of the seed bank by the following winter (Fig. 3).

In the Great Basin Desert, plant communities with numerous annuals have seed banks that fluctuate much like those of the hot deserts, as a result of the variable productivity of annuals (Young and Evans, 1975). Parmenter and MacMahon (1983) noted that soil seed densities more than tripled on shrub-removal plots that were invaded by the annual species *Bromus tectorum*. However, some Great Basin Desert communities have seed banks with a much greater proportion of seeds of perennial species than do the hot deserts, and the soil seed densities of these species are less variable than those of the annual species (Fig. 4; Parmenter and MacMahon, 1983). These characteristics appear to be the result of continuous additions to the seed bank by way of multiple or protracted flowering periods and/or extended seed release, rather than persistence of the seed bank.

Examination of habitat and plant community differences within a desert region (intermediate spatial scales) indicates that seed bank dynamics are governed by local plant distributions and local flowering and fruiting events. This scale is perhaps the one that we know least about in terms of either the processes that control vegetation or seed bank dynamics. Dye (1969) found significant differences in mean seed densities and species composition between two community types within 4 km of each other in the northern Chihuahua Desert. Both communities had *Larrea tridentata* and *Krameria parvifolia* as dominant species. The community with *Acacia constricta* as a codominant had approximately 13,000 seeds m^{-2}, whereas communities with *Ephedra trifurca* as a codominant had approximately 22,000 seeds m^{-2}, with significantly more annual grass species. Studies by P. R. Kemp (unpublished data) show even greater differences between two Chihuahua Desert sites separated by 40 km, but having similar dominant vegetation. At one site, mean seed densities ranged from 8800 to 24,500 seeds m^{-2} over 1 yr, but at the other site, which had coarser soils and fewer annuals, seed densities varied from 1300 to 6000 seeds m^{-2} during the same year.

At the smallest scales of spatial patterns, seed distributions are distinctly patchy (clumped). Figure 5 summarizes the variability of soil seed densities that occurs at one location within an Arizona Upland Sonora Desert community. The adult plants of annual species have patchy distributions to begin with, often related to shrub distribution (Went, 1942; Muller, 1953; Patten, 1978) or to edaphic factors (Beatley, 1969). Desert shrubs may have random, regular, or clumped distributions (Barbour, 1969, 1981; Woodell *et al.*, 1969; Fonteyn and Mahall, 1981). However, regardless of adult-plant distribution, seed distributions can emerge as clumped because of a lack of extensive dispersal. Generally, dispersal distances are relatively short, with little difference among seeds having different dispersal agents (Reynolds, 1958; Zohary, 1962; Burrows, 1973; Harper, 1977; Ellner and Schmida, 1981; Chew and Chew, 1970). The seeds that do achieve some dispersal by water or wind tend to accumu-

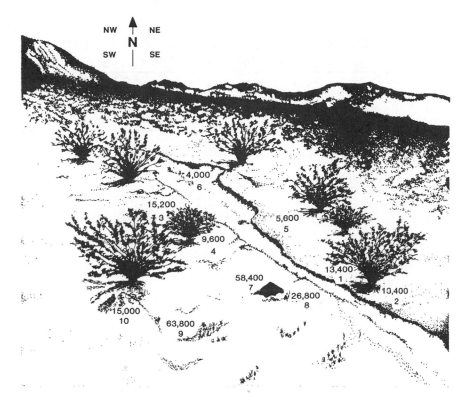

Figure 5. Average soil seed densities per square meter in 12 microhabitats (sampling times pooled) at an Arizona Upland Sonora Desert site (from Reichman, 1984). Microhabitats are (1) NW of *Larrea* bush, (2) SE of *Larrea* bush, (3) NW of *Ambrosia* bush, (4) SE of *Ambrosia* bush, (5) open area, (6) normally dry washes, (7) NW of obstruction (~25 × 100 mm), (8) SE of obstruction, (9) natural depression, and (10) extensive rodent digging.

late in depressions where water collects or in the wind shadows of obstructions (Fig. 5). Seeds dispersed by granivores are likely to be clumped in unrecovered caches (Price and Jenkins, 1987). Thus, patchiness in seed banks will tend to be amplified through time by seed deposit beneath successful parents and by dispersal into specific microsites or seed caches. Analysis of patchiness in a Sonora Desert community reveals that seed distributions are indeed more patchy than is the distribution of the plants in both space and time (Tables 1 and 2).

After seeds are dispersed into the soil, seed distributions will be affected by factors that cause depletion of seed reserves: germination, granivory, loss to deep soil, and senescence or other causes of death. Some of these processes can act in a nonrandom fashion and may cause

Table 1
Spatial and temporal heterogeneity of plant and soil seed densities (Arizona)[a]

Species	Patchiness in space[b]		Patchiness in time[c]	
	Plants	Seeds	Plants	Seeds
All species	3.7	16.7[d]	3.20	1200[d]
Perennials	0.8	5.4[d]	0.04	18.0[d]
Larrea tridentata	0.5	10.0[d]	0.01	16.2[d]
Ambrosia deltoidea	0.9	1.5	0.01	10.3[d]
Ambrosia dumosa	1.0	2.2	0.10	24.0[d]
Annuals	7.5	18.2	8.10	1011
Plantago	12.9	10.6	13.80	550[d]
Sphaeralcea	2.1	30.0	8.10	41.0
Cryptantha	2.5	—	2.80	—
Bouteloua	16.5	38.0	16.60	3546[d]
Oligomeris	—	7.1	—	0.1

[a]Data from species collected in the Sonora Desert near Ajo, Arizona (N. Scott, unpublished data).

[b]Average variance/mean ratio among samples at one time.

[c]Average variance/mean ratio of sample means from different times.

[d]Plant distribution significantly different ($p < 0.05$) than seed distribution.

Table 2
Spatial and temporal heterogeneity of plant and soil seed densities (New Mexico)[a]

Species	Patchiness in space[b]		Patchiness in time[c]	
	Plants	Seeds	Plants	Seeds
All species	4.0	30.4[d]	1.8	22.2[d]
Perennials	1.2	28.0[d]	0.1	12.2[d]
Oryzopsis hymenoides	0.8	4.0	0.2	0.3
Artemisia frigida	1.7	4.0	0.1	27.6[d]
Atriplex canescens	0.9	38.7[d]	0.1	1.4
Annuals	—	74.2	—	29.0
Descurainia	—	30.9	—	58.8
Chenopodium	40.0	154	3.2	16.4

[a]Data from species collected in the Great Basin Desert near Chaco, New Mexico (N. Scott, unpublished data).

[b]Average variance/mean ratio among samples at one time.

[c]Average variance/mean ratio of sample means from different times.

[d]Plant distribution significantly different ($p < 0.05$) than seed distribution.

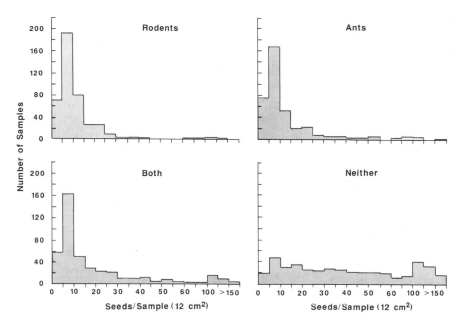

Figure 6.   Distribution of samples containing specific numbers of seeds from habitats that contained rodents (ants removed), ants (rodents removed), both ants and rodents, and neither ants nor rodents (from Reichman, 1979).

changes in patchiness. Of these processes, granivory has been studied most extensively.

Granivores have significant impacts on desert seed banks, consuming up to 70–95% of the seeds of some species in some localities (Chew and Chew, 1970; Soholt, 1973; Whitford, 1978; Brown et al., 1979b; Evenari et al., 1982). Experiments manipulating levels of granivores reveal that ants and rodents can affect both soil seed densities (Brown et al., 1979a; Reichman, 1977; Mehlhop, 1981) and plant densities (Inouye et al., 1980; Davidson et al., 1985). Granivores also appear to collect seeds from specific patch sizes, affecting the distribution of seeds in the soil (Davidson, 1977; Reichman and Oberstein, 1977; Price, 1978a,b; Reichman, 1979; Mehlhop, 1981). By selective foraging, ants and/or rodents may be the most important cause of local spatial heterogeneity in desert seed banks (Fig. 6).

C. Interspecific Variation in Seed Bank Patterns

Within seed banks, distributional patterns of seeds vary among individual species (see Figs. 7–10; Nelson and Chew, 1977). This variation may reflect differences among species in the degree to which the seed bank is part of an evolutionary strategy promoting population survival in the desert. Within hot deserts, most annual species have soil seed densities that are related to variability in plant production over time and from place to place (Beatley, 1969; Nelson and Chew, 1977; Patten, 1978). Their seeds are subject to continuous depletion through granivory and germination and, consequently, they do not have persistent seed banks. Most species probably achieve some production in most years, even in years with scant rainfall (Tevis, 1958; Beatley, 1974; Nelson and Chew, 1977). Their seed banks, however, may be subject to depletion by extreme droughts, leading to changes in plant distribution. Within this group of species, there are likely to be consistent species differences in seed distribution patterns related to plant phenology and morphology and to the differing effects of granivory. Those annual species with an indeterminate flowering may be able to disperse seeds for extended periods, resulting in less clumping and thus less risk of predation by rodents (Reichman, 1979). Similarly, plants that retain seeds on dead stalks (Gutterman, 1983) have prolonged release, decreasing clumping and predation. Granivores also selectively harvest certain species of seeds (Soholt, 1973; Reichman, 1975), sizes of seeds (Davidson, 1977), or patches of seeds (Reichman and Oberstein, 1977; Reichman, 1979), and they may avoid unpalatable or toxic seeds (N. Scott and R. G. Cates, unpublished data). Thus, some species would be expected to have more persistent seed banks because of characteristics that make their seeds less susceptible to granivory.

Toward one end of this continuum are species that maintain a persistent pool of seeds that only occasionally germinate or are replenished. They have small seeds with a somewhat uniform spatial distribution. This small seed size makes granivory rewarding only if seeds are in clumps, and the harvesting of these clumps tends to create a more constant density for the whole area (Reichman, 1979). Examples of such species are *Oligomeris linifolia* in the Sonora Desert (Fig. 7), *Mollugo cerviana* in the Chihuahua Desert (P. R. Kemp, unpublished data), and *Langloisia setosissima* in the Mojave Desert (Nelson and Chew, 1977). These species are closest to the "classical" desert ephemerals, and are most likely to possess adaptations or responses, such as innate dormancy and extensive dispersal, which promote the development of a persistent seed bank, as predicted by theoretical models (Cohen, 1966; Levins, 1969, Venable and Lawlor, 1980).

Shrubs and long-lived perennials in hot deserts have minimal dependence on seed banks for regeneration and protection against climatic uncertainty. Their strategy is one of producing a few seeds almost every year, most of which do not persist in the seed bank (Fig. 8; Boyd and Brum, 1983), probably because of predation by rodents (Soholt, 1973;

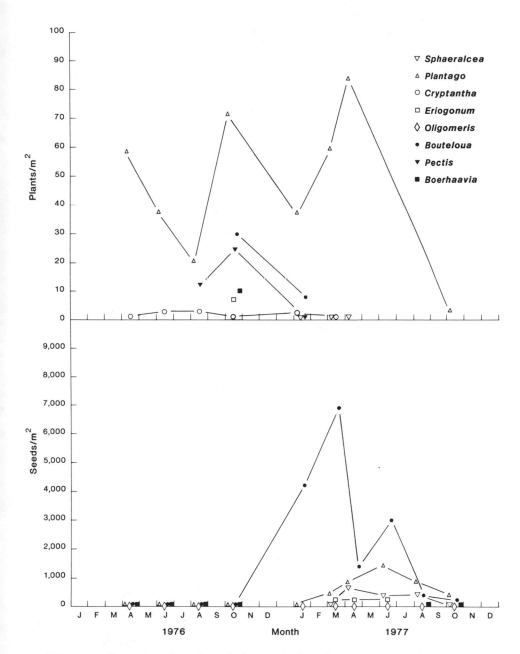

Figure 7.   Densities of seeds and plants of selected annual species from the Sonora Desert near Ajo, Arizona (N. Scott, unpublished data).

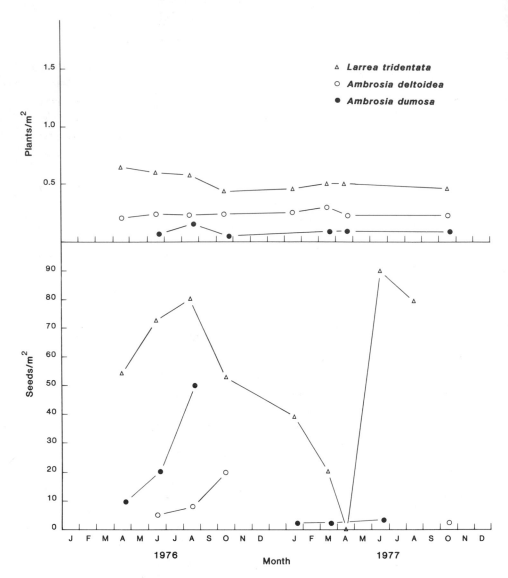

Figure 8. Densities of seeds and plants of selected perennial species
from the Sonora Desert near Ajo, Arizona (N. Scott, unpublished data).

Reichman, 1977; Inouye *et al.*, 1980). When conditions are favorable for
shrub establishment (perhaps a rare sequence of years), the seed source
will be primarily seeds produced during the previous season. Desert
perennials are protected against climatic uncertainty by a long life rather
than by a seed bank (Beatley, 1980).

The patterns described for the hot deserts also are found in the

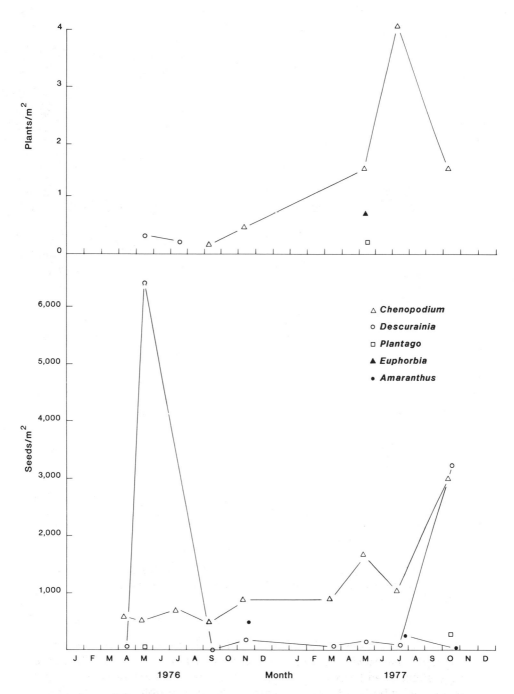

Figure 9. Densities of seeds and plants of selected annual species from the Great Basin Desert near Chaco, New Mexico (N. Scott, unpublished data).

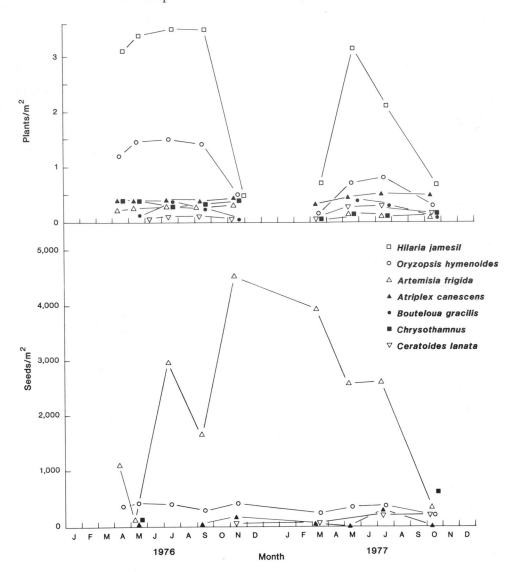

Figure 10. Densities of seeds and plants of selected perennial species from the Great Basin Desert near Chaco, New Mexico (N. Scott, unpublished data).

Great Basin cold desert, (see Figs. 9 and 10), but some differences occur. *Oryzopsis hymenoides*, a long-lived perennial grass (West, 1979), has a very stable seed bank (Fig. 10). However, its seeds are heavily foraged from soil by both ants and rodents (N. Scott, unpublished data), and it appears that the stability is the result of seed retention on old flowering culms that continuously replenish the seed bank. The seed bank for the perennial *Artemisia frigida* is uncharacteristically large and variable for a perennial species (Fig. 10). Among annual species, *Descurainia* sp. behaves similarly to the typical hot desert annual species, undergoing a more than 100-fold seasonal change in its seed bank, whereas *Chenopodium* sp. has more constant densities of soil seeds. Further studies of differences among species in seed characteristics and seed distribution patterns will be helpful in providing an understanding of the role of seed banks in population survival in desert ecosystems.

III. Relation of Seed Banks to Vegetation Dynamics

The interrelation between the seed bank and vegetation patterns in desert ecosystems is not well understood. The patterns discussed so far suggest that seed bank distributions are more patchy than plant distributions. Further, soil seed densities do not correlate well with plant densities, particularly at any one point in time (Figs. 7–10). This discrepancy is explained partly by the seed dormancy patterns of annual species and by the continually changing distributions of seeds and plants.

Studies of desert annual species showed that their seeds have a temperature-controlled dormancy that allows germination only in the appropriate season for a given species (Went, 1949; Juhren *et al.*, 1956). Summer annual species germinate in summer and complete their life cycle within about 4–10 wk (Kemp, 1983). Winter annual species germinate in autumn or winter and complete their life cycle in the following spring (Kemp, 1983). The germination of all seeds of a seasonal group of species (Went, 1973) would lead to a completely offset cycle: only plants would be present during the season, and only seeds during the off-season. However, the lack of seasonal rainfall could result in failure of the plant populations to set seed, and extinction if all seeds had germinated. These outcomes are prevented, to a large extent, by other dormancy mechanisms. Most seeds have a moisture-controlled dormancy; increasing rainfall or successive rainfall events induce germination during a period of more favorable soil moisture (Beatley, 1974; Freas and Kemp, 1983). After seedlings have become established, they may induce dormancy in the remaining seed bank, thus preventing competition among plants (Juhren *et al.*, 1956; Inouye, 1980). Some species also appear to have innate seed dormancy, with a fraction of the seeds remaining dormant in any one season even under optimal germination conditions (Freas and Kemp, 1983). All of these dormancy mechanisms act to couple germination to favorable growth periods and to prevent depletion of the seed bank as a result of exhaustive germination.

While knowledge of dormancy patterns helps explain some of the interrelationships between the plants and their seed banks, much more work is needed to understand the way in which seed banks determine vegetation patterns and vice versa. Simultaneous studies of vegetation, seed production, and the seed bank are needed to explain these relationships. Evolutionary and ecological modeling can be an important tool to help experimental scientists investigate processes that are potentially important as mechanisms of adaptation to desert environments. Models that incorporate evolutionary responses of seed bank traits along with adult-plant responses (Brown and Venable, 1986; Venable, Chapter 5, this volume) seem particularly useful for determining how the seed bank and the adult plants form an integrated survival unit in the uncertain desert environment. Future models must also incorporate the effects of granivory on the seed bank because of the important interrelationships between desert granivores and seed banks.

IV. Regional Comparisons of Desert Seed Banks

A. North American Deserts

Among North American deserts, the largest difference in seed banks is between the Great Basin cold desert and the hot deserts. Although most studies reveal that all deserts have similar average maximum soil seed densities (8000–30,000 seeds m^{-2}) following seasons of normal or above-average precipitation, there are some locations in the Great Basin Desert that appear to have very small seed banks (Hassan and West, 1986). Some Great Basin Desert seed banks also have an element of perennial species not found in the hot deserts (N. Scott, unpublished data). This difference is due partly to the low abundance of annual species in the Great Basin Desert communities, but mostly to the high seed production of certain perennial species. Seed production and dispersal patterns of these perennial species appear to lend stability to the seed bank.

Currently, we do not have sufficient information to ascertain consistent differences among seed banks of the North American hot deserts. Their seed banks all reflect the dynamics and patchiness of the annual species of which they primarily are composed. However, it is timely to explore more thoroughly the way in which individual species, particularly annuals, differ in adaptations and behavior of their individual seed banks. Numerous theoretical models predict that there would be different evolutionary responses by plants, including behavior of seed banks, to differing levels of environmental uncertainty (Cohen, 1966; Venable and Lawlor, 1980; Brown and Venable, 1986). Because precipitation patterns of hot deserts differ, it is reasonable to hypothesize that there are differences between species of different hot deserts. For example, we might expect to find differences in seed banks of summer annuals between the Mojave and Chihuahua deserts as a result of differences in summer moisture certainty. The relatively high certainty of summer rainfall in the Chihuahua Desert, combined with high evaporative demand, suggests that summer annuals of this desert would rely on morphological and physiological adaptations to assure success of and seed set by adults, rather than relying on adaptations that foster persistent seed banks (Brown and Venable, 1986). As summer rainfall becomes less certain, as in the Mojave Desert, annuals face a greater probability of unsuccessful seed set, and must rely more on a persistent seed bank for population survival (Brown and Venable, 1986). Thus, summer annuals of the Mojave Desert would be expected to have traits that foster a persistent seed bank, such as innate seed dormancy, small seed size, and antigranivory characteristics. While there have been no studies comparing these characteristics between species of different deserts, Freas and Kemp (1983) have tested a similar hypothesis that annuals within the Chihuahua Desert have evolved different strategies of population survival in response to the differences in relative certainty of moisture between seasons. Winter annuals face a greater chance of unsuccessful seed set than do summer annuals, because of the higher uncertainty of winter precipitation. The two winter annual species examined have innate seed dormancy (a requisite of persistent seed banks), while the summer annual species do not (Freas and Kemp, 1983). The finding of differences in innate dormancy among species of different seasons in the Chihuahua Desert suggests that comparisons of species from different deserts would be fruitful.

B. World Deserts

It is difficult to compare seed banks of the different desert regions of the world because of the paucity of information. In deserts with vegetation composed of a diversity of life-forms and a significant proportion of annuals, it is expected that seed banks would be occasionally large and always patchy, reflecting the dynamics and dispersal patterns of their annual species. A number of desert regions, such as the Sahara (Kassas and Batanouny, 1984), middle Asian (Walter and Box, 1983), and Negev (Gutterman, 1982) Deserts, have diverse life-forms and a significant proportion of annual species. Amelin (1947) reports seed production for central Asian semideserts to be in the range of 3000–30,000 seeds m^{-2}. If these seeds were incorporated into the soil, the soil seed density would closely match levels observed in North American deserts. In the Negev Desert, the abundance of annual species and their seed production is similar to that of North American deserts (Gutterman, 1982). In the

Negev Desert, the seed banks can be greatly affected by granivores (Evenari *et al.*, 1982).

However, in a desert region of western New South Wales, Australia, Westoby *et al.* (1982) reported that seed banks for several annual grass species were not greatly reduced over a 2-yr period, during which time there was no input into the seed bank because of drought, and ant granivores were apparently abundant and active. The seed densities were extremely high (83,000 seeds /m^2 for the 3 species reported) and Westoby *et al.* (1982) hypothesized that the seeds were able to escape predation through burial in the cracked clay soils during drought.

Some desert regions are distinctly drier than the North American deserts. An example is the coastal desert region of Peru and Chile. Much of this region is devoid of seed plants in most years (Rauh, 1985). In years with sufficient rainfall (20 mm), the ephemeral vegetation creates a flowering desert, revealing the presence of an extensive seed bank. However, paralleling the coast, about 30 km inland, is a narrow band of rock and debris desert that never has seed plants and apparently has no seed bank. Thus, there are limits to which even seed banks can provide a means for survival in desert plant populations.

V. Summary and Conclusions

Seed banks of North American deserts are spatially and temporally heterogeneous. This heterogeneity reflects the dynamics of the major processes that control additions to (seed production and dispersal) and deletions from (granivory and germination) the seed bank. As a whole, most desert seed banks have a turnover time of less than 1 yr, indicating little stability and persistence of the seed bank.

Soil seed densities are more variable from place to place within a desert region than between desert regions. All North American hot deserts can achieve maximum seed bank sizes (averaged over microhabitats) that are similar, approximately 8000–30,000 seeds m^{-2}. Seed banks of the Great Basin cold desert are perhaps not as large.

Seed banks of the hot deserts are composed primarily of annual species. The dynamics of the seed bank largely reflect the productivity and dispersal patterns of these species. The seed banks of the Great Basin cold desert have a significant proportion of seeds of perennial species. The behavior of this seed bank reflects, to a degree, the productivity and dispersal patterns of perennials.

Different species appear to have evolved different strategies with respect to the role of their seed bank in fostering survival in the desert.

All annual species are dependent upon seeds for population survival, but many do not appear to have a persistent seed bank. These species probably are subject to changes in distribution by external factors that can affect seed production or distribution, such as drought or granivory. Some annual species appear to have a seed bank that is relatively stable through time. These species would be expected to have adaptations, such as innate dormancy and granivory avoidance mechanisms, that foster persistent seed banks. Perennial species of hot deserts do not have persistent seed banks. The perennial species of the Great Basin cold desert that appear to have persistent seed banks probably have continuous replenishment of the seed pool rather than persistent soil populations.

Patchiness in desert seed banks may be one of the important causes of patchiness in the desert vegetation. However, densities of seeds in soil do not correlate well with densities of plants in time or space. Some of this lack of correspondence is related to constantly changing distributions of plants and seeds and to dormancy patterns in seeds. Much more work is needed to understand how seed banks determine vegetation pattern, and how seed banks and vegetation are interrelated in promoting population survival in the desert. Granivores must be included in these studies, as they play a fundamental role in the behavior of the seed bank.

Little is known about seed banks in desert regions outside North America.

Acknowledgments

I thank Dr. Norman Scott of the National Ecology Research Center, U.S. Fish and Wildlife Service, Albuquerque, New Mexico, for access to unpublished data on soil seed densities collected as part of a study of the use of natural seed crops by animal seed predators. Jean Hafner and Dr. Richard Smartt were primarily responsible for collecting these data, and Jean Hafner also contributed greatly to seed identification and analyses of soil samples from other unpublished data sets reported here. Drs. Jim Reichman and Kathleen Affholter provided many helpful comments on the manuscript.

CHAPTER 13

Wetland Seed Banks

Mary Allessio Leck

Department of Biology
Rider College
Lawrenceville, New Jersey

I. Introduction

Wetlands, which represent great habitat diversity, have three attributes in common: (1) flooded or saturated soils for at least part of the growing season, (2) vegetation adapted to a particular hydrological regime, and (3) hydric soils (Cowardin et al., 1979). Five major types of wetlands are identifiable: marine (open ocean shoreline), estuarine (salt and brackish waters of coastal rivers and bays), riverine (rivers and streams), lacustrine (lakes, reservoirs, and large ponds), and palustrine (marshes, bogs, swamps, and small shallow ponds). [See Cowardin et al. (1979) and Mitsch and Gosselink (1986) for discussion of wetland types.]

Seed banks, assayed by observing seeds germinating in soils, have been surveyed in all major wetland systems (Table 1), including marine

Table 1

Size and diversity of North American wetland seed banks

Wetland	Depth (cm)	Density ($\bar{x}$ m^{-2})	Range (m^{-2})	Species number	Location	Reference
Tidal marshes[a,b]						
Fresh (6)	0–10	9293	1620–13,670	52	New Jersey	Leck and Graveline (1979)
(4)	0–5	2430	1645–3620	35	New Jersey	Parker and Leck (1985)
(3)	0–10	26,957	14,805–41,010	53	New Jersey	Leck and Simpson (1987a)
Salt (7)	0–5	708	63–1375	17	California	Hopkins and Parker (1984)
(3)	0–10	691	259–1214	9	New Jersey	Engel (1983)
Nontidal marshes						
Fresh (1–3)[8]	0–~10	3203	696–9048	29	Iowa	van der Valk and Davis (1976)
Fresh (6)	0–5	29,753	10,875–36,230	45	Iowa	van der Valk and Davis (1978)
(6)	0–35	110,000	42,000–255,000	50	Iowa	van der Valk and Davis (1979)
Brackish (8)	0–5	3577	93–8253	34	Manitoba	Pederson (1981)
(6)	0–4	2455	70–6536	24	Utah	Smith and Kadlec (1983)
Salt (4)	0–4	191	50–430	3	Utah	Kadlec and Smith (1984)
(5)	0–10	8894	3036–20,182	4	Ohio	Ungar and Riehl (1980)

Other freshwater wetlands

Bogs (2)	0–10	165	0–330	~1	New Brunswick — Moore and Wein (1977)
(4)	0–45	171,830	12,874–377,041	12	West Virginia — McGraw (1987)
Floodplain (2)	0–3	11	0–21	1	Alaska — Walker et al. (1986)
Lake (5)c	0–10	315	0–542 (0–2335)d	5	Alberta — Haag (1983)
Lakeshore (6)	0–5	10,089	1862–19,798	41	Ontario — Keddy and Reznicek (1982)
(1)	0–21	36,639		25	Ontario — Nicholson and Keddy (1983)
Temporary ponds [2]	0–5	17,943	11,455–24,430	21	New Jersey — McCarthy (1987)
Swamp (2)	0–75/125	600	100–1100	6	Georgia — Gunther et al. (1984)
Riverine (2)[3]	0–10	2576	759–4392	59	South Carolina — Schneider and Sharitz (1986)
(1)e	0–10	276	76–611	10	Florida — Titus (1988)

[a]The number of sites (habitat types) is given in parentheses and the number of locations sampled in brackets. Data are for samples maintained under drawdown (moist) conditions and for surveys of the entire seed bank except as indicated. Nonseed plants are not included.

[b]Marshes are dominated by herbaceous macrophytes and swamps by woody species.

[c]Submersed samples only.

[d]Samples from heated areas near power plants.

[e]Woody component only (14 microhabitats).

(McMillan, 1981, 1983). Few studies have been conducted in tropical and subtropical habitats (Roberts, 1981), a problem long recognized for aquatic systems (Sculthorpe, 1967).

Studies of wetland soils have explored many questions related to seed biology. Among the earliest reports was that of Salter (1857), who noted that harbor mud contained seeds of agricultural and wild species, including several rare ones, that grew some distance from the site. Species of the local salt marsh vegetation were not present. His observations, corroborating a study by Darwin (1857), provided evidence for dispersal and longevity in salt water. Darwin (1859) used the presence of a pond seed bank (537 seedlings emerged from 3 tablespoons of soil) to support his thesis for the widespread dispersal of wetland species by birds, part of a larger consideration of migration, geographical distribution, isolation, and ultimately modification of species.

Between 1859 and the 1970s, the only wetland study was that of Milton (1939), who examined the relationship of the seed bank to the surface vegetation in a British salt marsh. Later studies also examined the relationship between seed banks and zonation patterns (e.g., van der Valk and Davis, 1978, 1979; Pederson, 1981; Keddy and Reznicek, 1982; Smith and Kadlec, 1983; Kadlec and Smith, 1984; McCarthy, 1987), succession (Wee, 1974; Oka, 1984), vegetation cycles (van der Valk and Davis, 1979), species diversity (many studies, e.g., Pederson, 1981; Leck et al., 1988, 1989), dynamics of species distribution (e.g., Ungar and Riehl, 1980; Hopkins and Parker, 1984; Parker and Leck, 1985), and establishment and recruitment to the vegetation (e.g., Meredith, 1985; Leck and Simpson, 1987a). Study of depth profiles provided information about longevity and vegetation history (e.g., van der Valk and Davis, 1979; Nicholson and Keddy, 1983; Gunther et al., 1984; Leck and Simpson, 1987a) and population history (McGraw, 1987). Soil samples from the same habitats collected at various times during the year showed seasonal variations in seed banks (Thompson and Grime, 1979; Ungar and Riehl, 1980; Hopkins and Parker, 1984; Schneider and Sharitz, 1986; Leck and Simpson, 1987a; Titus, 1988) and answered questions regarding seasonal depletion of seed banks due to germination (Leck and Simpson, 1987a), role of seed rain and flooding (Schneider and Sharitz, 1986), and the importance of persistent and transient seed bank strategies (Thompson and Grime, 1979). In addition, seed bank studies have led to models that predict vegetation changes based on water level fluctuations (van der Valk, 1981; Keddy and Reznicek, 1986) and processes contributing to seed burial and survival (McGraw, 1987).

This discussion examines the importance of environmental factors, especially the direct and indirect effects of an inundation regime on (1) the size, composition, and position of the seed bank; (2) recruitment from and renewal of the seed bank; (3) vegetation dynamics; and (4)

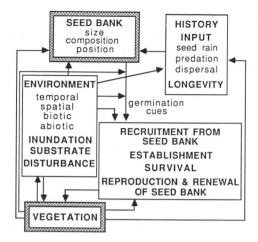

Figure 1. Relationship between wetland seed banks, vegetation, and environment.

germination cues (Fig. 1). Comparisons within and between wetlands are presented to document diversity and to establish, where possible, common principles. North American studies that cover a broad latitudinal range will be emphasized (Table 1), but reference will be made to other areas.

II. Seed Bank Dynamics

A. Establishment and Renewal of Seed Banks

Seed banks have been studied in a number of extant wetlands (Table 1). As a result, renewal processes are better understood than are colonization and establishment. Selection of successional stages at different sites may not provide the same level of insight gained from observing one site from the time of its origin.

Establishment of a wetland flora is dependent primarily on dispersal. Although some intertidal freshwater species may be adapted for limited dispersal or nondispersal (Sculthorpe, 1967; Ferren and Good, 1977; Ferren and Schuyler, 1980), many wetland taxa, as noted by Darwin (1859), have broad geographic ranges. These are the result of effective dispersal mechanisms, notably water and air (Sculthorpe, 1967; Moore, 1982), but also birds (deVlaming and Proctor, 1968) and other animals (e.g., fish; Gottsberger, 1978). Evidence for long-distance dispersal is most dramatically observed for species dispersed by ocean currents (D.M. Murray, 1986b) and in the amphitropical disjunctions of vernal pool plants (Zedler, 1987). Restricted ranges may be due to recent differentiation related to climatic changes and isolation (Zedler, 1987). The probabilty of a dispersal event occurring increases with time, coinciding with changes in the wetland that make it more conducive for growth of a larger number of species (Gaudet, 1977; Joenje, 1979). Accordingly, older wetlands (prairie glacial marshes) may have a larger number of seeds than do younger ones (Utah marsh; Smith and Kadlec, 1985b). From the limited number of sites surveyed (Table 1), it is tempting to suggest that wetlands from glaciated areas have less diversity than those in unglaciated areas, but this hypothesis requires testing.

Renewal of an established seed bank depends on both seed bank and vegetation dynamics, which in turn depend on recruitment from the seed bank, survival to subsequent reproduction, dispersal, predation, and viability (Fig. 1). The seed bank, depleted by germination flush and/or attrition, requires one growing season in a freshwater tidal wetland (Leck and Simpson, 1987a) for renewal, and variable, unknown lengths of time in other wetlands (Nicholson and Keddy, 1983). For species with transient seed banks or those with variable yearly population densities, lack of renewal can alter seed bank composition.

The seed rain in a wetland may be prodigous (e.g., Salisbury, 1970; van der Valk and Davis, 1979). However, species vary both in their reproductive capacity and dispersibility. Seed rain decreases with distance from seed source (Haag, 1983; Jerling, 1985). Species producing many seeds (coupled with prolonged viability and dispersibility) may be overrepresented in the seed bank (see van der Valk and Davis, 1979). Seed rain may be affected by the impact of herbivory on reproduction (Smith and Kadlec, 1985c; Cahoon and Stevenson, 1986) and by predispersal (e.g., Sickels and Simpson, 1985; Cahoon and Stevenson, 1986) or postdispersal seed predation (Ernst, 1985; T.J. Smith, 1987a), but relatively little is known about the importance of these factors.

B. Germination Strategies and Germination Traits

Four germination strategies, related to the persistence of seeds in soil and associated with morphological and physiological characteristics, have been identified by Thompson and Grime (1979) for temperate herbaceous seed banks. For the purposes of this discussion, Type I and Type II species have summer and winter transient seed banks, respectively; Type III species have a transient component and a relatively smaller persistent component; Type IV species have a large persistent seed bank relative to yearly seed input, and larger numbers of seeds in subsurface than in surface samples.

Among the 97 species with relevant data (based on depth distribution, exclosure studies, and/or seasonal changes; van der Valk and Davis, 1979; Ungar and Riehl, 1980; Leck and Simpson, 1987a; McCarthy, 1987) and adequate densities for evaluation, three species were Type II (*Ambrosia trifida*, *Impatiens capensis*, and *Peltandra virginica*), 66 species were Type III (e.g., *Bidens laevis*, *Acnida cannabina*, and *Sagittaria latifolia*), and 28 species were Type IV (e.g., *Dulicium arundinaceum*, *Juncus effusus*, and *Juncus pelocarpus*). The absence of Type I species reflects the kinds of habitats surveyed and/or the sampling regime. Wetland Type I species would include mangroves (e.g., *Avicennia germinans* and *Rhizophora mangle*) that usually germinate on the parent plant, and *Acer rubrum*, *Populus* spp., and *Salix* spp. that germinate soon after dispersal.

Most available data suggest that, as observed in the British flora (Thompson and Grime, 1979), specific morphological (Fig. 2) and phys-

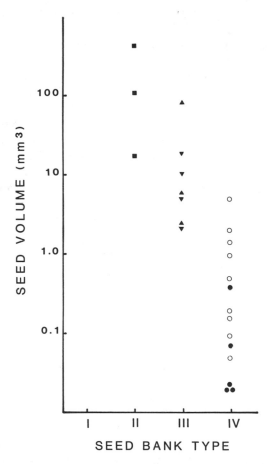

Figure 2. Relationship of seed size (estimated volume) to germination strategy (see text). Dimensions on which these estimates are based from Martin and Barkley (1961) and Montgomery (1977). ● Species found only in subsurface samples.

iological characteristics are associated with these strategies. As Type I species, the mangroves readily fit into the scheme because they have large seeds (>500 mm^3; sizes based on Gunn and Dennis, 1976) and germinate readily. However, the seeds of other species noted above are much smaller (~4–50 mm^3; sizes based on Schopmeyer, 1974), and some evidence exists (Canham and Marks, 1985) that seeds falling into an unfavorable light environment may germinate in response to disturbance during a second growing season. Type II species have relatively large seeds that escape burial by floating for extended periods or by having seed coat projections. They germinate in darkness or light. Although the actual dormancy mechanisms may vary (Leck, 1979; West and Whigham, 1975–1976), prolonged chilling results in complete spring germination.

In comparison, wetland species with persistent seed banks appear generally to have small seeds (Fig. 2), with several Type IV species having very tiny seeds (<0.1 mm^3). Species of Types III and IV appear to require light for germination, and this is often coupled to alternating temperatures (e.g., Gunther et al., 1984; Galinato and van der Valk, 1986). In addition, germination is affected by oxygen availability. Dormancy of these species with prolonged viability is often due to physical (mechanical) restrictions imposed by the seed coat (Sculthorpe, 1967). Germination of these seeds, as also noted by Baskin and Baskin (Chapter 4, this volume) requires dislodging of cuticularized or lignified plugs or rupture of the testa or endocarp, in some cases by drying. Drying, it should be noted, reduces the viability of some species (e.g., Zizania aquatica, Simpson, 1966; Impatiens capensis, Leck, 1979). Freshwater tidal wetland species, and presumably also species from other wetlands, differ in their capacity for enforced dormancy. Impatiens capensis, for example, over a 10-month period, loses viability under anaerobic conditions (Leck, 1979). Others, such as Bidens laevis, Acnida cannabina, and Polygonum arifolium, tolerate anaerobic conditions and germinate throughout the growing season following disturbance. Still others, such as Cuscuta gronovii and Mikania scandens, persist but germinate only in the spring (M.A. Leck, personal observation).

C. Size and Composition

Size, diversity, and composition of wetland seed banks vary considerably (Tables 1 and 2). Individual sites ranged from 0 to 59 species and 0 to 377,041 seeds m^{-2}. Generally, seed banks were smaller and less diverse as salinity increased. However, small size and low diversity were also found in some freshwater wetlands (Moore and Wein, 1977; Gunther et al., 1984). At the extremes were an Alaskan floodplain, which was the most depauperate, and a West Virginia bog, which had the largest seed

Table 2
Summary of wetland seed bank composition[a]

Wetland	Graminoids[b]	Percent Annuals	Woody	Most common	Most common species
Tidal marshes					
Fresh	<1	81	<1	26	*Acnida cannabina*
	3	86	<1	29	*Bidens laevis*
	3	31	<1	49	*Typha latifolia*
Salt	<1	1	<1	92	*Salicornia virginica*
Nontidal marshes					
Fresh	38	10	0	35	*Leersia oryzoides*
	44	26	0	42	*Scirpus validus*
	39		<<1	48	*Scirpus validus*
Brackish	58	21	<<1	50	*Scirpus validus*
	50		0	37	*Typha* spp.
Salt	0	100	0	98	*Salicornia rubra*
	53	47	0	56	*Juncus tenuis*
Other freshwater wetlands					
Lake	0	10	0	48	*Potomogeton pectinatus*
Lakeshore	43	<2	2	28	*Hypericum majus*
	54	1	<<1	15	*Eriocaulon septangularis* and
		(37 FA[c])		15	*Juncus effusus*
Temporary ponds	60	3	0	42	*Juncus pelocarpus*
Bogs	?	?	90	90	*Rubus strigosus*
Swamps	98	<<1	<1	90	*Juncus effusus*
	74	1	<1	61	*Rychospora inundata*
Riverine	48	16	11	35	*Cyperus* spp.

[a]Data are taken from the same sources as in Table 1.

[b]Graminoids include Gramineae, Cyperaceae, and Juncaceae.

[c]FA, Facultative annuals.

bank, with one site having the greatest density reported for any natural community (McGraw, 1987). The largest number of species (59) occurred in a South Carolina swamp (Schneider and Sharitz, 1986); however, 114 species were obtained from three freshwater tidal wetland studies over a 10-yr period (Leck *et al.*, 1988), more than double the number from any single study.

Seed bank composition varied within and among wetlands, as noted by Harper (1977), with one species often making up an overwhelming proportion of the seed bank. In wetlands these species comprised 15–90% of the seed bank, and were usually monocots, often graminoids (Table 2). Graminoids were not important in tidal wetlands, probably an artifact of studying wetlands not dominated by graminoids. Annual species were dominant in freshwater tidal wetlands (Leck and Grave-line, 1979; Parker and Leck, 1985; Leck and Simpson, 1987a), in some saline marshes (Ungar and Riehl, 1980; Kadlec and Smith, 1984), and along lakeshores (Keddy and Reznicek, 1982). The importance of annuals, often mud flat colonizers (Salisbury, 1970; van der Valk and Davis, 1978, 1979; van der Valk, 1981), can vary with frequency of drawdowns (Poiani, 1987), and may be confounded by the behavior of certain perennials that are facultative annuals (Nicholson and Keddy, 1983; Gunther *et al.*, 1984). These facultative annuals can alter considerably the potential importance of the annual component (e.g., from 1 to 37%; Nicholson and Keddy, 1983).

The woody component, even in swamps, was unimportant. The exception occurred in a bog (Moore and Wein, 1977) where *Rubus strigo-sus*, probably bird dispersed, was dominant. The only wetlands to have sizable numbers of viable seeds of woody species were swamps in South Carolina (Schneider and Sharitz, 1986) and Florida (Titus, 1988) that averaged 100–230 seeds m^{-2} and contained 5–15 woody species.

Few studies have noted the presence of a nonwetland component. Nonresident species in a tidal salt marsh were found in soils from near the bay front and along channels (Hopkins and Parker, 1984), and were particularly abundant when the salt marsh was adjacent to cultivated agricultural lands and meadows (Milton, 1939; Ungar and Riehl, 1980). Reproductive capacity of the local weed flora may influence drawdown succession (Gaudet, 1977), and perhaps also the seed bank. Such species may add considerable diversity to the potential flora (Leck *et al.*, 1988). Naturalized species, however, are not important (7%; based on Leck *et al.*, 1988b).

In addition to seeds and fruits, the soil contains vegetative propagules such as turions that may be more important than seeds for some species (e.g., Lemnaceae, van der Valk and Davis, 1979; *Potamogeton*, Rogers and Breen, 1980; Sastroutomo, 1981). Spores may also be present. *Chara*, a macroscopic alga, was an important species germinating from inland marsh soils (van der Valk and Davis, 1979; Kadlec and Smith, 1984). Fern spores of seven species exceeded seeds by 8–100 times, depending on time since clearing in Malaysia peat soils (Wee, 1974), and 23 species of bryophytes and ferns were recorded from freshwater tidal wetland soils (Leck and Simpson, 1987b). While vegetative propagules of seed plants may behave functionally as seeds (e.g., in overwintering), their behavior in soil (e.g., longevity and burial) remains unknown. In addition, little is known about how spore plants affect wetland vegetation dynamics (Leck and Simpson, 1987b), except possibly in bogs (Mitsch and Gosselink, 1986).

D. Longevity

Evidence for the viability and longevity of seeds may be gained by examining depth profiles (Fig. 3; see Fig. 5 in Keddy *et al.*, Chapter 16, this volume; McGraw, 1987). Sizable numbers of seeds have been observed below 25 cm (van der Valk and Davis, 1979; Gunther *et al.*, 1984; Leck and Simpson, 1987a; McGraw, 1987), with some occurring at 125 cm (Gunther *et al.*, 1984). Estimates of the ages of seeds in soil were ~45 yr in surface lakeshore sediments under 55 cm of water (Nicholson and Keddy, 1983), ~75 yr at 30–32 cm in a freshwater tidal wetland (Leck *et al.*, 1989), ~125 yr at 40 cm in a bog (McGraw, 1987), and >400 yr below 100 cm in a swamp (Gunther *et al.*, 1984). These observations and those where species appear at infrequent intervals when mud flats become exposed (Salisbury, 1970) appear to substantiate claims (Sculthorpe, 1967; Harper, 1977) that seeds of aquatic species have prolonged dormancy.

However, examination of seed bank profiles shows that, although decline with depth is often exponential, considerable variation occurs within and among wetlands (Fig. 3) and among species (Leck and Simpson, 1987a; McGraw, 1987). For lakeshores (Nicholson and Keddy, 1983), in temporary ponds, and in freshwater tidal wetlands, the seed bank was shallow, with >80% of the seeds occurring in the top 4–5 cm (Fig. 3A and B). In comparison, seeds in swamps, prairie marshes (Fig. 3C and D), and bogs (McGraw, 1987) were deeply buried, with only 20–50% occurring in the top 5 cm. These variable patterns may be due to past variations in seed rain, but may also be related to variable soil compression, to seed survival (McGraw, 1987), and/or to seed bank strategies (Leck and Simpson, 1987a).

Seed bank size, composition, and depth distribution are determined, in part, by longevity (Fig. 1). Factors selecting for longevity appear related to morphology, especially size and seed coat thickness; likewise, dispersibility is related to morphology (deVlaming and Proctor, 1968; Moore, 1982). Little is known about processes controlling lon-

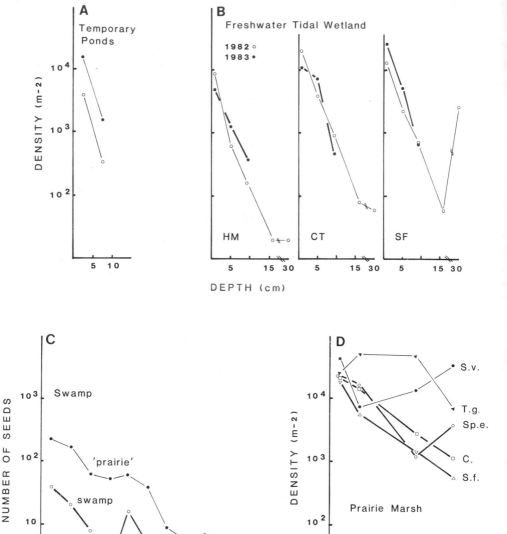

Figure 3. Depth profiles of wetland seed banks. (A) Two temporary ponds (data from McCarthy, 1987). (B) Three freshwater tidal wetland sites; HM, high marsh; CT, cattail; and SF, shrub forest (from Leck and Simpson, 1987a). (C) Okefenokee Swamp—prairie open and wooded swamp (data from Gunther *et al.*, 1984). A small number of seeds was found at 125 cm in the prairie site. (D) Prairie marsh communities; S.v., *Scirpus validus*; T.g., *Typha glauca*; Sp.e., *Sparganium eurycarpum*; C., *Carex*; and S.f., *Scirpus fluviatilis* (data from van der Valk and Davis, 1979).

gevity of wetland species, although imbibition and low oxygen levels reduce deterioration (Villiers, 1973).

E. Recruitment to the Vegetation

The importance of inundation, especially relative to recruitment from the seed bank, has been noted in freshwater tidal marshes (e.g., Parker and Leck, 1985), prairie marshes (e.g., van der Valk, 1981), inland saline marshes (e.g., Ungar and Riehl, 1980; Smith and Kadlec, 1983), riverine swamps (Schneider and Sharitz, 1986), and temporary wetlands (Gopal, 1986; McCarthy, 1987; Zedler, 1987). Many species appear to have broad tolerance limits for water level during the recruitment phase (Keddy and Ellis, 1985). Growth forms, however, vary in their response to flooding. Submersed species germinate almost exclusively under flooded conditions. The emergent perennials and mud flat annuals germinate under both flooded and drawdown conditions. Usually flooded samples produce fewer seedlings (van der Valk and Davis, 1978; Pederson, 1981; Smith and Kadlec, 1983; Leck and Simpson, 1987a). When this is not the case, it is due, at least in part, to species composition (Leck and Graveline, 1979; Kadlec and Smith, 1984).

Light, temperature, oxygen, and/or salinity may serve as germination cues (Fig. 1) (Ungar and Riehl, 1980; Gunther *et al.*, 1984; Galinato and van der Valk, 1986). These may be influenced by substrate characteristics, water flow, and water turbidity and color, and vary during drawdown cycles. Scouring and wave action may directly or indirectly determine recruitment and establishment (S.D. Wilson *et al.*, 1985; Keddy and Constabel, 1986). Some species may, in fact, require prolonged inundation (Gopal, 1986; Zedler, 1987) that can promote the growth of a fungus over the seeds (Griggs, 1981). Biotic factors, including allelopathy, shading, ingestion (Gottsberger, 1978), or specific nutritional requirements (e.g., of a hemiparasite, *Agalinis paupercula*; Keddy and Reznicek, 1982), may affect germination.

Furthermore, burial depth influences recruitment by affecting many of the factors noted above. As little as 1 cm of sand can substantially reduce germination (Galinato and van der Valk, 1986). Seedlings from buried large-seeded species are better able to reach the soil surface than are those from small-seeded species (Lee and Ignaciuk, 1985; Galinato and van der Valk, 1986).

Species differ in responses to germination cues. Because microenvironmental conditions vary from site to site during drawdown, different species may become established at sites with identical seed banks, as noted by Galinato and van der Valk (1986). Interactions between oxygen, light, and temperature appear particularly important. Our understanding of the correlation between germination traits of wetland species and field behavior is increasing (e.g., Conti and Gunther, 1984; Keddy and Ellis, 1985; S.D. Wilson *et al.*, 1985; Keddy and Constabel, 1986; Galinato and van der Valk, 1986). As germination traits are elucidated, the relationship of maternal effects, including inflorescence position, on responses to environmental cues and to survival (Berger, 1985; Ungar, 1987a) will be clearer.

F. Seed Bank and Vegetation Dynamics

1. Seed Bank Patterns

Factors that contribute to the distribution and maintenance of seeds in wetland soils include burial, inundation patterns (depth, duration, and timing of water fluctuation), physical and chemical characteristics of the substrate, and disturbance. These may directly affect the distribution of the seeds or may affect the availabilty of seeds by influencing reproductive output and/or recruitment.

Little is known about factors affecting vertical movement of seeds in the soil profile (McGraw, 1987; Leck and Simpson, 1987a), but they include earthworms (Grace, 1984), small mammals (McGraw, 1987), large mammals (cattle, hippopotamuses, Gaudet, 1977), or cracks which develop during drying (Gunther *et al.*, 1984; K.A. McCarthy, personal communication). Some movement by percolation may occur, with small seeds moving more rapidly than larger ones (McGraw, 1987). Most species that occur at considerable depth have small seeds (Harper, 1977; Type IV, Fig. 2), although exceptions are known (e.g., *Nelumbo nucifera*). Buoyant seeds are less likely to be deeply buried (Tyndall *et al.*, 1986). Sedimentation and soil compression also affect rates and depth of seed burial. Depth in the soil determines whether seeds will receive germination cues, and, even if cues are received, whether the shoot reaches the surface.

Fewer seeds are found in sites that are continually inundated (e.g., Pederson, 1981; Schneider and Sharitz, 1986; Poiani, 1987). Size and composition of the seed bank may vary along tidal channels (Parker and Leck, 1985) or with the presence of emergent sustrates where seeds may accumulate (Gunther *et al.*, 1984; Schneider and Sharitz, 1988; Titus, 1988). These patterns are coupled with the ability of many aquatic species to float (van der Valk and Davis, 1976; Keddy and Reznicek, 1982). Accumulation that occurs along drift lines (Pederson, 1981) may be enhanced by emergent vegetation (van der Valk and Davis, 1976, 1978; Smith and Kadlec, 1983, 1985b; Schneider and Sharitz, 1986, 1988). Where tides occur, tidal sorting of propagules by size may determine seed distribution (see Rabinowitz, 1978). Incorporation may be low if dispersal phenology needs to coincide with periods of low water, but if dispersal occurs over many months, chances for incorporation may be greater. Hydrological regime, however, appeared not to affect seed banks in a Utah salt marsh where seed bank size and composition of ridges and flats in a flood-irrigated area were similar (Kadlec and Smith, 1984).

Wetland seed bank composition may be affected by seasonal flooding. In a tidal salt marsh, increased species diversity and the proportion of annuals in the seed bank were related to high runoff from nearby rivers as well as to tidal dispersal patterns (Hopkins and Parker, 1984). Spring floods in a riverine swamp reduced both woody and herbaceous components of the seed bank in the site that was not continuously flooded, but caused no change in the one that was flooded (Schneider and Sharitz, 1986).

Size and composition of seed banks may also be influenced by substrate chemistry. Soil salinity can restrict germination of some species that then persist in the soil (Ungar and Riehl, 1980). Establishment of marsh plants on saline soils requires reduced salinity and an adequate seed source (Engel, 1983; Smith and Kadlec, 1983; Kadlec and Smith, 1984; Hopkins and Parker, 1984). Soil pH appears to play a role in the development and structure of some wetlands (Mitsch and Gosselink, 1986), but it is not known if it affects seeds present in the soil. Nutrients, which may vary during drawdown and with site, have been implicated in vegetation patterns (Dale, 1964; Gaudet, 1977).

Substrate texture may affect seed banks by influencing communities that develop (Dale, 1964, 1965). Germination, establishment, survival, and subsequent reproduction, for example, are lower on sand (Gaudet, 1977; Haag, 1983) or in sites with very finely textured or highly organic soils (Keddy, 1985a; Wilson and Keddy, 1985; Keddy and Constabel, 1986). Moreover, the presence of litter may cause allelopathic interactions (van der Valk and Davis, 1976), or more probably physical alteration of the environment (van der Valk, 1986).

Finally, various kinds of disturbance may influence seed bank and vegetation dynamics. For example, fire, superimposed on drought cycles in the Okefenokee Swamp, plays an important role in maintaining "prairies" and prairie lake holes in the peat sediments, retarding succession, and making the seed bank important to vegetation dynamics (Gunther et al., 1984). In contrast, in a young Utah marsh (Smith and Kadlec, 1985c), fire had little effect on seed banks.

Another element of disturbance is flotsam, which may inhibit recruitment from the seed bank (van der Valk, 1986) but may also be a source of seeds (van der Valk and Davis, 1976; Hopkins and Parker, 1984), providing microsites that may be preferred by certain species (M.A. Leck, personal observation). Masses of substrate, which float to the surface of the Okefenokee Swamp, allow recruitment of early successional species from the seed bank (Gunther *et al.*, 1984). Floating islands occur in other wetland systems, such as in the Amazon basin (Junk, 1970) and the sudd of Africa, but the importance of seed banks to their vegetation dynamics is poorly known.

Human activities, such as drainage, diking, damming, and irrigation, alter substrates and inundation regimes (Joenje, 1979; Tiner, 1984; Huenneke and Sharitz, 1986; Schneider and Sharitz, 1986, 1988; Gray and Bolen, 1987). Heated water from power-generating plants may interfere with natural stratification (Haag, 1983). In addition, agricultural activities may contribute to wetland disturbance. Reproductive success of a species and subsequent contribution to the seed bank depend on the species of grazers present (Gaudet, 1977). Recruitment from the seed bank and resulting vegetation may depend on if and when the canopy is removed (Meredith, 1985), and on size and composition of the seed bank (Pfadenhauer and Maas, 1987). Keddy *et al.* (Chapter 16, this volume) and van der Valk and Pederson (Chapter 15, this volume) discuss other issues related to human activities.

2. Correlation of Seed Bank and Vegetation Patterns

Similarity of the seed bank with the standing vegetation varied with wetland and often with the site within a wetland. In a freshwater tidal wetland, the composition of the seed bank reflected the standing vegetation, both as seedlings in the field and in the mature vegetation (Leck and Graveline, 1979; Parker and Leck, 1985; Leck and Simpson, 1987a); however, composition of both seed bank and vegetation varied with zone, and reflected the ability of species to survive (Parker and Leck, 1985). Similarly, in coastal salt marshes (Hopkins and Parker, 1984), along lakeshores (Keddy and Reznicek, 1982), and in inland marshes (van der Valk and Davis, 1978; Pederson, 1981), the seed bank composition reflected the vegetation, the latter two correlating best with the drawdown vegetation. In the Okefenokee Swamp, the seed bank of the open prairie site was similar to the vegetation, but the wooded swamp seed bank more closely resembled the vegetation of the prairie (Gunther *et al.*, 1984). In both a salt pan (Ungar and Riehl, 1980) and a bog (McGraw, 1987), the seed bank, overwhelmingly dominated by a *Juncus* sp. lacking or unimportant in the vegetation, did not resemble the standing vegetation.

Disparities between the floristic composition and the seed bank may also be caused by dominant species that contribute few seeds to the seed bank (e.g., *Acorus calamus*, Leck and Simpson, 1987a; *Phragmites australis*, Smith and Kadlec, 1983; *Peltandra virginica*, Whigham *et al.*, 1979), or have variable seed viability (e.g., *Spartina anglica*, Marks and Truscott,

Table 3
Comparison of species diversity in the seed bank, as field seedlings, and in standing vegetation

Wetland[a]	Number of species				Reference
	Seed bank	Seedlings	Vegetation	Total	
Tidal marshes					
Fresh (6)	52+	19	—	~54	Leck and Graveline (1979)
(4)	35+	20	24	37	Parker and Leck (1985)
(3)	55+	12	20	58	Leck and Simpson (1987a)
Salt (7)	17+	5	9	18	Hopkins and Parker (1984)
Nontidal marshes					
Fresh (6)	45+	—	34	48	van der Valk and Davis (1978)
Brackish (6)	29+	—	18	35	Smith and Kadlec (1983)
Salt (4)	9	—	14	15	Kadlec and Smith (1984)
(5)	4	—	7	8	Ungar and Riehl (1980)
Other freshwater wetlands					
Lake (5)	6	—	12	?	Haag (1983)
Lakeshore (6)	41	—	45	50	Keddy and Reznicek (1982)
Temporary ponds [2]	21	26	29	31	McCarthy (1987)
Riverine (2) [3]	59	—	49	73	Schneider and Sharitz (1986)

[a] The number of sites (habitat types) is given in parentheses and the number of locations sampled in brackets.

1985). It is possible that suitable germination conditions for specific species are not provided (e.g., van der Valk and Davis, 1976, 1978, 1979; Keddy and Reznicek, 1982; Smith and Kadlec, 1983), or that, once established, sóme species expand primarily by vegetative means (e.g., van der Valk, 1981). Although considerable *in situ* deposition of seeds may occur (Keddy and Reznicek, 1982; Haag, 1983; Parker and Leck, 1985), seeds of some species may disperse into communities where mature plants are lacking (e.g., van der Valk and Davis, 1976; Ungar and Riehl, 1980; Leck and Simpson, 1987a). In a salt pan a nonresident species formed a substantial part of the seed bank (Ungar and Riehl, 1980), whereas in a freshwater tidal wetland many did not persist (Leck and Simpson, 1987a). Where there is greater habitat and vegetation complexity there may be greater seed bank diversity (Schneider and Sharitz, 1986; Leck and Simpson, 1987a). Proximity to available seed sources and effective dispersal are also important.

Species diversity, however, varied with vegetation stage (seed bank, field seedlings, and standing vegetation) (Table 3). Because of the limited number of studies including all three stages, it is difficult to suggest whether any trend occurs. The importance of shifts in dominance with stage requires study.

Seed banks reveal clues to past vegetation. Along lakeshores, inundated soils contained a diverse flora, including rare species, present at a time when water levels were lower (Keddy and Reznicek, 1982). Distribution in the soil profile of a freshwater tidal wetland suggested different vegetation histories, and accumulation of species along the edge of the wetland compared with other areas (Leck and Simpson, 1987a). In fact, the depauperate seed banks in some sites may relate to vegetation history that, in turn, is related to inundation (e.g., Pederson, 1981; van der Valk, 1981; Haag, 1983) or to age of the wetland (Smith and Kadlec, 1985b).

G. Seed Banks and Populations

The importance of seed banks to population studies has been noted for several wetland species; four will be briefly considered here. However, for many wetlands little is known about the seed bank dynamics of individual species.

Few, if any, viable seeds of the annual *Impatiens capensis* remain after spring germination (Simpson *et al.*, 1985). Those that do may germinate following a second stratification period (M.A. Leck, unpublished data), but their importance to the population biology of the species is minimal. This species occurs in wetlands where the hydrological regime is predictable from year to year, and where there appears to be no selection for prolonged dormancy. Persistence is ensured by prolific seed production and effective dispersal (Leck, 1979).

Salicornia europaea, an annual halophyte, has variable seed bank dynamics. In a British salt marsh, its seed bank is essentially depleted by midsummer every year (Jefferies *et al.*, 1981), while Hudson Bay and Ohio populations have a persistent seed bank component (Jefferies *et al.*, 1983; Philipupillai and Ungar, 1984; Ungar, 1987b). In the latter population, field germination occurs over a prolonged period of time (February to June or later) and is related to seed dimorphism, with large seeds germinating more readily and earlier in the growing season and small seeds contributing to later recruitment. Establishment of those that germinate early is chancier because of early frost or increased salt stress. The small-seeded morph, more dormant under high salinity and requiring light and stratification to reach maximum germination, may persist for more than 1 yr, allowing for repopulation after years of reduced success. Correlations between seed bank strategy, germination ecophysiology, and morphology of the two *Salicornia europaea* morphs (and also *Atriplex triangularis*, Ungar, 1984; Khan and Ungar, 1986; Wertis and Ungar, 1986) parallel observations made for community seed banks by Thompson and Grime (1979).

Differences in soil seed reserves of two perennial sea grasses (*Halodule wrightii* and *Syringodium filiforme*) were related to seed ecology and reproductive biology (McMillan, 1981, 1983). Laboratory germination results, showing continuous low germination for >3 yr, suggested that both species should depend on persistent seed banks. However, under field conditions, *Syringodium filiforme* germinates readily and does not form a seed bank, while *Halodule wrightii* has sizable seed banks (26–3300 m^{-2}; $\bar{x}$ = 435 for three sites). Seeds of *Syringodium filiforme* are produced above the sediments on emergent inflorescences and are dispersed readily. Seeds of *Halodule wrightii* are produced at rhizome level and remain buried until disturbed. Reproductive biology of *Syringodium filiforme* favors more immediate colonization of new sites, and that of *Halodule wrightii* requires disturbance for dispersal and recruitment from the seed bank. Moreover, seeds of the seed bank species were smaller and seedlings did not produce adventitious roots as quickly.

H. Seed Banks and Communities

Inundation regimes differ with wetland (Table 1): in freshwater tidal wetlands, inundation occurs twice daily; in prairie marshes, along lakeshores, and in swamps, it is related to climatic cycles of 5–>30 yr; and in other wetlands (vernal pools, etc.), drawdowns are often yearly. These hydrological patterns establish the role seed banks play in vegetation dynamics.

Seed banks are central to the long-term survival of prairie marshes (van der Valk and Davis, 1976, 1978, 1979). During droughts, water levels drop so that mud flat species and emergents are recruited from the seed bank. With normal rainfall, standing water eliminates mud flat species, stops germination of emergent species, and triggers germination of submersed and floating species. If periods of high water continue, intolerant emergent species decline. The degenerating marsh and lake marsh have abundant submersed and free-floating plants. At each stage the seed bank contributes to the vegetation and, in turn, the vegetation contributes to renewal of the seed bank. Thus, in prairie marshes, at least in sites where drawdowns occur, the seed bank contains elements of each stage of the vegetative cycle.

Establishment in marshes depends, therefore, on germination of species relative to drawdown (e.g., Gaudet, 1977; van der Valk and Davis, 1978). Presence or absence of standing water is the "environmental sieve" that determines recruitment or extirpation of species in the seed bank model for vegetation change proposed by van der Valk (1981). This model was useful in predicting vegetation change in some wetlands (van der Valk, 1981), but not in another (Smith and Kadlec, 1985a). The lack of predictability was attributed to factors affecting germination, including light, soil moisture, temperature, competition, and salinity. Had these factors been considered, the accuracy of the model would have been improved (Smith and Kadlec, 1985a; Galinato and van der Valk, 1986).

In contrast to prairie marshes, the seed bank of a freshwater tidal wetland does not contain seeds of different (cyclic) successional stages. The seed bank closely resembles the surface vegetation. The high yearly turnover of many species in the surface layer and generally low densities in subsurface layers indicate that, except for very few species, long-term accumulation does not occur (Leck and Simpson, 1987a). This wetland is not affected by drought; although changes in water level would direct vegetation change, such changes would not be cyclic. Importance of a species may fluctuate in both the seed bank and in the vegetation over time (Leck *et al.*, 1989), but little is known about these population and community dynamics. Fluctuations in yearly recruitment from the seed bank may be determined by microclimatic differences (see Galinato and van der Valk, 1986) caused by tide cycles and/or weather patterns, or to differing tolerances of these microclimates by seedlings. Expansion of a species, therefore, may not be inferred from its presence in the seed bank, but may depend on survival of seedlings (Parker and Leck, 1985) or vegetative reproduction.

The relationships between vegetation types and fluctuations in water depth have been noted for the Great Lakes and other smaller lakes (Keddy and Reznicek, 1982, 1986). The periodic recruitment of species from the seed bank can only occur during specific water level phases, and boundaries between vegetation types shift as water levels change. Fluctuations in water level are necessary to maintain seed bank and floristic diversity.

Where the drawdown cycle is annual, as in temporary ponds, vernal pools, and monsoon climates, complex relationships between the seed bank and the vegetation are also observed. In temporary ponds, vegetation changes between periods of high and low water are primarily quantitative (McCarthy, 1987). Periods of high or low water are not long enough to eliminate the most abundant species, although some species occur only under specific conditions (e.g., flooding), as would be predicted by the van der Valk (1981) model. Species abundance, reflecting recruitment/survival differences, varies with precipitation during drawdown. The effects of extended drawdowns are unclear because of the greater potential for species interactions when vegetation density and species richness are greatest. For California vernal pools, Zedler (1987) hypothesizes that along the pool/nonpool boundary a tension exist between inundation-tolerant and -intolerant species caused by the varying effect of rainfall on community composition. Variation in pool level can contribute to the seed bank and vegetation diversity. If pools dry several times during a year, vegetation diversity may be reduced, but, because seeds persist in the soil, little evidence for extinction exists (Zedler, 1987). Finally, in monsoonal climates, two seasonally delimited communities may develop, one adapted to inundation and the other to drawdown (see Gopal, 1986), both contributing to and recruited from the seed bank.

When succession is linear, dispersal into the community of species tolerant of a specific hydrological regime determines the species composition and nature of the seed bank and the vegetation that develops. As rates of succession change, composition and dependence on the seed bank changes. For wetlands, rates of change can vary from less than a decade to several millenia (Windell *et al.*, 1986).

III. Summary and Conclusions

This discussion has highlighted relationships of wetland seed banks to vegetation and vegetation change (Fig. 1). The size and composition of seed banks vary within and between wetlands. These variations may be related to (1) seed longevity and germination traits; (2) composition of the surface vegetation; (3) seed rain, predation, and dispersal; (4) age and isolation of the wetland; (5) inundation regime; and/or (6) other variations in the physical and biotic environment. Size and diversity of freshwater wetland seed banks appear generally greater than those of saline wetlands, but there is considerable variability.

Seeds of most herbaceous species are capable of persisting >1 yr in soil. Persistent species often have small seeds that respond positively to light, increased aeration, and/or alternating temperature. Some, in fact, are seed bank fugitives, occurring in the vegetation only after environmental perturbation (e.g., drawdown or disturbance). Variable species response to microenvironmental conditions may alter recruitment and, therefore, seed bank and vegetation dynamics (Fig. 1).

Herbaceous species dominate wetland seed banks; annuals and perennials vary in importance with the wetland. Graminoids usually comprise >50% of the seed bank, while woody species are not common even in swamps. Lack of woody species may be related to high predation and decomposition rates, to delayed and variable reproduction rates (Harper, 1977), or perhaps to the lack of dependancy on long-lived seeds by long-lived species. The relationship between the wetland seed bank and vegetation dynamics varies with hydrological regime. Changes may involve daily, seasonal (annual), or cyclic (multiyear) fluctuations, or may be related to succession. Wetlands with daily tidal fluctuations contain seed banks that resemble the surface vegetation. In those with seasonal fluctuation, seed banks may contain seeds of two stages (e.g., inundated versus drawdown or dry versus wet season communities), the importance of each depending on vegetation tolerances to the inundation regime. Where cyclic changes occur, the seed bank contains components of various stages and, depending on water level, different communities can develop. Whether considering temporal or spatial vegetation patterns, distribution of species represents a continuum of tolerances to the physical, chemical, and biotic environment.

Predictions can be made about the successional status for a given wetland, but given the current level of knowledge, it is difficult to extrapolate between wetlands. What appears to be the case in one wetland (e.g., number of seed bank species > field seedling species > vegetation species; Table 3) may not occur in another. Comparative studies between wetlands are needed. In addition, wetlands are an important focus of conservation activity (Keddy et al., Chapter 16, this volume) and preservation legislation because human impact has been so great (Tiner, 1984); better understanding of wetland dynamics will aid management activities.

Information about a number of topics would be helpful. First, to understand recruitment and the ability to colonize available substrates, it is necessary to examine seed germination traits of component species and relate them to field studies. Some studies have begun to investigate such relationships (Conti and Gunther, 1984; Keddy and Ellis, 1985; Keddy and Constabel, 1986; Galinato and van der Valk, 1986). Second, although wetland productivity is well studied, relatively little appears known about seed production and pre- and postdispersal predation. Third, very small long-lived seeds can be deeply buried, but experimental studies are needed to determine the relationship of rate of burial to size, shape, and longevity. Interactions of selection pressures, as shown by the advantage of large-seededness in overcoming burial by sand being countered by heavy predation (Ernst, 1985), need to be examined. Fourth, the importance of the seed bank for a species, a community, or a wetland changes over time; how are diversity components (species richness, distribution of dominance, community structure, and genetic diversity; see Pickett and White, 1985b) involved? And finally, woody species are notably absent from wetland seed banks. More needs to be learned about the relationship between life history/growth form and dependence on seed banks.

Acknowledgments

I am indebted to John B. and Jean B. Allessio, who gave me free rein with Burpee seed catalogs; Robert B. Livingston and Erik K. Bonde for encouraging my professional interest in seeds; Robert L. Simpson and V. Thomas Parker for contributing to the development of this chapter; and Rider College for providing a research leave and grants that permitted completion of work contibuting to this discussion.

PART 4

Management and Soil Seed Banks

CHAPTER 14

Seed Banks in Arable Land

Paul B. Cavers

Department of Plant Sciences
University of Western Ontario
London, Ontario, Canada

Diane L. Benoit

Agriculture Canada
Research Station
St-Jean-sur-Richelieu
Québec, Canada

I. Introduction

Historically, interest in arable seed banks was aroused by the discovery that weed seeds could remain dormant and viable in the soil for many years although the vegetation had changed and there was no possibility of fresh seed input from weed species restricted to arable land. We cite three examples of such studies. The first is the work of Peter (1893; in Roberts, 1981), who collected soil samples from forests of different ages. The results suggest that seeds of arable weeds remain viable in soil for >50 yr, but not indefinitely.

The second study is of the famous example of the *Papaver* spp. that "blew in Flanders' fields" during World War I. Poppies sprang up on the disturbed soil of trenches, vehicle tracks, shell holes, and graves. Because they had not been seen before the war in places such as the Somme battlefield, it was concluded that they must have originated from dormant seeds in the soil (Hill, 1917). Some speculated that the seeds had probably been in the soil since the Franco–Prussian War of the early 1870s, the last time these fields had been cultivated and sown to grain crops. This speculation was supported by dormant seed floras of soil samples from certain field plots in England; as many as 28,000 seeds m^{-2} of *Papaver* spp. occurred compared with only 11,000 m^{-2} for the other 27 major weed species combined (Brenchley and Warington, 1930).

The final example is provided by a garden in Hertfordshire, England, created by Sir Edward Salisbury in 1928 in an area that had not been cultivated since the Napoleonic Wars of over a century before. Several species associated with arable land appeared in this garden, including *Anagallis foemina*, which was extremely rare in Hertfordshire in the 1920s but was common in France as a weed of cultivated land. Salisbury (1961) concluded that at least some of these *Anagallis foemina* plants were "doubtless arising from such seeds that had lain dormant for more than a century and which were stimulated into activity together with other cornfield weed seeds by the removal of the soil."

Although these and other reports did not prove conclusively that the seeds germinating on recently disturbed soil had been dormant for 50–150 yr, they did stimulate great interest in seed banks of arable weeds. Two classic studies, the Beal experiment, which has now run for >100 yr (Kivilaan and Bandurski, 1981), and the Duvel experiment, which ran for 39 yr (Toole and Brown, 1946), demonstrated that weed seeds could remain dormant for 30–100 yr. Both experiments used seeds placed in sterilized soil within glass bottles or similar containers; the containers were buried and then excavated at regular intervals and the seeds tested for germinability.

These early studies spawned a vast literature on agricultural seed banks. Because it is so extensive, we have decided to limit our discussion to seed banks of arable land in temperate regions. Studies of other agricultural communities (e.g., rangelands and long-term pastures) will be considered elsewhere (Garwood, Chapter 9; Rice, Chapter 10; van der Valk and Pederson, Chapter 15, this volume).

II. Inputs to Arable Seed Banks

Huge numbers of seeds are produced by both crop and weed species. In most cultivated land, the species with the greatest density and usually the highest seed yield per unit area are the crop species. However, few of their seeds are deposited in the seed bank because most are harvested and removed from the field. Cereal species, for example, have been bred to resist shattering and to ripen synchronously, thereby facilitating their harvest (de Wet, 1975). Crop seeds that are deposited on the soil contribute little to the seed bank. They usually have short life-spans because of predation, susceptibility to microbial degradation, and rapid germination through lack of dormancy.

The vast majority (often over 95%) of seeds entering the seed bank in arable land come from annual weeds growing on that land (Roberts, 1981; Hume and Archibold, 1986). Only 4% of the weed species in arable land are described as perennial (Kropáč, 1966). Biennial species can produce very high seed yields per plant, but these species are usually restricted to intermittently cultivated areas.

Limited data exist on the total input of all seeds into arable soil. Differential dates of ripening between and within species (Stevens, 1932), rapid release of ripe seeds, and damage to ripening seeds during collection have made such studies difficult. Nevertheless, Chancellor (in Roberts, 1970) estimated that a dense stand of weeds produced more than 1,235,000 seeds m^{-2}.

Estimates of seed yields from individual species are more common than those for total yield. A selection of recent estimates is given in Table 1. In general, the input from annual species is much greater than that from perennials. Also, yields from the same weed species vary greatly with different associated crops.

Seed yields for individual weed species are more consistent from year to year than seed yields of comparable nonweedy species. Weed species are often apomictic or self-pollinated, and thus do not suffer the almost total failures of seed production that befall some insect-pollinated species. Even primarily wind-pollinated plants (e.g., *Rumex crispus*) seem to have consistently high proportions of flowers producing viable seeds (P.B. Cavers, personal observation). When annual seed output does fall dramatically for weeds of cultivation, such as *Chondrilla juncea* (Cuthbertson, 1970) or *Sinapis arvensis* (Edwards, 1980), the cause is usually unfavorable weather, which can act directly (e.g., drought during flowering) or indirectly (e.g., the spread of diseases among seeds under wet conditions). Herbicide use can greatly reduce total seed input and can drastically alter the relative input of different species. This will be considered in a later section.

Table 1

Examples of seed production for weeds of arable land

Species	Crop and/or cultural treatment	Seeds m^{-2}	Location	Reference
Agropyron repens	Highly cultivated	Up to 634	Michigan	Werner and Rioux (1977)
	Low cultivation	Up to 886		
Alliaria petiolata	Arable land, unculti-vated for 1 yr	73,420	Ontario	Cavers et al. (1979)
	Garden (partial shade)	107,580	Ontario	Cavers et al. (1979)
Alopecurus myosuroides	Winter cereals	6500	England	Naylor (1972)
Amaranthus palmeri	Vegetable fields	110,000	Texas	Menges (1987)
Amaranthus powellii	Fertilized plots	709,500	California	Hauptli and Jain (1978)
	Unfertilized plots	424,800	California	Hauptli and Jain (1978)
Amaranthus retroflexus	Fertilized plots	1,038,000	California	Hauptli and Jain (1978)
	Unfertilized plots	415,800	California	Hauptli and Jain (1978)
Asclepias syriaca	Not specified	Up to 8685	Ontario	Bhowmik and Bandeen (1976)
Bromus tectorum	Not specified	14,850	Western North America	Upadhyaya et al. (1986)
Centaurea maculosa	Irrigated plots	Up to 40,000	British Columbia	Watson and Renney (1974)

Species	Habitat	Number	Location	Reference
Convolvulus arvensis	Various	5–2000	Various	Weaver and Riley (1982)
Euphorbia esula	Crop land	2500	Saskatchewan	Best et al. (1980)
Lotus corniculatus	Not specified	47,000–79,000	Not specified	Turkington and Franko (1980)
Medicago lupulina	Fallow land	Up to 6600	Ontario	Sidhu (1971)
Panicum miliaceum	White beans	42,600	Ontario	O'Toole and Cavers (1983)
	Corn	3400	Ontario	O'Toole and Cavers (1983)
	Barley	150	Ontario	O'Toole and Cavers (1983)
Polygonum convolvulus	Various	~543	Finland	Hume et al. (1983)
Portulaca oleracea	Not specified	78,600	Texas	Menges (1987)
Setaria viridis	Field plots	100,000–200,000	Alberta	Vanden Born (1971)
Sonchus asper	Field plots, low density, unfertilized	18,800	British Columbia	Hutchinson et al. (1984)
	Field plots, low density, fertilized	5800	British Columbia	Hutchinson et al. (1984)
	Field plots, high density, unfertilized	14,500	British Columbia	Hutchinson et al. (1984)
	Field plots, high density, fertilized	35,400	British Columbia	Hutchinson et al. (1984)
Thlaspi arvense	Small grains	168,125	Saskatchewan	Best and McIntyre (1975)
Trifolium repens	Sown hay/pasture	910–109,000	Various	Turkington and Burdon (1983)
Xanthium strumarium	Fallow	200–300	India	Weaver and Lechowicz (1982)

The pattern of seed distribution within a field generally follows the direction of crop rows (Benoit, 1986; McCanny and Cavers, 1988). Weed seeds from an initial infestation are seldom moved far by wind, but they are carried along the rows by farm machinery during planting, cultivation, and harvest. McCanny and Cavers (1988) found that 3.3% or fewer of the seeds on plants of *Panicum miliaceum* were carried more than 50 m by combines during crop harvest. Nevertheless, transport by farm machinery changed an infestation of isolated patches of *Panicum miliaceum* to almost complete coverage of a field within 2 yr (McCanny and Cavers, 1988).

We have noted from many weed surveys in southern Ontario that the greatest diversity of weed species in a field is found at the edges, and particularly in the headlands. Similarly, Hume and Archibold (1986) reported that most seeds, dispersed from a weedy pasture into an adjacent fallow field, were found in the seed bank within 3 m of the edge of that field. There are four main reasons for the greater diversity at field edges. First, most seeds generally land within a few meters of the parent plant (McCanny, 1986); wind dispersal is not effective for most weeds of arable land. Second, there are many species that grow only in uncultivated fence rows at the field edge; their propagules are dispersed for short distances into the field. Aerodynamically efficient propagules blown into an arable field (e.g., from species of composites) are often trapped on crop plants near the edge of the field and deposited there. Third, farm equipment always enters a field along the edge or in the headland; seeds transported from other fields are deposited there during cleaning, attaching of implements, etc. Fourth, the light intensity at ground level is usually higher at the edge of a crop. It is also higher in many headlands where crop seeds are either unplanted or incompletely planted. In such places there can be germination and establishment of shade-intolerant weeds that would never emerge under dense crop cover.

III. Numbers of Seeds in the Seed Bank

The size of a seed bank in an arable field reflects the past and present weed management in that field, as well as the crops grown. In vegetable crops, where intensive management by herbicides, cultivation, hand weeding, crop rotation, and the like is evident, the total banks of viable weed seeds can range between 250 and 46,819 seeds m^{-2} (Table 2). In small grain cereal crops the range is 4742–73,350 seeds m^{-2} (Table 2). Data collected before chemical herbicides came into general use are partially responsible for the higher values. In corn (maize) fields, which are

Table 2
Numbers of viable weed seeds in arable soils

Country	Crop or cropping sequence	Herbicide application[a]	Range in number of viable seeds m^{-2}	Depth sampled (cm)	Reference
Poland	Onions	±	11,197–20,800	20	Lewandowska and Skapski (1979)
United Kingdom	Cabbage/leeks/brussels sprouts/peas	−	2773–46,819	15	Roberts (1963a)
	Vegetables	+	1386–3240	15	Roberts (1968)
	Corn	+	7000	15	Roberts and Neilson (1981)
	Carrots	+	5025	15	
	Vegetables	+	250–24,330	15	Roberts and Neilson (1982b)
	Cereals	−	8329–73,350	15	Brenchley and Warington (1930)
	Wheat	−	28,709	15	Brenchley and Warington (1933)
	Barley	−	29,952	15	
	Cereals	−	12,831–43,950	15	Brenchley and Warington (1945)
Canada	Fallow/wheat/wheat	−	4742–22,939	15	Budd et al. (1954)
Germany	Cereals	+	17,712	25	Hurle (1974)
	Cereals (no weed control)	−	43,778	25	Hurle (1974)
United States	Corn	+	2080–130,300	25	Schweizer and Zimdahl (1984a)
Hungary	Corn	−	5503–14,908	40	Fekete (1975)
	Corn	−	21,353	40	Fekete (1975)
	Corn	+	19,210	40	Fekete (1975)

[a] +, With herbicide; −, without herbicide.

intensively managed monocultures, total seed banks have been estimated from direct counts at between 2080 and 130,300 seeds m^{-2} (Table 2). If Kropáč's (1966) assumption that the viable seed bank represents 20% of the total seed bank is true, then the viable seed banks in corn and soybean fields are slightly smaller than those of vegetable crops (416–26,060 seeds m^{-2}).

A few species are prominent in seed banks throughout the temperate regions. *Chenopodium album* and *Stellaria media* are major contributors to seed banks in cool temperate regions, and *Amaranthus* spp., *Echinochloa crusgalli*, and *Portulaca oleracea* are dominant in warm-summer temperate areas (Roberts, 1981).

Populations of weed seeds in the soil can increase very rapidly from one year to the next. Edwards (1980) found that an initial input of 200 seeds m^{-2} of *Sinapis arvensis* into a clean field was followed by an input of nearly 5000 seeds m^{-2} the next year. Similarly, seeds of *Amaranthus palmeri* introduced in irrigation water produced large plants in a green pepper crop, and within 1 yr there were 17,300 seeds m^{-2} in the soil (Menges, 1987).

The numbers of seeds of a black-seeded biotype of *Panicum miliaceum* in the seed bank at different times of the year from fields of barley, corn, and white beans reflected the seed input the previous year. The largest seed bank was found in a field that had had white beans (O'Toole, 1982). Further studies in the same area (P.B. Cavers and J.J. O'Toole, unpublished data) have confirmed that seed banks of black-seeded *Panicum miliaceum* following white bean or corn crops are almost always several times as large as those following barley.

IV. Seed Banks in Different Crops

Each crop has a particular suite of weeds that are associated with it (e.g., tea weeds, sugar beet weeds, lowbush blueberry weeds, soybean weeds, etc., Holzner, 1982). Most weeds that appear in a crop are also represented in the seed bank. Roberts (1981) noted that some of the greatest differences in weed floras are between crops that have weed seedlings appearing in the fall and winter (e.g., winter cereals) and crops that have weed seedlings appearing in the spring and summer (e.g., corn, soybeans, and spring cereals).

Few crop species are weedy, primarily because they are unimportant in the seed bank. Duvel's buried seed experiment (Toole and Brown, 1946) showed that most crop seeds, especially those of large-seeded species, died very quickly. A few small-seeded crops such as tobacco, red clover, celery, and Kentucky bluegrass (*Poa pratensis*) had some viable seeds after 39 yr (Toole and Brown, 1946). Some crop varieties are particularly weedy; for example, annual bolting types of sugar beets can form a persistent seed bank that is difficult to eradicate (Roberts, 1983).

Among the worst arable weeds are the weed races of crops (Harlan, 1982; Egley and Elmore, 1987); they are similar to the crops in a multitude of ways. Some have such similar seeds that they are harvested with the crop and cannot be removed by seed-cleaning processes. An interesting example is provided by sorghum (*Sorghum bicolor*), in which a number of weed races have arisen in different parts of the world (Harlan, 1982). A center of wild *Sorghum bicolor* is in Ethiopia and Sudan. There the *verticilliflorum* race, a truly wild plant, crosses with cultivated varieties to form fully fertile weed races that infest the local sorghum crops. A completely separate race, called shattercane, arose in the United States and now is a serious weed of American sorghum crops (Harlan, 1982). Many of these weed races have very dormant seeds and can form large and persistent seed banks.

The potential of different populations of proso millet (*Panicum miliaceum*) to form seed banks has been investigated (Cavers and Bough, 1986). Generally, seed banks of crop biotypes are not large, because most crop seeds have little dormancy and are vigorously attacked by soil fungi (M. Bough, personal communication). Nevertheless, several crop biotypes have become serious weeds in Canada. The most extensive of these infestations occur in the corn-growing areas of southern Manitoba, where a biotype of proso millet, apparently identical to the licensed crop variety "crown," has become an abundant and difficult weed. Fewer than 5% of "crown" seeds survive the winter in most sites. We believe that this infestation continues because of enormous annual input into the soil coupled with a cropping pattern of continuous corn that precludes meaningful crop or weed competition for proso millet.

Several infestations of proso millet in Ontario and Québec are caused by biotypes resembling crop varieties. Biotypes with white and orange–red seeds have little dormancy and do not cause serious infestations. One population with golden seeds and the dark-red-seeded biotype are a greater problem because they have much stronger dormancy and can establish persistent seed banks (M.A. Bough and P.B. Cavers, unpublished data).

Another weedy biotype with black seeds has strong dormancy and large, persistent seed banks. It causes the most extensive and rapidly spreading infestations of proso millet in the midwestern United States (Harvey, 1979) and southern Ontario (Cavers and Bough, 1986) and is probably the original weed race of this species.

Many crops are infested with unrelated mimics that are the result of hand weeding, which, over the centuries, has selected weeds that resemble the crop (Harlan, 1982). One of the most costly is *Echinochloa crusgalli* var. *oryzicola*, which in the vegetative stage is a very close mimic of rice. Although the inflorescences of the crop and weed are entirely different, effective removal of the weed from the crop after flowering is impossible (Harlan, 1982). Many mimic weeds survive because they germinate synchronously with the crop and have a seed bank that is persistent enough to outlast short-lived crop rotations.

V. Changes over Time in Arable Seed Banks

Most seeds are incorporated into arable soils from the surface by mechanical cultivation, but other forces can be important. Soil eroded by water can cover seeds whenever the soil is not frozen. In addition, freezing and thawing cycles can open up large cracks into which seeds can be washed from the surface by melted snow or early spring rains. Cracks formed by excessive drying also permit entry of seeds into the soil. Earthworms, burrowing mammals, and other animals can also incorporate seeds into the soil (Harper, 1977).

After incorporation into the soil, viable seeds decline continuously in numbers because of germination or rotting. Observations by Roberts (1970) suggest that the rate of loss of all seeds in the soil profile follows an exponential decay curve; i.e., a constant percentage is lost each year. However, the rates vary for different species (Roberts and Feast, 1972). Loss by germination is largely restricted to the upper layers of the soil profile (Chepil, 1946).

Viability is greatly affected by the intensity of microbial infection and the microbial species involved (Pitty *et al.*, 1987). Associated with these effects are the changes in seeds caused by biological or mechanical wounding. For example, in *Avena fatua* wounding often overcomes primary dormancy, with the result that there is more rapid germination and a less persistent seed bank (Foley, 1987).

Several studies have shown that populations of the same species can differ profoundly in seed longevity in the soil. These variations are attributable primarily to differences in dormancy caused by a complex interaction of genetic and environmental influences. These act during seed formation or after dispersal, before and after burial (Chadoeuf-Hannel, 1985). For example, seeds of *Amaranthus retroflexus*, *Barbarea vulgaris*, and *Echinochloa crusgalli* are either nondormant or dormant after collection, depending on the date of collection and the population (Taylorson, 1970). Nondormant seeds of *Amaranthus retroflexus* and *Barbarea vulgaris* are shorter lived in the soil than the dormant ones are. This difference is not apparent for *Echinochloa crusgalli* (Taylorson, 1970). Other factors that affect seed dormancy and longevity are the microclimate during ripening (Grant Lipp and Ballard, 1963; Kigel *et al.*, 1977; Sidhu and Cavers, 1977) and wetting and drying in the soil (Vincent and Cavers, 1978).

Courtney (1968) described annual dormancy cycles for seeds of *Polygonum aviculare* stored in soil. Regardless of when during the year they were excavated, virtually all seeds germinated at 4°C. However, at constant higher temperatures (8–23°C), seeds removed from February to May usually exceeded 50% germination, while those removed from June to November never exceeded 20%. Similar annual cycles in dormancy have been reported for many other species (Taylorson, 1970, 1972; Karssen, 1980–1981; Baskin and Baskin, 1985a; Baskin and Baskin, Chapter 4, this volume). Some species do not show regular cycles (e.g., several legume species, P.B. Cavers and S.S. Sidhu, unpublished data), and there are other species, such as *Stellaria media*, where one population exhibits cycles in dormancy but another does not (van der Vegte, 1978).

VI. Seed Banks in Different Soils

Soil characteristics influence the density and species richness of seed banks. Pawlowski (1963) examined the more important soils of the Lublin district of Poland and found an average of 43,292 seeds m^{-2} of 45 species in the 0 to 20-cm-deep layer of sandy soils, 21,632 seeds m^{-2} of 45 species in the same layer in chernozems, and 14,888 m^{-2} of 70 species in rendzinas. Later, more extensive studies throughout Poland gave similar results (Wesolowski, 1979a,b). Likewise, the relative abundance of individual species is greatly dependent on soil type (Brenchley and Warington, 1930; Paatela and Erviö, 1971; Fekete, 1975). In clay soils seeds of *Stellaria media* and *Lapsana communis* were more frequent and *Polygonum lapathifolium*, *Rumex acetosella*, *Anthoxanthum odoratum*, and *Potentilla* spp. were less frequent than in other soils; in peaty soils *Carex* spp. were more frequent and *Viola arvensis*, *Polygonum aviculare*, *Polygonum convolvulus*, *Taraxacum vulgare*, and *Scleranthus annuus* were less frequent (Paatela and Erviö, 1971). Such differences in seed banks undoubtedly reflect the fact that some weeds are only common on particular kinds of soil.

Some weed species are more or less restricted to either basic or acidic soils (King, 1966). In fact, many more weed species were associated with particular soils in the past (Hilbig, 1982); for example, there were weeds of dry acidic soils and weeds of chronically flooded limestone soils. With soil improvements for agriculture many of these agrestal weeds became liable to extinction. Reduced crop rotations, widespread and increasing fertilizer and herbicide use, liming, irrigation, drainage, and improved seed-cleaning methods accelerated the trend toward fewer weed species. Coincidentally, several species of shade-tolerant nitrophiles have expanded in importance (Hilbig, 1982). These trends have alarmed European botanists, and several European countries have set up "arable land reserves" with nineteenth century farming methods to preserve rare arable weeds.

The moisture level in the soil has a major influence on the size and nature of the seed bank. Seeds of a weedy biotype of *Panicum miliaceum* germinated more rapidly in a well-drained, sandy soil, regardless of depth, than in medium or poorly drained clay loams. Thus, the seed bank in the sandy soil was depleted more quickly (Colosi *et al.*, 1988). Likewise, wetter soils had less seedling emergence, and thus a more persistent seed bank (Pareja and Staniforth, 1985). A probable reason is that greater fluctuations in soil moisture in the drier sandy soils cause hydration–dehydration cycles in weed seeds. This process results in increased germination of grasses (e.g., *Digitaria* spp. and *Setaria faberi*; King, 1966) as well as dicots (e.g., *Rumex crispus*; Vincent and Cavers, 1978).

Very wet soils, however, show smaller diurnal fluctuations in temperature than do drier ones (Daubenmire, 1974), thus maintaining dormancy and viability in many weed seeds that require diurnally fluctuating temperatures for germination. In very dry soils, of course, weed seeds of certain species remain dormant and viable until sufficient moisture for germination is available (e.g., *Chaenorrhinum minus*; P.B. Cavers, unpublished data).

Other soil attributes that affect the size and composition of the seed bank are the percentage of plant cover, soil density, carbon dioxide and oxygen levels (soil aeration), and the presence of volatile germination inhibitors or allelochemicals (Egley, 1986). The effects of fertility levels will be discussed in the following section.

VII. Effects of Fertilizers and Manures

Weeds are generally more productive on fertile soils, adding more seeds to the seed bank (Roberts, 1981). For example, *Amaranthus powellii* and *Amaranthus retroflexus* responded to increased soil fertility with increased seed production (Table 1; Hauptli and Jain, 1978). *Sonchus asper*, in contrast, produced only one-third as many seeds on low-density fertilized plots as compared to unfertilized ones (Table 1; Hutchinson *et al.*, 1984).

However, this latter result was reversed at high density, suggesting that the relationship between weed seed production and fertility may be complex.

The addition of fertilizers can lead to an overall depletion of the weed seed bank because fertilizers containing nitrates or nitrites can stimulate the germination of dormant seeds. Egley (1986) noted that four studies reported increased seedling emergence in response to the application of nitrate fertilizers, whereas two other studies reported no increase in emergence. However, the latter two studies are suspect; one was not a field investigation, and, in the second, seeds of *Chenopodium album* collected from the soil of treated plots were less dormant and contained more nitrate than did seeds from untreated plots. Egley (1986) also noted that loss from the seed bank after fertilizer treatment may be underestimated if only emergence counts are used, because many seeds germinate but do not emerge. Nitrate is most effective in stimulating germination when combined with other stimuli such as light, alternating temperatures, and ethylene. A combination of ethephon (a source of ethylene), potassium nitrate, and afterripening caused virtually all dormant seeds of *Avena fatua* to germinate (Saini *et al.*, 1986). This approach to depleting the seed bank, and thereby reducing weed problems, is apparently not yet practical in the field (Roberts, 1983). The difficulty is getting a combination of chemicals, each in effective concentrations, to the individual seed, because breakdown or immobilization in the soil may be rapid. This topic is presently under study (e.g., Donald, 1985). One successful application has been the use of ethylene to stimulate germination of dormant seeds of *Striga* spp., greatly decreasing the populations of this serious parasitic weed in North and South Carolina (Egley, 1986).

Farm manure, spread as fertilizer, can be an important source of seeds for seed banks of arable soils. Seeds of many weed species can pass through the digestive tracts of cattle, sheep, horses, pigs, and goats to be deposited in a viable state in the feces (Dore and Raymond, 1942; Salisbury, 1961) although they are usually destroyed by passage through chickens (Harmon and Keim, 1934). The feces are mixed with straw, which may also contain seeds, to form manure that is left to rot for several months. Many seeds or fruits lose their viability under the high interior temperatures of the manure pile, but seeds can remain viable under the much lower temperatures near the periphery (Salisbury, 1961). The seed bank of *Chenopodium* spp. was significantly higher in a field receiving a manure treatment than in a comparable neighboring field receiving none (Benoit, 1986). The seed bank of dormant weeds appears largest immediately after the application of manure, and decreases with time (Lewandowska and Skapski, 1979).

VIII. Effects of Cultivation

Cultivation is used to destroy weed seedlings, break up the soil crust, aerate the soil, and reduce the size of the weed seed bank in the soil by stimulating seed germination. Regardless of differences in the depth and frequency of cultivation, seed banks decrease exponentially (Rice, Chapter 10, this volume). However, the longer a cohort of seeds remains in cultivated soil, the slower it loses viability (Chadoeuf *et al.*, 1984; Warnes and Andersen, 1984).

In the absence of input from seed production, the seed bank declines more rapidly with cultivation than without it (Roberts, 1966, 1970; Roberts and Dawkins, 1967; Roberts and Feast, 1973b; Froud-Williams *et al.*, 1983; Warnes and Andersen, 1984; Bridges and Walker, 1985). The total loss of viable seeds in the soil increases with the number of cultivations per annum (Roberts, 1966, 1970; Roberts and Dawkins, 1967; Roberts and Feast, 1973b; Cook, 1980). However, the annual cycle in dormancy found in many weed seeds may influence the rate of loss. If cultivation brings a nondormant seed to or near the surface, it will probably germinate, but, if it is dormant, it may be returned to a lower depth by subsequent cultivation.

In England, seedling emergence on undisturbed soil accounted for only 1% of the viable seed bank (Roberts and Dawkins, 1967; Froud-Williams *et al.*, 1983). After a single cultivation, seedling emergence represented, on average, 5% of the viable seed bank in the top 10 cm, but varied for each species depending on the timing of cultivation (Roberts and Ricketts, 1979). Seedling emergence on plots cultivated twice annually accounted for 6–7% of the viable seed bank (Roberts and Dawkins, 1967; Roberts and Feast, 1973b) while emergence on plots cultivated four times represented 9% (Roberts and Dawkins, 1967). In contrast, in France, seedling emergence from uncultivated soil averaged 8% of the total seed bank (viable plus nonviable seeds), and after one shallow cultivation (10 cm) or one deep ploughing (30 cm), averaged 12% of the viable seed bank (Barralis and Chadoeuf, 1980).

Changes in primary tillage operations can cause changes in weed species composition (Knab and Hurle, 1986). For example, when cultivation is intensified, weed species associated with intensive cropping (e.g., vegetable production) tend to increase in importance (Roberts, 1962; Roberts and Stokes, 1965).

Comparison of four types of cultivation after two cycles of vegetable crop rotation showed the lowest density of weed seeds for deep ploughing (30 cm) and the highest density on plots that received shallow rotary cultivation (7.5–10 cm) (Roberts, 1963a,b; Roberts and Stokes, 1965) (Table 3). Differences in seed density after the three ploughing treatments

Table 3
Effects of cultivation procedures[a]

Cultivation procedure	Depth of cultivation (cm)	Number of viable seeds m^{-2} in the top 10 cm of soil[b]		
		Site A	Site B	Site C
Ploughing	18–23	9500	5100	2800
Ploughing	30	13,400	5700	2000
Ploughing + subsoiling[c]	18 + 13	11,000	4300	2600
Rotary	8–10	16,300	21,400	6900

[a]From Roberts (1963b).

[b]The three sites were farms in Hampshire (A), Warwickshire (B), and Yorkshire (C) counties in England.

[c]Subsoiling involves the use of a cultivation tine to a depth lower than the plough. Shallow ploughing to 18 cm plus subsoiling would give less soil inversion than would deep ploughing.

were small and not statistically significant (Table 3). When differences between cultivation treatments occurred, they usually reflected different seed distributions in different layers of soil rather than total number of seeds present throughout the working depth (Roberts, 1963a; Roberts and Stokes, 1965).

Considerable interest exists in reducing seed banks by crop rotations, fallowing, and other crop management methods involving cultivation. Schweizer and Zimdahl (1984b), summarizing 50 yr of crop management studies, found one common theme: If input of weed seeds was prevented or was minimal, most viable seeds were lost from the seed bank within 1–4 yr, regardless of the agricultural practice (fallowing, reduced tillage, monoculture and associated tillage routines, crop rotation, and herbicide combinations).

Perhaps the most popular technique for reducing weed seed banks has been some form of fallowing (e.g., Brenchley and Warington, 1936, 1945; Budd *et al.*, 1954; Roberts and Dawkins, 1967; Roberts and Feast, 1973b; Archibold and Hume, 1983). The most successful fallows combine prevention of further seed input with cultivation that forces dormant weed seeds to germinate. A fallow with at least two cultivations per year reduced the seed bank to a greater extent than did a "chemical fallow," where there was no cultivation but herbicides prevented seed input (Roberts and Dawkins, 1967; Roberts and Feast, 1973b; Bridges and Walker, 1985).

It is often assumed that the distribution of seeds in arable soils is relatively uniform throughout the working depth (Kropáč, 1966; Roberts, 1981). However, most available data contradict this assumption. Pawlowski and Malicki (1968) reported that all methods of plough-ing tested moved the majority of seeds from the surface into the top 0- to 15-cm-deep layer, and the deeper the ploughing the further downward was the placement of some weed seeds. The first ploughing moved some seeds downward, but subsequent ploughing moved seeds up-ward or downward (Pawlowski and Malicki, 1968). Shallow tillage, such as skimming (5–7 cm) and shallow ploughing (15 cm), moved more seeds than medium (20 cm) and deep (25 cm) ploughing (Pawlowski and Malicki, 1968). Moldboard ploughing buried about 80% of surface-sown seeds, but returned about 38% of them toward the surface in a subse-quent operation (Van Esso *et al.*, 1986). After 9 yr of cultivation, the distribution of viable seeds with depth varied with treatment (Fig. 1; Roberts, 1963a; Roberts and Stokes, 1965; Knab and Hurle, 1986). Rotary or tine cultivation resulted in the highest proportion of seeds in the surface layer, whereas most ploughing methods tended to concentrate seeds at intermediate depths.

Ploughing depth, soil humidity and texture, and working speed of the tractor modify the amount of inversion of the plough layer, introduc-ing small-scale variability in the vertical seed distribution pattern (Van Esso *et al.*, 1986). Pareja *et al.* (1985) found that conventional tillage not only incorporated seeds into deeper layers of the soil, but also into larger soil aggregates (clods), an effect more pronounced with increasing soil depth (Table 4). Under reduced-tillage conditions, most seeds were con-centrated in the unaggregated fraction of the soil (particles <0.31 cm in diameter). Terpstra (1986) examined the effects of clod size and hardness on germination and seed longevity for five weed species. There was significantly greater germination and emergence from the smaller and

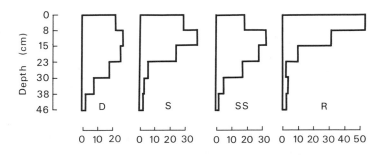

Number of seeds as % of totals

Figure 1. Depth distribution of viable weed seeds after 9 yr using four cultivation treatments. D, Deep ploughing (36–41 cm); S, shallow ploughing (15–18 cm); SS, shallow ploughing plus subsoiling (41–46 cm); R, rotary cultivation (15–18 cm) (from Roberts and Stokes, 1965).

Table 4
Number of weed seeds for three soil aggregate size classes[a,b]

Soil aggregate size class (cm)	Soil depth (cm)	Tillage regime	
		Conventional (seeds 100 g^{-1})	Reduced (seeds 100 g^{-1})
>7.49	0–5	2	70
	5–10	6	2
	10–20	11	2
5.00–7.49	0–5	4	48
	5–10	4	8
	10–20	9	—
<0.31	0–5	3	118
	5–10	4	12
	10–20	8	8

[a]From Pareja et al. (1985).

[b]As a function of soil depth and tillage regime. Data were collected in the spring. Conventional tillage was fall moldboard ploughing to 20 cm deep plus spring disking and harrowing. Reduced tillage was "slot" planting in the row of the previous crop with no preplanting soil disturbance.

softer clods. Larger seeds were less affected by clod condition. In the unaggregated fraction of a soil (particles <0.31 cm in diameter), seeds are likely to occupy spaces between the particles, while aggregates >5.0 cm in diameter may contain fissures or macropores large enough for seeds to enter (Pareja et al., 1985). Seeds inside soil aggregates may be exposed to high moisture and low oxygen, conditions conducive to maintenance of seed dormancy but not to germination (Currie, 1972; Pareja et al., 1985). Twice as many seedlings may appear on a fine, firm seedbed as on a rough one (Roberts, 1981).

Some properties of seeds (hairy, ridged, or sticky coats) may facilitate their association with soil particles (Pareja et al., 1985). Harper et al. (1965) showed that the relative abundance of seedlings of two species of Bromus could be changed by altering the particle size of the soil. The shape of the seed, including whether the awn was curved or straight, was important.

IX. Effects of Herbicides

The use of herbicides to complement standard cultivation practices can drastically reduce the population of weed seeds in the soil (Roberts and

Neilson, 1981). For example, atrazine used in a corn monoculture reduced the seed bank by 98% after 6 yr (Schweizer and Zimdahl, 1984a).
Nevertheless, 3 yr after atrazine use had been discontinued in plots
treated for 3 yr, the seed bank had rebounded to half of its original
density (Schweizer and Zimdahl, 1984a). No significant differences in
weed seed density were reported between crop production with standard, as compared with intensive (more frequent and/or higher applications) herbicide treatment (Schweizer and Zimdahl, 1984b; Bridges and
Walker, 1985). However, intensive herbicide treatments kept seed densities to low levels in all cases, whereas standard (recommended) treatments were more affected by unfavorable climatic conditions (Bridges
and Walker, 1985).

Norris (1985; in Menges, 1987) estimated that weed control must
exceed 99.99% efficiency to allow the seed bank of an arable field to
remain static. However, in evaluating herbicides, scientists rate a herbicide as good if it gives 95% control in test plots; under field conditions,
licensed herbicides rarely give much more than 80% control (C.J. Swan-
ton, personal communication). Some seeds of many weed species will
be produced after most herbicide applications. Nevertheless, because
they do not affect all species equally, herbicides can change the weed
flora dramatically (Hurle, 1974; Chancellor, 1980; Dvořák and Krejčíř,
1980). Across the Canadian prairies the use of 2,4-D and related compounds for the past 40 yr has been accompanied by a rapid increase in
the abundance of resistant weed species. Such a species is *Setaria viridis*,
which has become the most widely distributed weed in the area
(Thomas, 1979a,b). In southern Ontario, the continued use of triazine
herbicides in cornfields has resulted in great increases in triazine-
resistant populations of *Chenopodium album* (Benoit, 1986). Long-term
use of residual herbicides in crop production has decreased seed banks
(Roberts and Neilson, 1981; Roberts, 1983), but few, if any, weed species
have been eliminated and no new species have appeared.

X. Summary and Conclusions

The seed bank is now recognized as the prime source of new plants of
annual species that cause the main weed problems in arable land. The
particular management regime used in crop production governs the size
of the weed seed infestation in the soil and its species composition
(Holzner, 1982). Three steps must be followed to reduce or eliminate the
potential of seed banks for causing new or continuing weed infestations.

   1.   *Determine the size and species composition of the seed bank.* Population ecologists sometimes refer to the seed bank as a black box, because

it is very difficult or impossible to determine how many viable seeds are present on a plot of land at a given time. At least 100 individual samples of soil totaling 800–1000 cm^{-2} are needed to give a reliable estimate of the size and species composition of the seed bank (Benoit, 1986). Ideally, all viable seeds must be extracted from the soil and identified because simple emergence tests often do not lead to the appearance of seedlings from all dormant seeds.

2. *Determine dormancy/longevity for major seed bank species and select appropriate germination procedures.* Detailed studies of the dormancy and longevity of individual species have been made (see references in Table 1). Also, good general comparisons between species are available. However, intraspecific differences in genotype and the effects of environment on the seed during and following ripening can have major effects on dormancy and longevity in the soil. These are not fully understood.

A few studies have provided models of seed bank systems where actual numbers of seeds entering and leaving the seed bank are given (Naylor, 1972; Sagar, 1982; Mortimer, 1983). These models, made for individual species, have been surprisingly useful in making practical predictions about the number of weed seedlings that may be expected in a particular field. A basic model for such studies is illustrated in Fig. 2.

3. *Invent and practice new crop management systems to take advantage of current knowledge.* Management that gives long-term prevention of any new seed input would be useful on land that already has persistent seed banks. Models have proved most helpful (e.g., Schweizer *et al.*, 1986). In the prechemical era of weed control, farmers recognized that if they

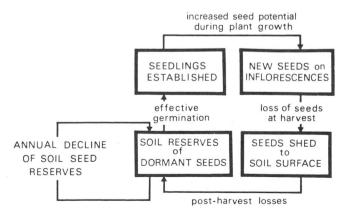

Figure 2.   A model of a weed seed cycle (from Cussans, 1987).

prevented weed seeds from entering their fields, they could control their weed problems. They used a variety of techniques, including cultivation throughout the season, mowing, pulling of single plants of dangerous species, crop rotation, and use of smother crops in headlands, for control. Now that farmers are dependent on herbicides, they spend far less time on the traditional means of weed control; populations of resistant or nonsusceptible weeds have established huge seed banks in many fields. A combination of nonchemical techniques and appropriate herbicides must be used in future attempts to reduce weed seed banks on arable land.

Acknowledgments

We thank Agriculture Canada, NSERC (Operating Grant to PBC), and CSIRO for financial support. We would like to express our appreciation to Claire Ménard, Donna Irwin, and Susan Kelo for typing the manuscript and to Richard Groves for useful comments.

CHAPTER 15

Seed Banks and the Management and Restoration of Natural Vegetation

A. G. van der Valk

Department of Botany
Iowa State University
Ames, Iowa

Roger L. Pederson*

Delta Waterfowl
and Wetlands
Research Station
Portage La Prairie
Manitoba, Canada

*Present address: Ducks Unlimited, Inc., Western Regional Office, 9823 Old Winery Place, Suite 16, Sacramento, California 95827.

I. Introduction

Whenever it becomes necessary to establish or reestablish native species as food or cover for animals, to prevent soil erosion, or for esthetic reasons, managers are faced with choices. Should they rely on natural dispersal, should they sow or transplant the required species, should they alter environmental conditions to make the site more favorable for establishment, or should they leave the site as is? Because of the increasing isolation of many managed sites from areas where native species are still extant, natural dispersal is often slow and unreliable (Bauer, 1973; Johnson and Bradshaw, 1979). The vagaries of natural dispersal also can result in a random assemblage of species at a site that may bear little resemblance to any native plant community in the same region (Bauer, 1973). Consequently, the seeds of native species routinely are sown or individual plants are transplanted to sites where they are wanted (Johnson and Bradshaw, 1979; American Society of Agricultural Engineers, 1981; Lewis, 1982; Brooker and Corder, 1986). Sowing and transplanting, however, are expensive, and sometimes are impossible because commercial sources of seeds or plants may not be available.

Alternatively, there is the option to recruit from the seed bank if seeds of required or preferred species are present. The presence of species in a seed bank disposes of many of the problems associated with collecting, storing, and sowing seeds or transplanting individuals, but it does not eliminate uncertainties associated with seed germination and seedling survival. Environmental influences on recruitment, not dispersal, in these situations becomes the major determinant of the success or failure of a revegetation project. The impact of environmental conditions on recruitment from seed banks is a phenomenon whose significance sometimes has been inadequately appreciated, and whose management potential has not been fully realized.

Seed banks have been exploited in two contexts: to manage the composition and structure of existing vegetation, and to restore or establish native vegetation. Our objectives in this chapter are (1) to review attempts to exploit *in situ* seed banks to manage vegetation, (2) to examine the potential of relict and donor seed banks to restore vegetation, and (3) to assess the influence of environmental conditions on recruitment from the seed bank and its management implications. Although wetland examples will be emphasized in this chapter, the techniques used and lessons learned can be applied directly, or are easily adaptable, to terrestrial situations.

II. Vegetation Management

After the destruction or disturbance of vegetation by fire, overgrazing, drought, flooding, etc., seed banks often play a role in natural regeneration (secondary succession) (Grime, 1981; Roberts, 1981). Nevertheless, the management of vegetation by deliberately altering environmental conditions to recruit preferred species from the seed bank is uncommon except for freshwater wetlands. Many freshwater wetlands are routinely managed by lowering their water levels (drawdowns) to recruit species from their seed bank (van der Valk, 1981).

The use of drawdowns as a management tool was developed largely in the American midwest to reestablish emergent vegetation after it had been eliminated by high water or overgrazing (Kadlec, 1962; Harris and Marshall, 1963; Weller and Spatcher, 1965; Meeks, 1969; Weller and Fredrickson, 1974; Bishop *et al.*, 1979). In larger, midwestern wetlands, species with long-lived seed predominate in the flora because they can survive years of high water as seeds and become reestablished immediately during the next drawdown, or because they can survive the drawdown as seeds and become reestablished when standing water returns (van der Valk, 1981). Emergent species with short-lived seed are at a disadvantage in these wetlands because their reestablishment during periodic drawdowns that last only 1 or 2 yr is dependent on the uncertainties of seed dispersal. Another group of species found in these wetlands are mud flat annuals that become established during drawdowns because they also have long-lived seed (van der Valk and Davis, 1978). However, species in all wetlands are not subject to this type of selection pressure. In wetlands with more stable water levels, seed banks may play little or no role in the recruitment of emergents.

An example from a wetland in Texas illustrates why an examination of seed banks prior to the initiation, or even planning, of management procedures is always recommended. Eagle Lake, Texas, formerly was covered with emergent vegetation, but because of high water levels during the summer (the lake is used as a reservoir for rice farms) and the introduction of nutria, the emergent vegetation has disappeared from most of the lake. To restore the emergent vegetation, a drawdown was proposed, but a study was first conducted to determine if there were still viable seeds of emergents in the seed bank. Other than seeds of two annuals, a *Cyperus* and a *Polygonum*, no viable seeds remained (A.G. van der Valk, unpublished data). Over the years that Eagle Lake had been used as a reservoir, the lake bottom had become covered with a thick layer of silt and clay that buried its seed banks. A drawdown would not restore emergent vegetation because emergent seed was no longer present. Thus, the planned drawdown was abandoned.

Not all wetland and terrestrial vegetation types have exploitable seed banks. Determining the presence of a seed bank and its composition is an essential first step before a new management plan that relies on a seed bank is attempted. Where previous experience has established the presence and composition of a seed bank, an initial seed bank study is not necessary.

A. Predicting Vegetation Composition

Predictive models that make use of seed bank information, plus information about other characteristics of the reproduction and growth of species, have been developed for forests by Noble and Slatyer (1980) and for freshwater wetlands by van der Valk (1981). Currently these types of models are available mostly for situations in which a major disturbance (fire or flooding) has removed the previous vegetation. In other words, they are models of secondary allogenic succession. Predicting recruitment of species in a vegetated area (secondary autogenic succession) or in an area that has never been vegetated (primary succession) is much more difficult.

An examination of the composition of the seed bank makes it possible to predict the initial composition of the postrecruitment vegetation, particularly on exposed drawdown surfaces in wetlands or other substrates cleared of vegetation in terrestrial situations. Seed bank data can yield information on three features of new vegetation: (1) its species composition, (2) the relative abundance of recently recruited species, and (3) the potential distribution of each species (Welling *et al.*, 1988). An analysis of the composition data can reveal which desirable or undesirable species may become established, and which desirable species are missing from the seed bank. The reliability and accuracy of such predictions about the future composition of postrecruitment vegetation have rarely been investigated.

Carney and Chabreck (1977) examined the drawdown vegetation and the seed bank of an impounded Louisiana wetland that was managed for waterfowl. The most abundant genera (in descending order) during the drawdown were *Hydrocotyle*, *Eleocharis*, *Paspalum*, *Cyperus*, *Echinochloa*, and *Sacciolepis*. The most abundant genera in the seed bank included *Cyperus*, *Hydrocotyle*, *Eleocharis*, *Scirpus*, *Sacciolepis*, and *Polygonum*. Carney and Chabreck (1977) concluded that it was possible to predict which genera would be abundant, but that it was impossible to predict their order of abundance.

Smith and Kadlec (1985a) used seed bank information from five vegetation types (*Distichlis spicata*, *Scirpus lacustris*, *Scirpus maritimus*, *Typha* spp., and open water) in conjunction with plant life history characteristics to predict the postfire vegetation composition in Utah in a

Great Salt Lake marsh. While predictions for specific vegetation types were often not very accurate, a fairly reasonable prediction of the composition of drawdown vegetation was made (seven of the nine species that were found in the field were predicted to occur from the seed bank data) for the whole study area. Additional information on the impact of light, salinity, and soil moisture on seed germination for most species would have allowed more accurate predictions.

Similarly, predictions made from a comparable set of data about the composition of drawdown vegetation in an experimental marsh complex located in the Delta Marsh, Manitoba, Canada, were generally correct qualitatively, but were often inaccurate quantitatively (Table 1). Nevertheless, all but one uncommon species found during the first year of

Table 1
Predicted (seed density) and actual
seedling densities[a,b]

Species	Seed density (m^{-2})	Seedling density (m^{-2})
Emergent species		
Scirpus lacustris	610	400
Typha glauca	220	30
Scolochloa festucacea	30	100
Phragmites australis	14	1
Carex atherodes	30	30
Puccinellia nuttalliana	0	<1
Scirpus maritimus	14	20
Hordeum jubatum	1	20
Mudflat annuals		
Atriplex patula	16	700
Aster laurentius	12	20
Chenopodium rubrum	180	50
Ranunculus sceleratus	60	6
Rumex maritimus	28	2
Wet meadow perennials		
Stachys palustris	3	4
Mentha arvensis	16	1
Teucrium occidentale	2	<1
Lycopus asper	10	<1
Sonchus arvensis	4	3
Cirsium arvense	7	1
Urtica dioica	3	6

[a]Data from Pederson (1983) and Welling (1987).

[b]During the first year of a drawdown in the 10 cells of the experimental marsh complex of the Marsh Ecology Research Program, Delta Marsh, Manitoba, Canada.

the drawdown were predicted to occur from seed bank data. Some emergent and annual species (e.g., *Scirpus lacustris*, *Typha glauca*, and *Chenopodium rubrum*) were predicted to occur at much higher densities in the field than they were later found (Table 1). Other species (e.g., *Scolochloa festucacea*, *Atriplex patula*, and *Aster laurentius*) had higher field densities than predicted (Table 1). Soil moisture and temperature conditions for seed bank samples in the greenhouse were different from those in the field during the drawdown. These environmental differences seem to be a major cause for the discrepancies between predicted and field densities (see Section IV). As in the previous study, predictions for large areas were much more reliable than were those for small areas (van der Valk *et al.*, 1989).

An examination of recruitment at different elevations at Delta Marsh indicated that some common species were found in the field at elevations lower than predicted by the seed bank data (van der Valk *et al.*, 1989). This suggests a movement of seeds downslope as water levels are lowered during a drawdown, introducing another potential error into predictions of drawdown vegetation that are made from seed bank data.

The composition of the seed bank can be used to predict the overall composition of postdisturbance vegetation. Because of differences between field conditions and those in the greenhouse, predictions based on greenhouse studies are often quantitatively unreliable.

B. Managing Seed Banks for Preferred Species

The composition of the seed bank is a function of the composition of the seed production of the present and previous vegetation and the longevity of seed of each species under local conditions. If there is a seed bank, the postdisturbance vegetation recruited from the seed bank will resemble that of the previous vegetation to some extent. This relationship between the composition of the seed bank and vegetation is particularly important in managed vegetation dominated by annuals. If management procedures are the same from year to year, areas with a stand of desirable annual species will have similar vegetation the next year. The same applies to areas in which undesirable species have become established. Consequently, control of the germination, growth, and seed production of undesirable species is as essential as that for preferred species (Fredrickson and Taylor, 1982).

Moist-soil plants are species, primarily annuals (e.g., *Echinochloa* spp., *Polygonum* spp., *Bidens* spp., *Cyperus* spp., *Panicum* spp., and *Paspalum* spp.), that grow on mud flats (Bellrose, 1941). Moist-soil management involves diking areas to create impoundments, establishing water-control capabilities, and instigating summer drawdowns (Fredrickson and Taylor, 1982). During the first year of a drawdown, large populations of annual plants typically become established, providing cover and food for wildlife. However, with successive years of summer draw-downs, there may be a gradual increase in unwanted plant species, such as *Typha* spp., vines, and woody vegetation.

Management techniques to control problem vegetation involve herbicides, flooding, irrigation, drying, disking, plowing, rotovating, burning, or combinations of these treatments. Once an area has been successfully managed for moist-soil plants for 5 to 7 yr, there is less need for control techniques, presumably because the seed bank now is dominated by the seeds of desirable species (Fredrickson and Taylor, 1982). In some cases, however, moist-soil management attempts have led to severe infestations of problem plant species such as *Lythrum salicaria*, *Typha* spp., or woody vegetation (Givens and Atkeson, 1957; Steenis and Warren, 1959; Bednarik, 1963; Harris and Marshall, 1963; Kadlec and Wentz, 1974).

III. Vegetation Restoration

A. Reestablishing Vegetation from Relict Seed Banks

Where natural vegetation has been eliminated, the potential for re-establishing it from a relict seed bank is an intriguing possibility that is currently being investigated. How long seeds of native species can survive in cultivated or pasture soils is poorly documented. It is known that in some forest plantations seeds of certain species from previous vegetation types can survive for at least 30 yr and possibly 45 yr (Hill and Stevens, 1981), and a recent experimental study revealed that viable seed of 9 of 14 species that were stored in forest soils could survive at least 5 yr (Granström, 1987). There are very few studies, however, that have examined the fate of seeds of native species when an area is cultivated or when one vegetation type is replaced by another.

In Alaska the seed banks of undisturbed *Picea* forests and of cultivated areas that were formerly *Picea* forest (Fig. 1) were investigated by Conn *et al.* (1984). Seeds of *Picea* spp. and *Alnus crispa*, the dominant tree species, were restricted almost completely to undisturbed forest seed banks. After 5 yr or more, the seed banks of agricultural fields were dominated by seeds of introduced colonizing species (weeds), but the total density of native colonizer species had not begun to decline even after 20 yr. Abandoning these agricultural fields would result in an "old field" vegetation that would be dominated by introduced colonizers and also contain many native colonizer species, but no *Picea* or *Alnus*. This is

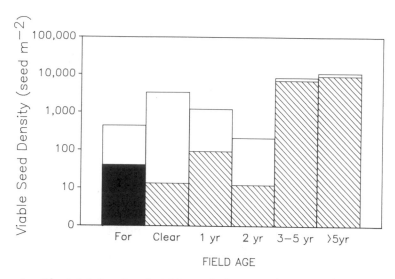

Figure 1.   The total density of viable seed of mature forest species (black), native colonizers (white), and introduced colonizers (hatched) found in the seed bank of forested (For), cleared (Clear), and cultivated (1– >5 yr) sites in Alaska. From Conn *et al.* (1984).

similar to patterns of old field succession that have been reported in many parts of United States.

The drainage of wetlands to create tillable agricultural land has been the single most important cause of wetland loss in the United States (Tiner, 1984). Unlike losses due to filling, losses due to drainage are theoretically reversible. By plugging drainage tiles or ditches, the hydrology of a former wetland basin can be restored to something resembling predrainage conditions (Madsen, 1986). If the drained wetland to be restored still has a seed bank containing wetland species, wetland vegetation will become reestablished rapidly. Erlandson (1987) investigated the potential use of relict seed banks to restore wetland vegetation to drained basins in the prairie pothole region. In all the drained wetlands examined, there were remnants of wetland seed banks, but the number of wetland species declined with time (Fig. 2).

In The Netherlands, the restoration of quaking fens from floating forests is being attempted by removing trees from these forests. An investigation of the seed banks and understory vegetation of floating forests in the Westbroek polder near Utrecht (Table 2) indicated that of the 23 rare fen species 11 were present in the seed bank and that 7 of these were also still present in the understory vegetation. Similarly,

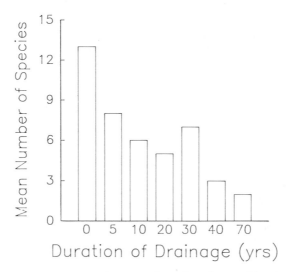

Figure 2. The mean number of species found in the seed banks of existing wetlands (0 yr) and wetlands drained 5–70 yr in Iowa, Minnesota, and North Dakota. Data from Erlandson (1987).

Pfadenhauer and Maas (1987) have demonstrated that seeds of fen species still are found in hay meadows in Germany.

Relict seed banks can play a role in the restoration of native vegetation, but their utility diminishes with time because seeds of desirable species lose their viability and those of undesirable species accumulate. Sites where native vegetation has only recently been eliminated make the best candidates for restoration.

B. Establishing Vegetation Using Donor Seed Banks

Potentially, one of the least expensive and easiest ways to establish native vegetation in an area where it has previously not existed is to use donor soil from a nearby site with the appropriate vegetation. Many important questions must be considered whenever this method is used. What is the species composition of the donor seed bank? What time of the year should donor soils be collected? How long can donor soil be stockpiled before there is a significant loss in seed viability? How deep a layer of donor soil is needed? What environmental conditions must be established and maintained to ensure that adequate recruitment from the seed bank occurs? And finally, what can be done to prevent the establishment of unwanted species?

Table 2
Status of 23 fen indicator species in floating forests[a,b]

Species	Absent	Veg	SB	Both
Carex lasiocarpa	X			
Cicuta virosa	X			
Drosera rotundifolia	X			
Eriophorum gracile	X			
Pedicularis palustris	X			
Ranunculus lingua	X			
Scirpus lacustris	X			
Utricularia minor	X			
Carex acutiformis		X		
Carex rostrata		X		
Juncus subnodulosus		X		
Menyanthes trifoliata		X		
Carex diandra			X	
Carex disticha			X	
Epilobium palustre			X	
Stellaria palustris			X	
Calamagrostis canescens				X
Carex curta				X
Carex pseudocyperus				X
Lysimachia thyrsiflora				X
Potentilla palustris				X
Thelypteris palustris				X
Viola palustris				X

[a]Data from van der Valk and Verhoeven (1988).

[b]Species from forests in Westbroek polder, The Netherlands. Absent, Not present in vegetation or seed bank; Veg, present only in the vegetation; SB, present only in the seed bank; Both, present in both vegetation and seed bank.

This approach to vegetation establishment on unvegetated sites was pioneered by workers interested in mine spoil reclamation (Bradshaw and Chadwick, 1980), and it has been investigated with a view to creating both wetland (Dunn and Best, 1984) and terrestrial vegetation (Beauchamp *et al.*, 1975; Tacey and Glossop, 1980; Farmer *et al.*, 1982). It involves the removal of the topsoil prior to mining, storing it, and putting it back after the ore has been removed, or, alternatively, removing topsoil from an area to be mined and immediately placing it on an area to be revegetated. Beauchamp *et al.* (1975) in Wyoming, Tacey and Glossop (1980) in Western Australia, and Farmer *et al.* (1982) in Tennessee have examined the seed banks of local native vegetation to determine the feasibility of this approach, and all of them concluded that it has potential. The way topsoil is handled and stored, however, can have a significant impact on the success of a revegetation project (Johnson and Bradshaw, 1979). Removal of the upper 5 cm of topsoil from an unmined area and placing it immediately on a mined area gives the best results (Tacey and Glossop, 1980). Donor seed banks will not produce vegetation identical to the local native vegetation from which they were taken because some native species may not be present in the seed banks or their seeds may lose viability rapidly when topsoil is stockpiled. Nevertheless, donor soils can be used to establish rapidly a species-rich vegetation dominated by native species that are adapted to local conditions.

An experimental study of wetland creation in which donor soils were used to revegetate plots in a recontoured Florida phosphate strip-mining site illustrates some of the advantages, limitations, and practical difficulties associated with this method. In this study, different thicknesses of peat from a forested wetland were deposited on experimental plots at the old mine site (Brown and Ødum, 1985). Within 1 yr, many more herbaceous species became established on plots with donor soils than on the control plots. However, no woody species, one of the principal aims of this study, were recruited. The biomass of wetland species on plots with donor soils also was much higher than on control plots (Fig. 3). Microenvironmental conditions in plots with donor soils re-

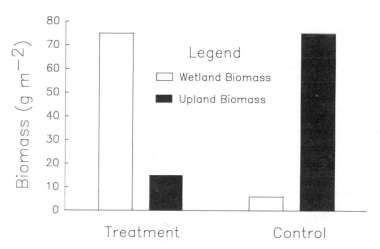

Figure 3.   Mean October biomass of wetland and upland species on control sites and treatment sites with donor soils in a Florida wetland creation project. Data from Brown and Ødum (1985).

sulted in different species dominating the tops of hummocks and depressions. This study also encountered many practical problems, including difficulties applying specified depths of peat, differences in the composition of the seed banks from place to place in the donor swamp, low water levels that hampered recruitment during most of the study period, and the elimination of seedlings in areas where water hyacinths were deposited during a period of high water. In spite of the practical problems encountered and the lack of woody species establishment, this method of wetland creation was judged to be effective and economically competitive with the more traditional method of establishing new wetlands by transplantation (Brown and Ødum 1985; Robertson, 1985).

Many of the practical questions regarding the use of donor soils to establish wetlands have been investigated by Wilhelm et al. (1987) as part of a wetland creation project along the Des Plaines River in Illinois (Hey, 1987). These included the time of year that donor soils should be collected, the depth to which viable propagules are found in donor soils, and the impact that soil moisture has on recruitment from the seed bank. Two recommendations came from this study: (1) donor soil should be collected at the beginning of the growing season, and (2) donor soils should be collected to a depth of no more than 25 cm. Soil moisture levels had a significant impact on the number of species recruited from donor soil, but the responses to the moisture treatments varied among donor soil types (Fig. 4).

Donor seed banks are potentially an economically and ecologically sound way to establish vegetation on mine sites where an adequate supply of donor soil exists. The donor soil not only supplies seeds and other propagules, but also provides a supply of nutrients and a surface soil structure and chemistry somewhat similar to that found in nearby undisturbed areas. Various soil amendments are used in most reclamation projects to improve the germination and growth of species (Bradshaw and Chadwick, 1980; Lyle, 1987). Such amendments (e.g., fertilizers, organic matter, and irrigation) may also be needed to ensure adequate recruitment of species from donor seed banks and their survival. The lack of a supply of donor soil, whose efficient collection with earth-moving equipment normally requires the prior removal of the existing native vegetation, limits the application of this technique to other types of restoration projects.

"In the final analysis, seed banks are not a panacea for restoration of native . . . vegetation" (Dunn and Best, 1984) because often they lack seeds of required species. The use of donor soils to restore native vegetation on site, however, does have many advantages, such as enabling the rapid development of a diverse vegetation that can impede the establishment of unwanted species, and it immediately provides the newly created vegetation with a seed bank that contains a diversity of species that may buffer the new vegetation against disturbances.

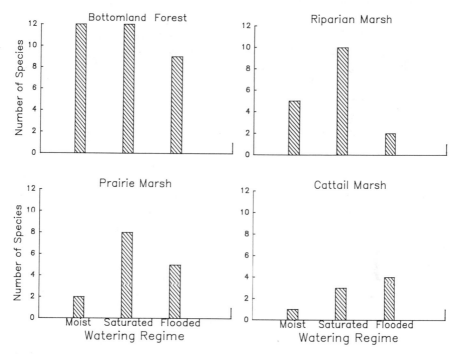

Figure 4.   The number of species that became established in seed bank samples taken from bottomland forest, riparian marsh, prairie marsh, and cattail marsh at the Des Plaines River, Illinois, demonstration project kept under moist, saturated, and flooded conditions. Data from Wilhelm *et al.* (1987).

IV. Environmental Control of Recruitment

Seed banks normally contain seeds of a number of species, each of which has different seed germination characteristics (Thompson and Grime, 1979; Keddy and Ellis, 1985; Galinato and van der Valk, 1986) and seedling survival characteristics (van der Valk and Welling, 1988). A plant canopy (van der Valk and Davis, 1978; Meredith, 1985) and standing or fallen litter (van der Valk, 1986) can reduce significantly, or prevent completely, the recruitment of species from the seed bank. When these are absent, soil moisture, as is illustrated by the seed bank study at Des Plaines River site (Fig. 4), and other environmental conditions such as soil temperatures (Galinato and van der Valk, 1986), soil salinity (Ungar, 1978, 1987a), and photoperiod seem to be the major factors regulating recruitment. Because recruitment is dependent on environmental conditions (Welling *et al.*, 1988), it may be feasible for managers to optimize the recruitment of desirable species or, at least, reduce the recruitment of undesirable ones by manipulating environmental conditions (Harris and Marshall, 1963; Meeks, 1969; Fredrickson and Taylor, 1982; Naim, 1987). The potential for doing this has been investigated in a number of recent wetland studies.

How soil moisture affects recruitment from wetland seed banks has been studied by applying different watering treatments to subsamples of a seed bank sample (R.L. Pederson, unpublished data). Although some seeds of most species could germinate in most of the treatments, the treatment in which there was optimal germination varied from species to species. For example, among annuals, *Aster laurentius* seed germinated best in the watered-every-day and every-other-day treatments, while *Chenopodium rubrum* seeds germinated best in the saturated and watered-every-day treatments (Table 3). Optimum germination of *Hordeum jubatum* seed occurred in the watered-daily and every-other-day treatments, while for *Typha glauca* it occurred in the flooded and saturated treatments (Table 3). Differences in seed germination due to soil moisture conditions, plus those due to soil temperature (Galinato and van der Valk, 1986) are enough to cause changes in the dominant species during drawdowns from year to year in the Delta Marsh (Welling, 1987).

Naim (1987) examined the impact of soil moisture conditions in both the greenhouse and field on seed germination for four species found in moist-soil units at the Mingo National Wildlife Refuge in southern Missouri. In the greenhouse, seeds of these species were sown on flats maintained at 10 different soil water matrix potentials. In the field, soil water matrix potentials also were monitored for 4 weeks after the onset

Table 3
Relative percentage germination of seed of four species from a composite seed bank sample[a]

Moisture treatment	Aster laurentius		Chenopodium rubrum		Hordeum jubatum		Typha glauca	
	F	B	F	B	F	B	F	B
FL	7	0	0	0	0	0	33	32
SA	25	22	35	27	10	11	41	34
W1	35	42	28	46	30	22	22	31
W2	27	29	20	14	40	56	4	3
W3	16	7	15	13	2	1	0	0

[a]Species from the Delta Marsh, Manitoba, Canada, in five different watering treatments. The treatments were flooded (FL), saturated soil (SA), watered daily (W1), watered every second day (W2), and watered every third day (W3). There were both freshwater (F) and brackish water (B) watering treatments.

of a drawdown in an area were all four species grew. For each species, seed germination was highest at a particular soil matrix potential. In the greenhouse, seeds of *Polygonum hydropiperoides*, *Eleocharis obtusa*, and *Echinochloa crusgalli*, all desirable species, germinated best at 0, −10, and −20 kPa, respectively, and seeds of *Xanthium pensylvanicum*, an undesirable species, germinated best at −60 kPa. In the field, the emergence of these species during a drawdown was as predicted: *Polygonum* germinated first, followed by *Eleocharis* and *Echinochloa*. *Xanthium* did not begin to germinate until soils were much drier (Table 4). Naim's data make it possible for managers to develop drawdown procedures that can reduce the recruitment of *Xanthium*.

Attempts to establish vegetation in saline inland wetlands illustrate another type of environmental constraint on recruitment during drawdowns. Seed germination is adversely affected by salinity (Ungar, 1978; Galinato and van der Valk, 1986). In saline wetlands, soil salinity levels during low-water periods often restrict vegetation colonization and seed production (Christiansen and Low, 1970; Ungar and Riehl, 1980; Lieffers and Shay, 1981; Smith and Kadlec, 1983). Seed densities in the seed bank of an impounded brackish wetland in Manitoba, Canada, were extremely low, with *Scirpus maritimus* dominating its scant seed bank (Table 5). Shallow flooding produced over twice the seedling density of simulated drawdown conditions (Table 5) due to the reduction in salinity. Simulated drawdown samples in which no seedlings grew had double the soil conductivity of those with seedlings. Therefore, flushing or irrigation would be necessary to reduce the soil salinity to levels that would permit the establishment of emergent vegetation at this and other saline sites (Kadlec and Smith, 1984).

Table 4
Seed germination and soil moisture[a,b]

Species	Soil water matrix potential (kPa)	
	Laboratory	Field
Polygonum hydropiperoides	0	− 5
Eleocharis obtusa	−10	−10
Echinochloa crusgalli	−20	−40
Xanthium pensylvanicum	−60	−55

[a]Data from Naim (1987).

[b]Potentials are those at which maximum percentage germination of seeds occurred in the greenhouse and at which 50% emergence occurred in the field at the Mingo National Wildlife Refuge in southeastern Missouri.

Table 5
Number of seedlings from
a saline wetland seed bank[a]

	Watering treatment	
Species	Watered daily	Shallowly flooded
Atriplex patula	5	0
Hordeum jubatum	1	0
Scirpus maritimus	12	34
Typha spp.	0	2
Zannichellia palustris	0	1
Potamogeton pectinatus	0	2
Total	17	39

[a]Seed bank samples ($n = 25$) were taken from an impounded saline wetland in southwestern Manitoba, Canada. Samples were either watered daily or kept under 5 cm of water (shallowly flooded).

Manipulating environmental conditions during drawdowns to favor the establishment of desirable species has great potential for controlling the composition of drawdown vegetation, and deserves further study. This approach would make the results of wetland vegetation management more predictable. To do this, however, a better understanding of the germination requirements of seeds of both desirable and undesirable species is needed.

In wetlands, conditions for seed germination and seedling growth on exposed mud flats are usually close to ideal when no salinity problems exist and environmental conditions can be manipulated by controlling the timing and duration of drawdowns. In terrestrial situations, the potential for manipulating environmental conditions to ensure the establishment of desirable species also exists. For example, nitrogen fertilization triggers the germination of seeds of *Prunus pensylvanica* from the seed bank of hardwood forests in Pennsylvania (Auchmoody, 1979). Fertilization, liming, irrigation, and incorporation of organic matter are all amendments that can be used to alter site conditions to foster the recruitment of desirable species (Auchmoody, 1979; Johnson and Bradshaw, 1979; Lyle, 1987).

V. Summary and Conclusions

Vegetation management, based on the exploitation of seed banks, will be successful only when (1) the seeds of the required or preferred species are present in the seed bank, (2) the seeds of unwanted species are not present or, at least, are uncommon, and (3) conditions suitable for the germination of the seeds of preferred species can be established and maintained. Where these conditions are met, managers have successfully exploited seed banks to establish desirable vegetation. It is not possible to determine *a priori* which types of wetlands or terrestrial vegetation have exploitable seed banks and which do not. The seed banks of only a few vegetation types, scattered across North America, have been investigated (see, however, the appropriate chapters in this volume). Consequently, it is prudent to establish the composition of the seed bank prior to the initiation of any management plan that presupposes its existence.

Besides a basic knowledge of the composition of the seed bank, more detailed knowledge is needed of both the seed germination requirements of preferred and undesirable species and how to create conditions in the field that favor the germination of preferred species. Recruitment from the seed bank is greatly influenced by soil moisture and other environmental conditions (e.g., soil temperature, soil salinity, light, nutrients, etc.). Wetland management plans should consider ways to influence soil moisture to maximize recruitment of desirable species. The timing, the duration, the rate at which water levels are lowered and raised, and the reflooding of drawdown surfaces periodically to maintain high soil moisture levels must be considered. When it is feasible, dikes, commonly used for water control in wetland management, should be built with the capability to manipulate water levels, permitting establishment of ideal soil moisture conditions during drawdown. Terrestrial management plans should consider various soil amendments to improve recruitment from the seed bank, such as irrigation, incorporation of organic matter, and fertilization.

Egler (1954) first pointed out that the recruitment of species from seed is restricted to periods when specific environmental conditions occur and that this had profound management implications. Noble and Slatyer (1980) have extended Egler's concepts to develop a general model of fire-induced forest succession that explicitly includes a consideration of the behavior of species with seeds in the seed bank. A comparable model has been developed by van der Valk (1981) for freshwater wetlands subject to water level fluctuations. To understand and predict the course of both natural and management-induced vegetation change, these models require information about the characteristics of individual species, including seed longevity, seed germination requirements, and age to sexual reproduction. For example, better comparative information about seed germination for many species under different environmental conditions would greatly improve predictions about the composition of postrecruitment vegetation. Recruitment of species from the seed bank is, of course, only the first step in establishing any vegetation. Seedling and adult mortality due to drought, flooding, herbivory, disease, and/or competition will also play a role in determining the composition and structure of vegetation (Pickett *et al.*, 1987; van der Valk and Welling, 1988). As our understanding of life histories of species improves, vegetation management will become more precise.

Acknowledgments

The preparation of this manuscript was supported by the Marsh Ecology Research Project of the Delta Waterfowl and Wetlands Research Station, Delta, Manitoba, Canada. The initial drafts of this manuscript were written while the senior author was on leave in the Department of Plant Ecology, University of Utrecht, The Netherlands, where he was supported by Grant 84-249 from The Netherland's Organization for the Advancement of Pure Research (ZWO) and by a faculty improvement leave from Iowa State University. We would particularly like to acknowledge the many contributions of Carol Erlandson, Parvaiz Naim, Louisa Squires, and Chip Welling, whose ongoing or recently completed studies have improved significantly our understanding of wetland seed banks and their potential exploitation for wetland management. This is paper Number 40 of the Marsh Ecology Research Program, a joint project of the Delta Waterfowl and Wetlands Research Station and Ducks Unlimited Canada.

CHAPTER 16

Seed Banks and Vegetation Management for Conservation: Toward Predictive Community Ecology

Paul A. Keddy,
Irene C. Wisheu,
Bill Shipley,
and Connie Gaudet

Department of Biology, University of Ottawa,
Ottawa, Ontario, Canada

I. Applying Ecological Research

The ultimate test of our scientific understanding of an ecological phenomenon is our ability to use this understanding to manage ecosystems for conservation. Our objective in this chapter is to consider how a knowledge of seed banks can be used to manage vegetation for conservation purposes. We will begin by considering general principles, and then conclude with three case studies where rare plant communities are being managed based on knowledge of their seed banks.

There are three components in conservation management: determining the characteristics of the system of interest, selecting the goals of management, and developing the techniques for attaining these goals. The preceding chapters have summarized what is known about the ecology of seed banks for a variety of vegetation types. For any particular vegetation type, we would need to answer the following questions to document resource characteristics: Are there buried seeds? What natural processes lead to establishment of seedlings from these seeds? Which species in the vegetation do not have buried seeds? Which species of buried seeds are not present as adults? What natural processes regulate which seeds produce seedlings, and which seedlings produce adults? Because vast amounts of descriptive data could be collected from any system, a theoretical framework is needed to help decide which observations deserve highest priority. Such a framework is provided by management models designed to predict community responses to perturbation (Starfield and Bleloch, 1986).

The second component is determining goals. We will assume that the goal of vegetation management is to maintain the biological diversity of the planet. Scientists can contribute to the setting of these goals by documenting which species are rare or endangered, and whether the rarity is local or global. It is up to the professional ecologist to judge which goals are feasible based on site constraints. While this may seem straightforward, we must remember that interest groups will have different goals, many of which will be incompatible with those we have assumed to be important. Even within the conservation movement itself, strong constituencies exist for certain groups of plants (e.g., orchids) and weak ones for others (e.g., grasses). Sometimes a conflict exists between maintaining diversity and maintaining the vigor of a rare plant at a local scale. As a consequence, the choice of management goals is rarely based solely upon scientific criteria. In general, the goal can be specified by stating the desired abundances of certain target species; they may be rare species which the management is designed to enhance or alien species which are targeted for elimination.

Once the goal has been determined, management techniques to

attain this goal are needed. In the simplest case, management will consist only of protecting the site from human interference. Even in this case, it is necessary to define carefully what is meant by interference, and how protection will actually accomplish this. In many other cases, some form of manipulation of the disturbance agents may be necessary, and choosing the appropriate management techniques becomes more difficult. The professional ecologist can contribute at this point by designing and testing the different possible techniques which exist. Because there are major constraints imposed by the time, budgets, and abilities of the managers who will implement these techniques, close cooperation between the managers and ecologists is needed at this stage. Much of this management is carried out on a case by case basis in an *ad hoc* manner. There is a need for general predictive models of vegetation response to perturbation that can be adapted to suit individual situations.

In this context, we will consider the role of seed banks in the management of vegetation for conservation. We will first sketch a general framework for accomplishing this goal; the specific details will have to be provided for the particular system of interest. We will illustrate this framework with examples from shoreline plant communities, because these are the systems with which we have the greatest familiarity, but the primary emphasis will always be upon the general principles.

II. Manipulating Regeneration Processes

In the broadest sense, a general management model would need only a few essential elements: (1) a vector of species abundances describing the current state of the system, (2) a matrix of species traits, and (3) a function which describes the perturbation. The function would use the information in the matrix to predict the future state of the vegetation, a new vector of species abundances. In practice we have few usable matrices of species traits, but they could be constructed by screening (e.g., Grime and Hunt, 1975). In the context of this chapter, the traits might be simply densities of buried seeds in the soil, estimates of vegetative regeneration rates, or the sensitivities of adults to different kinds of disturbances. Such a general model would not only lead toward predictive community ecology, but would simultaneously guide both managers and scientists in their choice of which aspects of the system need to be measured. We will first consider the essential aspects of regeneration from seed banks that such a model could include. We then present a simple model as a first step toward predictive community ecology.

The preceding chapters have documented the importance of regeneration processes in natural vegetation types. Fundamentally, we know that all individuals must eventually die. The problem for the manager is to control when death occurs, and what finally replaces each individual. This is conveniently divided into two phases—manipulating the disturbance regime to reduce existing biomass, and manipulating recruitment patterns to favor selected species (Fig. 1).

A. Manipulating Disturbance

As part of defining the characteristics of the system, it is necessary to determine the size, frequency, and intensity of disturbance produced by specified agents (Sousa, 1984), and how the target species respond to these events. Let us define disturbance as any event which permits regeneration from buried seeds. This is a slightly broader definition than that of Grime (1979), who defines disturbance as any event that removes biomass from a plant community. While the two definitions are similar,

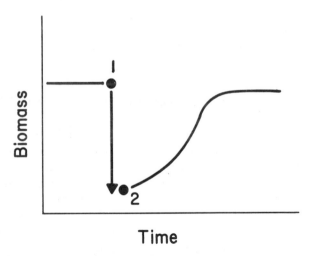

Time

Figure 1.   The two kinds of intervention which the manager can use to control regeneration from buried seeds. In case 1, the disturbance regime is controlled by varying timing, area, or intensity. Examples would include manipulating the density of burrowing animals or frequency of fire. The *y* axis refers only to disturbance leading to a reduction in existing biomass. The model could easily be expanded to include any event which permits regeneration from buried seed. In case 2, recruitment is carefully manipulated to select the most desired species. Examples would include controlling the amount and species composition of litter or manipulating soil moisture to enhance recruitment of an endangered species.

there may be cases where disturbances turn over the substrate and expose seeds without actually killing adult plants or removing litter. Target species may be rare species which we wish to maintain or increase, or aliens which we wish to eliminate. In a prairie, for example, cattle, badgers, and fires are principal disturbance agents. Each of these will create different sizes, areas, and intensities of disturbance, and will therefore create different regeneration niches for target species.

Once this information is available, it becomes possible to make qualitative predictions of the kinds of disturbance that are best suited to produce the desired goal. In some cases the manager may decide to maintain the natural regime, but in others there may be enhancement or reduction of particular disturbances. In practice, simple models should predict vegetation associations expected to result from different strategies and combinations of strategies. For example, Platt and Weiss (1977, 1985) examined the species that colonize badger mounds on prairies. Based on life history characteristics such as seed dispersal distances and establishment requirements, they predicted the changes in abundance that would occur in different badger disturbance regimes.

In summary, the information required is (1) natural disturbance agents; (2) area, frequency, and intensity of each disturbance; and (3) performance of target species in these regeneration sites. There is a definite need for quantitative models that managers could use to project species composition under different scenarios.

B. Manipulating Recruitment after Disturbance

A disturbance may create a gap in the vegetation, but the specific characteristics of the gap (Sousa, 1984) will control what grows there. Subtle changes in the characteristics of this gap may produce radically different recruitment rates of target species. Moisture availability, litter loading, and substrate texture are but three characteristics of gaps which managers could manipulate. Harper *et al.* (1965) is the classic study which showed how minor differences in microtopography can control the germination of different species; small holes, small boxes, glass slides, and even worm casts produced differences in the germination of three *Plantago* species. Such manipulations may be difficult to perform in nature reserves covering many hectares, but others are possible. For example, Keddy and Constabel (1986) created a soil texture gradient and sowed different shoreline plants along it. Species showed a significant shared preference for the fine-textured end of the gradient, but differed in their ability to occupy the coarse substrates. Van der Valk (1986) showed that the presence of litter reduced germination of wetland plants during low-water years. Not only the presence of litter, but also the kind of litter can produce differential germination (Fig. 2). *Bidens cernua* was less sensitive

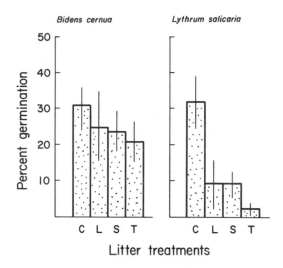

Figure 2.   The effects of three kinds of litter (270 g m⁻²) on the germination of *Bidens cernua* and *Lythrum salicaria* (F. Terillon, unpublished data). C, Control (no litter); L, *Lythrum salicaria*; S, *Scirpus americanus*; T, *Typha latifolia*.

to the effects of litter loading than was *Lythrum salicaria*, presumably due to its larger seeds (0.01 g as opposed to 0.0005 g). *Typha* litter produced the lowest germination for each species, probably because the large flat leaves produce the most impenetrable layer, although allelopathic effects cannot be excluded. When seeds of shoreline plants are sown at different water levels, differences in water level can have a major impact on which seeds germinate and which kind of communities establish (Keddy and Ellis, 1985; Moore and Keddy, 1988). These are factors managers could control, in turn controlling the kinds of species which are recruited from seed banks. If a seed bank contained both a rare species and a common one, it is easy to imagine a series of experiments to determine the kind of postdisturbance environment that would enhance the performance of the rare species.

   In summary, information is required to answer three questions: (1) What environmental factors regulate germination in the vegetation gaps? (2) How do the target species differ in their response to these factors? (3) Which factor or factors are most easily manipulated to obtain the desired effect?

C. The Management Plan

The simplest management plan could be produced by specifying the following parameters:

1. Current abundance of target species.

2. Desired abundance of target species.

3. Major natural disturbance agents.

4. Regeneration of target species in each kind of natural disturbance.

5. Disturbance regime to be established and methods for attaining this regime.

6. Postdisturbance manipulations (if any) to be performed.

7. Monitoring system to assess progress toward goals and provide information on refinements which may be needed in steps (5) and (6).

III. A Quantitative Model: Toward Predictive Community Ecology

A. Making Simple Predictions

The ideas discussed above could be put into a more quantitative form to clarify the problems which a manager faces, and to provide an opportunity to explore the consequences of different management options so that the best options are chosen. Let us express the current abundance of target species as the vector V_0 $(x_1, \ldots, x_s)$ where x_i is the abundance of species i and there are s species of interest. The objective of the management is to convert V_0 to a new vector of desired abundances V^* $(x_1^*, \ldots, x_s^*)$. The target species may be desirable rare species, undesirable alien species, or simply a group of convenient indicator species for a particular habitat type.

The major disturbance agents are then identified and experimental manipulations are conducted to measure the effects of each kind of disturbance upon the vector of target species. At this time any potential postdisturbance manipulations could also be explored. This yields a set of n statements as follows: V_0 with perturbation 1 becomes V_1, V_0 with perturbation 2 becomes V_2, etc.

Based upon this information the manager must then chose a management strategy (that is, a proposed set of manipulations) which will change V_0 as close as possible to V^*. The problem is to solve for the various combinations of a_i in the following equation:

$$a_1 V_1 + a_2 V_2 + \cdots + a_n V_n = V'$$

where a_i is the area of disturbance i and V' is the resulting vector of species abundances. The objective is to minimize the distance between

V' (the actual vegetation type produced) and V^* (the desired vegetation type). That is, the objective is to minimize, for s species,

$$\left(\sum_{j=1}^{s} (x'_j - x^*_j)^2 \right)^{0.5}$$

This can be solved using the standard techniques of nonlinear programming.

This problem is illustrated graphically in Fig. 3. There is rarely a single solution. As illustrated in Fig. 3, the manager has the option of increasing the area of disturbance type 3, or else trying two kinds of disturbance in different patches using agents 2 and 4.

B. Uncertainty

The above treatment is completely deterministic. In reality, there will be many sources of uncertainty. There are a number of important elements to be included in this measure of uncertainty. Management plans should minimize extrapolation from the initial field experiments. Changing the area of the treatments (even if patch size is kept constant) could produce a wide range of unknown effects resulting from factors such as pollinators or seed dispersers being influenced by the total area of the

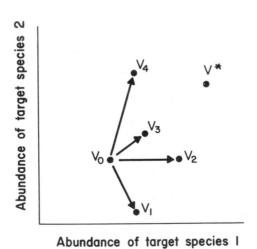

Figure 3. A simple management model. The vegetation is currently in state V_0, and the goal is V^*. Field experiments using four different disturbance regimes produce response vectors V_1–V_4.

manipulation. The greater the deviation from the areas used in the experiments, the greater the uncertainty of the model.

The vector addition assumes there is no interaction between treatments. In fact, such interactions are highly probable. The more perturbations used, the greater the number of possible interactions, and, therefore, the greater the uncertainty of the outcome. Also, the outcome vectors of the experiments will all have variances, although they have not been used above. The greater the variance in a particular vector, the greater the uncertainty of the outcome when applying that manipulation.

While we have not done so, it is possible to cast these measures of uncertainty in quantitative form. The manager's objective would then be to convert V_0 to V^* while minimizing the uncertainty of the outcome. Minimizing uncertainty might be weighted more heavily where one is managing very small populations of an endangered species.

C. Incorporating Constraints

At least three constraints on the manager are likely, including size of area that can be disturbed, timing of disturbance, and the budget. These constraints can be incorporated by adding the following specifications to the model: Where k_1 and k_2 are the minimum and maximum area practical to disturb using a given method, that disturbance would be expressed as $k_1 < a < k_2$. For the program as a whole, the total cost must be less than the amount specified. That is, where c is the cost per unit area of each kind of perturbation, $\Sigma\, c_j a_j$ must be less than T, the total budget.

The above treatment is a preliminary attempt at a quantitative presentation of intuitive ideas on the management of rare species and/or vegetation types using disturbance regimes. The model could be made more sophisticated and rigorous, but this simple approach illustrates the basic concepts.

IV. Case Histories: Fire and Water

A. Disturbance by Fire

In principle, any species which regenerates after disturbance can be managed using this conceptual approach. The importance of the interaction between disturbance and regeneration in the maintenance of many species and ecosystems is widely recognized (Watt, 1974; Grubb, 1977b; Grime, 1980; Vogl, 1980; Risser *et al.*, 1981; Bazzaz, 1983; Gillon, 1983; Mooney and Godron, 1983; Sousa, 1985; Pickett and White, 1985). Although agents of disturbance range from the activity of small mammals (Watt, 1974) and grazing (Risser *et al.*, 1981) to drought and flooding (Vogl, 1980), disturbance by fire has received the most attention in vegetation management. Fire was initially identified as a destructive force incompatible with conservation objectives, but it soon became apparent that strict control of fires often threatened the very systems resource managers were trying to protect (Hartesveldt *et al.*, 1969; Owen, 1980; Gillon, 1983). Disturbance by fire is now recognized as essential to the perpetuation of natural systems from prairie and savanna (Daubenmire, 1968; Gillon, 1983; White 1983) to forests (Hartesveldt *et al.*, 1969; Kozlowski and Ahlgren, 1974; Kelsall *et al.*, 1977; Wright and Bailey, 1982; Alexander and Dubé, 1983) and heathlands (Mallick and Gimingham, 1985).

The ability of component species to survive and regenerate after fire is important for predicting system responses. Fire generally promotes recruitment from seed banks through removal of accumulated litter and elimination of competing species. Many seeds are tolerant to heat (Daubenmire, 1968) and, if covered even slightly with soil, can survive a relatively intense fire. In contrast, aboveground vascular plant tissue is easily killed by heat (Wright and Bailey, 1982). Therefore, available seed can provide important information in predicting the effects of fire. Several other aspects of regeneration, however, can be involved in response to fire and these may be highly specialized and species specific. Some species are dependent on fire for seed dispersal (Kozlowski and Ahlgren, 1974; O'Dowd and Gill, 1984), germination and establishment (Hartesveldt *et al.*, 1969), and even seed bank formation (e.g., Wellington and Noble, 1985). Vegetative regeneration from rhizomes, root crowns, or protected meristems can account for a substantial proportion of postfire recruitment (e.g., Lyon and Stickney, 1976; Mallick and Gimingham, 1985). Seasonal timing and frequency of fire further affects regeneration of component species (e.g., Wright and Klemmedson, 1965). Obviously, realistic prediction of responses to fire must be based on a consideration of several aspects of regeneration.

B. South African Renosterveld: Case Study 1

The interplay of regenerative strategies and their importance to the design of effective management programs is exemplified by a proposal for the conservation of rare South African renosterveld. Renosterveld is an endangered South African shrub–grassland with an estimated 188 threatened taxa (Hall and Veldhuis, 1985). This land is being converted to pasture and cropland at an alarming rate and conservation of remaining tracts has become critical. However, effective management is complicated by several factors as outlined by Cowling *et al.* (1986).

Selective grazing of grasses by domestic cattle and creation of gaps for shrub seedling establishment leads to an increase in shrub cover. Once an abundance of shrubs renders the land unsuitable for grazing, there is a tendency to convert the land to crops to maintain economic viability. Therefore, protection of the rare vegetation type depends on keeping it in a productive grassland state. Although fire seems an appropriate management tool, the situation is further complicated by the fact that both the undesirable shrubs and the desirable dominant grasses are fire adapted. The solution to this conservation problem lay in a detailed analysis of the seasonal patterns of regeneration and growth of the component species.

Phenodiagrams depicting seasonal patterns of budding, flowering, immature seed, mature seed, and growth maxima were prepared for key indicator species, two undesirable shrubs, and the desirable dominant grass. These data indicated that a fall burn would coincide with the end of the shrub growth period when carbohydrate stores were minimal and with the dormant period of the dominant grass. In addition, a fall burn would coincide with seed production in the shrub species, preventing further contributions to the seed bank. An understanding of the seed bank dynamics proved particularly important to effective management in this system. Research indicated that shrub seed germination was stimulated by fire and that seed viability in the soil lasted about 7 yr. Thus, plans for follow-up burns to exhaust seed stores were incorporated into management recommendations.

The renosterveld example elegantly illustrates the importance of considering several aspects of postfire regeneration for both desirable and undesirable species in predicting system responses. Clearly, a poorly timed burn could have aggravated, not solved, the problem. The potential to exhaust seed bank resources through repeated burns has particular relevance to rare species conservation. Emphasis of the renosterveld program was on depleting seed stores of an undesirable species. If instead of undesirable species rare or desirable species were represented in the seed bank, the effect of exhausting seed stores through repeated and ill-timed fires could seriously endanger the long-term survival of these species.

C. Disturbance by Flooding

High water levels are also a source of disturbance to vegetation (Salisbury, 1970; Raup, 1975; van der Valk and Davis, 1978; van der Valk, 1981; Keddy and Reznicek, 1982, 1986; Pederson and van der Valk, 1984; McCarthy, 1987). The relationship between disturbance by high water levels and regeneration from buried seeds has been investigated in prairie potholes (van der Valk, 1981; Pederson and van der Valk, 1984) wherein incorporation of species life history traits was used to create a predictive model (van der Valk, 1981). This work has been applied primarily in management of waterfowl habitat (Pederson and van der Valk, 1984). Much less attention has been paid to the component plant species or to the importance of flooding for maintaining the biological diversity of natural areas. The following examples emphasize conservation of rare and endangered plant species in natural areas.

D. Matchedash Lake, Ontario: Case Study 2

In Ontario, many rare Atlantic coastal plain species occur on sandy lakeshores within the old shorelines of glacial Lake Algonquin (Keddy, 1981a), where some form of natural disturbance (e.g., fluctuating water levels, Keddy and Reznicek, 1982; or waves, Keddy, 1985a) appears necessary to prevent the encroachment of competitive dominants. One of the largest concentrations of rare species occurs in Matchedash Lake, located on the extreme southern edge of the precambrian shield near Georgian Bay. Keddy and Reznicek (1982) discuss the role of fluctuating water levels in the persistence of Atlantic coastal plain vegetation in this lake. Water levels were controlled by a beaver dam on a small outlet stream which flowed over solid rock. This beaver dam raised water levels approximately 1.5 m in some years, but when the beaver dam washed out, levels dropped dramatically. During low-water years, a rich coastal plain flora emerged from buried seeds, and during high-water periods these species persisted as buried seeds. In the absence of water level fluctuations, shoreline dominants such as *Myrica gale* and *Calamagrostis canadensis* covered the entire shoreline down to the waterline, eliminating any sites for the growth of coastal plain species. A similar relationship between water level fluctuations and rare coastal plain plants has been described in New Jersey (McCarthy, 1987).

The future of the rare plant community of Matchedash Lake was threatened when the Ontario Ministry of Natural Resources built a dam on the outlet stream to maintain the lake permanently at a high water level (Keddy, 1981b; Reid, 1987). The objective was to protect a species of sport fish introduced the previous decade. The dam was built despite the evidence of botanical significance, leading to several years of controversy. Eventually the Ministry agreed to reinitiate water level fluctuations in the lake by replacing the existing dam with one capable of manipulating water levels. They also agreed to fund a 7-yr study of the responses of rare plants to this manipulation. The experiment began in 1983 when water levels were lowered 30 cm. Water levels were lowered 30 cm more in each of the following 2 yr and were maintained at that low level until 1987. Rare coastal plain plants increased in abundance as water levels fell (Fig. 4).

A number of important questions had to be addressed in planning

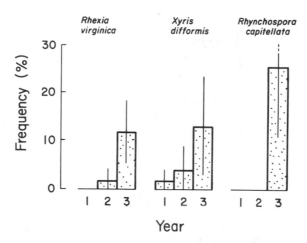

Figure 4. The response of three rare coastal plain plants to a low-water period in Matchedash Lake. ($n = 20$ quadrats, years are 1983–1985).

this first manipulation. The frequency of natural cycles was unknown, so a 7-yr cycle was adopted as a first approximation. High water levels that lasted too long might eliminate the seeds, whereas low levels that lasted too long might have threatened the fishery. The amplitude of past fluctuations could be judged from historical observations, from the presence of drowned stumps, and from the distribution of buried seeds (Keddy and Reznicek, 1982). The foundation of the dam set a practical lower water limit somewhat higher than desirable, but the restrictions, if any, placed on the germination of rare plants will only be known after the study is completed. The greatest threat was that the low-water period would be too short. Unlike marshes where seed banks can be deep, the seed bank at Matchedash Lake was extremely shallow (Fig. 5). Therefore, there was a risk that germination during the low-water period would exhaust the seed bank and that water levels would rise before it was replenished. This reinforces the need for a thorough understanding of the characteristics of the resource. Lacking a thorough understanding in this case, management decisions had to be based upon available data from previous studies (Keddy and Reznicek, 1982; Nicholson and Keddy, 1983). The ongoing monitoring will undoubtedly permit refinement of the program.

Questions remain that need to be answered: What is the optimal frequency and amplitude of fluctuations? What is the life-span of buried seeds? What time of year should water levels be changed? How do the answers to the foregoing questions vary among species? Which are the most appropriate indicator species for the success of the program?

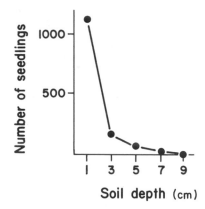

Figure 5. The depth of buried seeds in Matchedash Lake estimated by the number of seedlings emerging from different depths in soil cores (after Nicholson and Keddy, 1983).

The situation is complicated by the need to consider the sport fishery as well; the timing of drawdowns desirable for rare plants may be very different from those tolerated by fish.

E. Wilsons Lake, Nova Scotia: Case Study 3

The Atlantic coastal plain flora in Nova Scotia (Fernald, 1921, 1922; Roland and Smith, 1969) includes many species rare in Nova Scotia (Maher *et al.*, 1978) and endangered (*Coreopsis rosea* and *Hydrocotyle umbellata*) or threatened (*Sabatia kennedyana*, *Lophiola aurea*, and *Clethra alnifolia*) in Canada. The highest concentration of these species occurs along the Tusket River in Yarmouth County, Nova Scotia (Roland and Smith, 1969; Keddy, 1985b; Wisheu, 1987). A combination of infertile glacial till and dramatic water level fluctuations apparently prevents domination of shorelines by ericaceous shrubs and large grasses such as *Calamagrostis canadensis*. The flora is thus composed of species that tolerate low fertility and repeated natural disturbance. The major threat to these species is not a reduction in natural disturbance as at Matchedash Lake, but an increase in disturbance due to human activities.

To protect an intact example of the rare coastal plain flora, a key section of shoreline was purchased and has been designated as an ecological reserve. The major cottage and shoreline developments that are eliminating populations of these species along the Tusket River no longer threaten this particular site. However, other human activities, especially all-terrain vehicle (ATV) use, continue to damage coastal plain vegetation at this and other privately owned sites. Conservation of this unique flora depends on the ability to predict the impact of this type of disturbance on the long-term survival of the species and to make appropriate management recommendations. We conducted a research program with the goal of predicting the capacity of these species to recover from damage by ATVs based on characteristics of the seed bank.

The first step was to document the distribution and abundance of adult plants (Wisheu, 1987). The next question was whether there were buried seeds of the target species to allow reestablishment when the adults were killed. Densities of buried seed ranged from 1500 seeds m^{-2} in 30 cm of water, to 30,000 seeds m^{-2} in the seasonally flooded zone 75 cm above the August water level. At the elevation where many of the target species occurred (15–30 cm above the August water level), seed densities were approximately 10,000 seeds m^{-2}. Rare species were not well represented in the seed bank compared to common species (Fig. 6). This suggested that repeated disturbance by ATVs would lead to an increase in the common species and a decrease in the populations of rare and endangered ones.

Preliminary studies of seed depth profiles showed that the seeds are

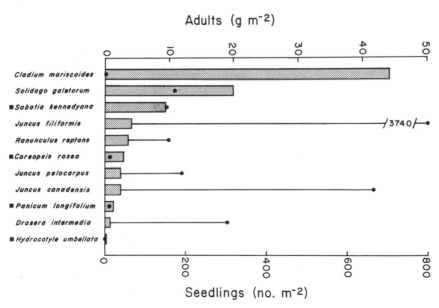

Figure 6.   Biomass and buried seeds of selected species at Wilsons Lake, Tusket River Valley, Nova Scotia. Adult weights (bars) and seedling counts (circles) are given for the five most abundant angiosperms, the five most abundant types of seedlings, and the four target species (squares).

restricted to the top few centimeters of substrate as in Matchedash Lake (Fig. 5). This could mean that repeated damage would lead to exhaustion of the seed bank. This has occurred at nearby Gillfillan Lake, where one of the largest populations of *Sabatia kennedyana* in Canada (Keddy, 1985b) had been subjected to several years of repeated damage from ATVs. The adult populations have dramatically declined and samples taken from this site showed that the total density of buried seeds was more than an order of magnitude lower than in similar vegetation types on Wilsons Lake (Gillfillan Lake, 550 seeds m^{-2}; Wilsons Lake, 10,000 seeds m^{-2}).

This study illustrates the importance of understanding seed bank characteristics in predicting response of threatened species to disturbance. However, other plant attributes need to be considered as well. Accurate predictions of response would also depend on the capacity for vegetative regeneration. Although several of the rare species at this site propagate from rhizomes, the low fertility of the site and the presence of insectivorous and evergreen species suggest that they are stress tolerators (*sensu* Grime, 1977) with a high investment in leaves and low rates of growth (see also Grime and Hunt, 1975; Boston and Adams, 1987). A low adult recovery rate coupled with the low density of buried seed would make these species particularly sensitive to damage from ATVs.

The foregoing examples deal with comparatively small areas. The problem becomes orders of magnitude more difficult when dealing with large heterogeneous areas. For example, there are a variety of engineering proposals that would significantly alter the natural water level fluctuations of the Great Lakes, yet there is minimal information on the potential impact of such projects upon shoreline floras. Keddy and Reznicek (1986) reviewed the existing data on seed banks and water level fluctuations in wetlands, and produced a qualitative model for existing vegetation–seed bank relationships in the Great Lakes. They concluded that stabilization of water levels would greatly reduce the areas of marsh, fen, and wet prairie on the shorelines of the Great Lakes. The potential scale of the damage from such engineering projects emphasizes the importance of understanding the role of seed banks in maintaining such vegetation types.

V. Conclusion

In the first part of this chapter we proposed some general management principles and ideal procedures for managing vegetation using seed banks. These include characterization of the system, management goals, and techniques that can be used to achieve these goals. Although application of science for conservation requires us to accept the messiness of reality and the inevitable compromises, the case histories show the feasibility of applying these basic principles to actual management problems. In each case, species responses to disturbance were predicted from seed bank and regeneration data, and specific disturbance regimes could be recommended for conservation of target species. Managers may face situations where it is desirable to increase natural disturbance or decrease unnatural disturbances. Seed banks may be the major source of regeneration of the desirable species or undesirable species. The timing of the disturbance and its duration may be critical to successful implementation of the management goal. Because management decisions are often made in the absence of adequate data on the characteristics of the system, it is essential to have ongoing monitoring to allow refinement.

For an ecosystem the size of the Great Lakes, the data are so sparse that it is necessary to extrapolate from freshwater tidal marshes, prairie marshes, and small lakes. However, such extrapolation may be a source of inspiration and encouragement: if generalizations on this scale are valid, it suggests that we are approaching a level of ecological understanding where vegetation management can be based upon general principles. Although the efficacy of these principles can only be fully validated through more site-specific studies, it is clear that the control of vegetation composition by manipulating regeneration from buried seeds provides a powerful tool for conservation management.

Acknowledgments

We gratefully acknowledge support from the Natural Sciences and Engineering Research Council of Canada and the World Wildlife Fund. We thank the Ontario Ministry of Natural Resources, Scott Wilson, and Mirek Sharp for access to the unpublished germination data from Matchedash Lake. The Nature Conservancy of Canada and Wildlife Habitat Canada also deserve mention for providing the funding to acquire the nationally significant site in Nova Scotia. We also thank Anita Payne for her assistance with manuscript preparation.

PART 5

Synthesis

CHAPTER 17

Pattern and Process in the Dynamics of Seed Banks

V. Thomas Parker

Department of Biology
San Francisco State University
San Francisco, California

Robert L. Simpson

School of Science
William Paterson College
Wayne, New Jersey

Mary Allessio Leck

Biology Department
Rider College
Lawrenceville, New Jersey

I. Introduction

The array of seed bank patterns that exist reveals the diversity, complexity, and interactions of environmental influences on plant life histories.

Seed banks reflect the filtering of plant life histories by establishment environments distributed along spatial and temporal scales followed by local selection. The dynamics of populations and communities fundamentally depend upon seed and seedling stages (Mayer, 1980–1981). Probabilities of mortality are greater at these stages (e.g., Harper, 1977; Werner, 1979; Angevine and Chabot, 1979; Cavers, 1983; Fenner, 1987; Louda, Chapter 3, this volume) and selection has resulted in numerous successful "designs" (Williams, 1966).

We will consider seed bank dynamics from the perspective of individual populations and communities. At both levels of organization, seed banks integrate past and current selective pressures on seed and seedling stages and, consequently, influence the life history characteristics of species and the dynamics of communities. Spatial and temporal changes in seed banks related to the timing or type of dispersal, survival, longevity and germination are often adaptive (Venable, Chapter 5, this volume). Several generalizations emerge about how these underlying processes influence population patterns. Other generalizations emerge at the community level where the coincidence as well as the diversity of response reflect environmental opportunities and constraints. The following discussion attempts to place these generalizations in perspective.

II. Population Seed Bank Dynamics

Seed bank dynamics of individual populations are primarily influenced by inputs and losses (Simpson *et al.*, Chapter 1, this volume). Factors controlling these vary on spatial and temporal scales, providing the impetus on which ecological and evolutionary processes can act. Some factors may influence seed banks more than others. Seed production, for example, can overwhelmingly influence the rate of seed input compared to other processes, such as long-range dispersal within a habitat. If local populations are small, on the other hand, or if environmental conditions have changed so that reproduction is limited, then processes like long-range dispersal can be of far greater significance. For certain events, such as initial colonization, long-range dispersal may always be most important (e.g., Leck, Chapter 13, this volume). Similarly, as some processes decline in importance or magnitude, others, such as loss of viability during long-term burial, may increase proportionately in importance.

The environment fluctuates constantly on both spatial and temporal scales. Variations in microenvironments affect germination (Keddy and Ellis, 1985; Galinato and van der Valk, 1986), with seeds of some species able to tolerate wide ranges of environmental conditions. Large-magnitude changes in environmental conditions, such as from precipitation fluctuations or disturbance, also occur and result in considerable seed bank response.

The shifting in importance of processes and conditions must be evaluated when considering the ecological status of populations. On ecological and evolutionary scales, the shifting importance of such factors has undoubtedly been integrated into the overall characteristics of species populations. In this section we first consider spatial heterogeneity of seed banks in the context of dispersal, predation, germination, and burial; we then evaluate temporal heterogeneity, especially as determined by persistence of seeds in soil; finally, we focus on the implications of these processes at the population level.

A. Spatial Heterogeneity

Even though spatial distribution patterns will be principally influenced by source individuals, with seed density usually declining exponentially with distance from the source (Howe, 1986), a variety of factors can influence dispersal patterns and result in seeds being nonrandomly distributed. For example, seeds of *Acacia* species distributed by ants are clumped near ant nests (Auld, 1986b). Other aspects of seed characteristics integrated with dispersal vectors will have a similar impact. Fruit quality results in nonrandom and clumped distribution patterns for species with similar fruit quality (e.g., Howe and Smallwood, 1982; Howe, 1986). The timing of dispersal, size of the seed crop, dispersal characteristics such as bouyancy or fruit quality, and variations in environmental conditions such as inundation patterns or gaps can all influence spatial distribution of seeds and, thus, the habitats available for seedling establishment (Rabinowitz, 1978; Leck, Chapter 13, this volume). Moreover, because dispersal influences the shape of the seed and seedling shadow curves, seed and seedling demography are linked to it (Howe, 1986).

Other processes can cause variation even with initial homogeneous postdispersal distribution of seeds. Predation of seeds and related activities of predators can be of great importance (e.g., in tropical forests, Garwood, Chapter 9; Mediterranean climate shrublands, Parker and Kelly, Chapter 11; deserts, Kemp, Chapter 12; grasslands, Rice, Chapter 10, this volume). Predation intensity varies considerably within habitats (Louda, Chapter 3, this volume); seed bank depletion can be sizable and differential.

Germination may often be the principal process by which seeds are lost from the seed bank. It is tightly coupled with spatial variations in microclimate that cue germination. Seeds of pioneer tree species in tropical forests, for example, respond sensitively to variation in light levels (Garwood, Chapter 9, this volume). Seed bank density of such pioneer species is low in recently disturbed areas where seeds have germinated, and increases with distance from the disturbance into the forest. *Erodium botrys*, a grassland species, can be induced to germinate by large diurnal temperature fluctuations, so that within grasslands germination occurs readily on bare ground but not where seeds are buried by gopher mounds (Rice, Chapter 10, this volume). A number of environmental factors can vary within a habitat and result in differential seed bank decay patterns (e.g., Baker, Chapter 2; Baskin and Baskin, Chapter 4, this volume).

Burial is one aspect of seed bank dynamics about which little is known (e.g., Garwood, Chapter 9; McGraw and Vavrek, Chapter 6; Leck, Chapter 13, this volume). For some habitats, such as those where fire is an important factor, distribution of seeds within the soil is critical for survival (e.g., Auld, 1986b; Parker and Kelly, Chapter 11, this volume). In such habitats, soil temperatures may result in high seed mortality near the surface, may cue seeds to germinate below that, and may be ineffectual for stimulation at still lower levels (Auld, 1986b). Several factors can influence burial rates, although their relative importance will vary among habitats. These include soil texture and structure, soil deposition and compression, seed size and other morphological characteristics, length of seed longevity, earthworm and other animal activity, and local environmental conditions that relate to seed longevity, such as cold or low pH. Longevity of seeds seems to be enhanced by conditions well below the soil surface (Karssen, 1980–1981).

Burial processes may be important for the long-term survival of some species, such as colonizers of past disturbances (Baker, Chapter 2, this volume) as well as fugitive species (Leck, Chapter 13, this volume). Little is known about the depth from which buried seeds can germinate, but the response is probably species specific and dependent on seed size (e.g., Garwood, Chapter 9, this volume). Interpretations of seed depth profiles are confounded by the interaction of seed survival and other factors, such as soil compression (McGraw, 1987).

Because habitats are spatially diverse, the spatial heterogeneity of a population's seed bank, both horizontally and vertically, can be critical for establishment. Processes such as dispersal, predation, germination, and burial determine the spatial patterns of seeds in the soil.

B. Temporal Heterogeneity

The interrelationships among environmental fluctuations, seed production, dispersal, habitat quality, and germination are responsible for most temporal patterns. Of these processes, germination has a clearly predictable temporal dimension that, in most cases, is directly related to successful transition from seed to establishment in the vegetation.

Some species lack any form of seed dormancy. Seeds that do not germinate at the first opportunity have little chance of survival; later germinants are at greater risk from competitors (Cook, 1979; Werner, 1979; Angevine and Chabot, 1979). For some that lack seed dormancy, dispersal must be timed to favorable germination periods, the case with many riparian species (e.g., Canham and Marks, 1985). For others, dispersal may occur when environmental conditions inhibit germination, and germination occurs when favorable conditions return (Garwood, Chapter 9, this volume). The ability to germinate at any time would seem to be advantageous, but it appears to be common only under certain circumstances, such as in habitats having favorable growing conditions throughout the year (Garwood, Chapter 9, this volume). Even in species lacking dormancy, germination may have a temporal component that results from temporal differences in seed production and dispersal (Terborgh, 1986).

Most species have some type of primary dormancy at the time of dispersal (Karssen, 1980–1981, Evenari, 1980–1981; Baskin and Baskin, Chapter 4, this volume). Although delays in germination can be risky, dormancy and the processes involved in breaking dormancy are usually adaptive (Cohen, 1966, 1967; Marks, 1974; Angevine and Chabot, 1979). Seasonality, for example, imposes a strong influence on the availability of favorable germination sites. Germination timing typically depends on seasonal variation in environmental conditions that are relatively predictable, such as winter in temperate forests, or seasonal droughts in tropical forests, Mediterranean climate grasslands, or deserts. Species may cue germination by either intrinsic or extrinsic mechanisms to periods favorable for seedling establishment. The ability of populations to respond to cues, which become in effect habitat filters, determines at the point of seed germination which species will compose the vegetation.

Between-year variation in habitat conditions exerts a strong selection pressure which leads to long-term persistence of a fraction of each year's seed production (Cohen, 1966, 1967). Habitat unpredictability such as unusual droughts in deserts (Kemp, Chapter 12, this volume), drawdowns in wetlands (Leck, Chapter 13, this volume), and fire in conifer forests or chaparral (Archibold, Chapter 7; Parker and Kelly, Chapter 11, this volume) are examples. On a smaller scale, the quality of the habitat may fluctuate relative to various factors, such as gopher activity in grasslands (Rice, Chapter 10, this volume), gap formation in deciduous forest (Collins *et al.*, 1985; Baskin and Baskin, Chapter 4; Pickett and McDonnell, Chapter 8, this volume), or soil disturbance in tundra (McGraw and Vavrek, Chapter 6, this volume). These impacts favor persistence.

Species may disperse to temporarily unfavorable locations and remain dormant in the soil until the appropriate cues occur. Arable weed seeds, for example, can be transported to the surface of the soil or buried by agricultural practices. These species may have complex secondary dormancy and are excellent examples of adaptive responses to unpredictable and potentially catastrophic temporal variation (Cavers and Benoit, Chapter 14, this volume).

C. Implications for Populations

The ability of plants to produce, through intrinsic or extrinsic mechanisms, a persistent seed population in effect becomes an additional habitat filter controlling species composition of the vegetation. Species with transient seed banks may compensate with alternative mechanisms, such as by long-lived adults or persistent bulbs or rhizomes (Harper, 1977; Baker, Chapter 2; Pickett and McDonnell, Chapter 8; Parker and Kelly, Chapter 11, this volume).

Persistent seed banks range greatly in size even in a given vegetation type. In part the rate of accumulation depends on the fraction of each year's seed production that remains dormant. Factors that influence longevity have been little explored (e.g., Garwood, Chapter 9; Leck, Chapter 13; McGraw and Vavrek, Chapter 6; Rice, Chapter 10, this volume). We can speculate, however, that the dormant fraction will depend on the probability of seedling establishment. As the probability of seedling establishment declines, the dormant fraction would increase. Seed longevity is also important; the greater it is, the more rapidly a dormant seed bank will accumulate. Predation directly reduces the probability of seedling establishment and selects for increases in the fraction remaining dormant. Predation thus acts in a manner opposite to seed longevity in influencing the fraction remaining dormant. Because the size of the surviving seed bank may be critical for many species, natural selection would act to maintain a particular seed bank density and therefore integrate the influences of seed production, dispersal, predation, seed longevity, and other factors illustrated in Fig. 1 of Chapter 1 (Simpson *et al.*; see also Venable, Chapter 5, this volume). Cohen (1966, 1967) has developed models that pursue these ideas in greater depth.

These processes, in turn, are affected by environmental factors. Temporally, seed bank patterns are determined, in large part, by seed longevity. Selection pressures determine the transient/persistent nature of seed banks, with the constancy of the environment and alternate survival strategies helping shape the seed bank response. For any particular species the importance of dispersal, predation, and other processes can shift along spatial and temporal scales in response to environmental variation.

Although the relationship of plant vigor to seed age is well documented for crops (Barton, 1961), how reproductive and competitive abilities of different aged seeds influence both seed bank and vegetation dynamics of natural populations require study. Moreover, little is known about physical and biotic characteristics of the soil that influence the seed bank (Mayer and Poljakoff-Mayber, 1975). Mechanisms involving the loss of dormancy notwithstanding, physiological and biochemical changes that occur while seeds are buried are not well understood. Except for a few species, it is unclear how dispersal, predation, dormancy, and other processes interact to determine seed bank size and distribution. Several authors (e.g. Garwood, Chapter 9; Kemp, Chapter 12; McGraw and Vavrek, Chapter 6; and Rice, Chapter 10, this volume) continue to emphasize the need for population studies to determine the relative importance of seed bank processes, and how differences in vegetation composition and disturbance regimes affect the relative importance of a species seed bank. Perhaps study of recalcitrant seeds, which are common in the tropics (Garwood, Chapter 9, this volume) and some wetlands (Leck, Chapter 13, this volume), coupled with population manipulations may provide insights. Furthermore, study of seed banks in contrasting environments would help interpret their role in vegetation dynamics and determine survival value of germination-regulating mechanisms, which, as noted by Mayer and Poljakoff-Mayber (1975), are difficult to establish.

III. Community Seed Bank Dynamics

The same processes operating at the population level are important in determining community dynamics. In addition, at this level the diversity of successful regeneration syndromes becomes clear. Opportunities for regeneration are not limitless; the patterns observed reflect physical and biotic constraints. Community comparisons provide insights regarding the importance of persistent seed banks and their ecologically equivalent alternatives.

A. Regeneration Patterns

A variety of patterns is observed in seed bank dynamics among species within a community and among communities. These patterns are based principally on the timing of germination and are indicative of regeneration opportunities. Two different ecological patterns occur. One pattern involves the age of germinating seeds. Seeds germinating within a year of dispersal are transient with respect to the seed bank, whereas those germinating later emerge from the persistent seed bank. The second pattern reflects external physical and biotic environmental variation influencing germination.

Within a year, germination timing in a community is often controlled by several factors (e.g., light, temperature, and/or rainfall) which restrict germination to one (e.g., Parker and Leck, 1985) or two periods (e.g., Thompson and Grime, 1979). If strong spatial patterns of establishment environments exist within a community, then the vegetation composition will reflect those patterns (e.g., Parker and Leck, 1985). Within a vegetation dominated by perennials, there may be only a small number of seeds that germinate and establish. When annuals are a large or dominant proportion as in some deserts (Kemp, Chapter 12, this volume), annual grasslands (Rice, Chapter 10, this volume), or certain wetlands (Leck, Chapter 13, this volume), the transient component will disproportionately determine the yearly fluctuation in seed bank composition and dominance.

On a longer time scale, patterns are also discernible. Within-year patterns discussed above generally reflect local environmental constraints on all species, but among-year patterns reflect life history differences in the frequency of species-specific safe sites. Persistent seed banks are usually found in populations whose environment lacks safe sites for a sufficiently long period that the population would otherwise be at risk. Reasons for the lack of safe sites vary greatly among species and communities, but are often related to climate. The climatic fluctuations of North American deserts and the complex germination patterns of annuals cued to them (Beatley, 1974) are sufficient to select for persistent seed banks (Cohen, 1966, 1967). Persistence may also be selected by biotic restrictions on safe sites. Often disturbances at some scale, involving single plants or larger areas, are required for successful establishment and reproduction (Grubb, 1977b). Persistent seed banks, cued to disturbance, allow a population to disperse temporally to sites where resources are temporarily available.

Any stage of the life cycle may act as an ecologically equivalent alternative to a persistent soil seed bank (Fig. 1). These may be serotinous aboveground seed banks, seedling banks, sapling banks, bud banks (Harper, 1977) or long-lived adults. All represent a persistent stage that can survive some environmental extremes and respond to disturbance. The initial floristic composition concept (Egler, 1954) in which most species are present before or soon after a disturbance indicates the importance of a persistent stage at or near a site at the time of disturbance. Initial floristic composition generally reflects preexisting regeneration potential found in seed or sapling banks, vegetative reproduction, or long-range dispersal (e.g., Bormann and Likens, 1979; Aplet *et al.*, 1988). Recent work has reemphasized the importance of disturbance to vegetation dynamics (Pickett and White, 1985a). Seed banks

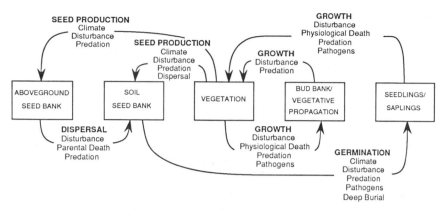

Figure 1.   Generalized model of vegetation dynamics. Processes and factors affecting rates of change are indicated between pairs of ecologically equivalent persistent stages.

and their ecologically functional alternatives are the key to regeneration following disturbance, the "regeneration niche" (Grubb, 1977b). Each regenerative alternative differentially responds to a variety of influences (Fig. 1). A brief comparison follows.

Soil seed banks may persist during many habitat disturbances and environmental fluctuations. Through the seed bank, successional species, which dominate for only a short time after severe disturbances, can be a part of future vegetation dynamics at a site. Soil seeds banks can cue to a variety of disturbances, and thus in a temporal sense can be at the right place at the right time. Larger disturbances, however, actually may be required to break dormancy and so limit the conditions in which seedlings establish. Seed banks must survive long-term exposure to possible predation and must maintain viability; responses can vary with age or isolation of sites.

Aboveground seed banks represent a specialized type of bank related to fire (Archibold, Chapter 7; Parker and Kelly, Chapter 11, this volume). Serotinous cones or fruits provide a method of releasing germinable seed following a fire. Recent studies have shown the degree of serotiny varies with frequency of fire (Wellington and Noble, 1985; Zammit and Westoby, 1987a). The relative disadvantage of aboveground seed banks depends on sensitivity to other influences, including predation and variation in fire frequency.

Seedling and sapling banks can be transient or persistent, with sapling banks being persistent seedling banks of woody species. If seeds suffer high losses from predation, the case for many large-seeded species, then having persistent seedlings rather than a seed bank may be a preferred alternative. Seedling and sapling banks must tolerate the understory environment, principally limited by shade, water, and nutrients. Physiological adaptation to these stresses may limit speed of response (Grime, 1977; Bormann and Likens, 1979). A significant advantage is responsiveness to small changes in the environment. A small opening in a canopy may be sufficient for sapling response, but may not be detected by dormant seeds in the soil. A disadvantage is that seedlings and saplings may not survive certain disturbances, e.g., fire and browsing.

Vegetative propagation is common among some life-forms (e.g., herbaceous perennials) and in certain habitats (e.g., wetlands or chaparral). This is an effective means of surviving disturbances or environmental fluctuations. Spread of a population can be rapid, but there may be limits to colonizing potential and genetic diversity.

Long-range dispersal, the spatial analog to persistent seed banks, is temporally limited and requires sufficient dispersal distance and numbers of seeds to invade disturbances. Dispersal must be appropriately timed. When disturbances are large relative to dispersal ranges, then dispersal may be ineffective in covering a site.

These regeneration alternatives vary in tolerance to environmental stresses and sensitivity to small environmental changes. Under different conditions one alternative may be favored over another, reflecting its relative advantage both spatially and temporally.

B. Patterns at the Community Level

In environments with a continuous vegetative cover, germination from the seed bank seems to follow similar patterns for both dominant and subordinate life-forms. One pattern is a transient seed bank for species whose seedlings are generally tolerant of the conditions imposed on the habitat by the dominant vegetation. Dominant species may require an opening later for entrance to the canopy while subordinate life-forms generally lack that requirement. For species intolerant of environmental conditions in undisturbed vegetation, spatial or temporal dispersal to a disturbance is required. Species that rely on spatial dispersal have seedlings that last several weeks to several years, but growth is rapid if they land in suitably disturbed areas. Persistent seed banks contain species that in some examples may survive several hundred years, as seeds await environmental cues associated with disturbance.

The importance of persistent seed banks within a community will generally reflect possible seed sources and evolutionary constraints, as well as the relative advantages of other regeneration alternatives. Between-year variation in environmental fluctuations, as might be experienced in the desert, favors persistent seed banks for annuals, but gen-

erally not for perennials (Kemp, Chapter 12, this volume). Climatic fluctuations result in large changes in rates of germination from persistent
seed banks of annual species as well as subsequent additions to the seed
bank (Fig. 2). The diversity of species with similar life forms, cues, and
responses becomes apparent at the community level as expressed by
yearly variation.

That disturbances are important for regeneration is well recognized
(e.g., Watt, 1947; Grubb, 1977a,b; Oliver, 1981; Pickett and White,
1985a). Within closed vegetation, the relative importance of persistent
seed banks usually depends on the type and timing of disturbances.
Small disturbances in forests tend to favor seedlings already present in
the opening. After a large disturbance, most forest regeneration occurs
quickly, often from a variety of sources (Oliver, 1981; Aplet et al., 1988).
The size and type of disturbance fundamentally determines the species
response (Grubb, 1977b; Whitmore, 1978; Denslow, 1980). The importance and magnitude of regenerative alternatives differ greatly before
and after a particular disturbance. In the case of a fire-prone vegetation,
such as chaparral, transient seed banks and their persistent seedlings
may be the primary means of regeneration before a fire, while persistent
soil seed banks, aboveground serotinous seed banks, and vegetative
regeneration predominate after a fire (Fig. 3; Parker and Kelly, Chapter
11, this volume).

The predominance of disturbance-oriented syndromes indicates the
importance of release of resources for regeneration. If disturbance of the

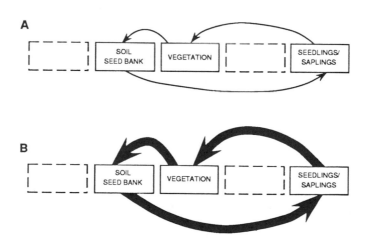

Figure 2.   Effect of 2 different climatic years on the dynamics of desert
annuals. The two years illustrated are a dry or low precipitation year (A)
and a relatively high precipitation year (B). Thickness of the arrow lines
reflects relative rates of change among stages.

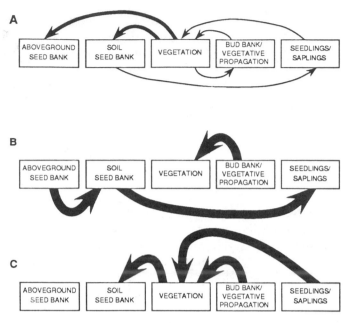

Figure 3. Effect of a fire on chaparral dynamics. The prefire condition (A) shows net accumulations into aboveground and soil seed banks with some small changes in other stages. After the fire (B), large numbers of seedlings are produced from aboveground and soil seed banks, and vegetative regeneration from bulbs, woody burls, or other structures occurs. Later in the first postfire year (C), vegetative regeneration continues, while annuals and many perennials flower and add to the soil seed bank. Thickness of arrow lines reflects relative rates of change among stages.

vegetative cover is sufficiently infrequent, then species specialize by surviving and reproducing in areas such as vegetation gaps. The eastern deciduous forest of North America contains many species specialized to disturbances. Most tend to be short-lived with respect to disturbance frequency (Bormann and Likens, 1979) and are considered successional. When disturbance is frequent, successional species may not die between disturbances, and then they are considered dominant species (e.g., in chaparral). Redwoods (*Sequoia sempervirens*), although long-lived, require some type of disturbance for seedling establishment, but are dominants because their life span is longer than the interval between disturbances. In some cases, whether dominant species have persistent seed banks is cued to catastrophic disturbance (e.g., fire; Archibold, Chapter 7; Parker and Kelly, Chapter 11, this volume) and its frequency relative to the juvenile period or vegetative life-span. Alternatives, such as seedling or sapling banks, may not survive the disturbance, and dispersal may be ineffectual if the disturbance is large (McMaster and Zedler, 1981).

Within a life-form, some distinct community variations in seed bank dynamics suggest significant differences between habitats. For example, in the case of shade-intolerant trees (pioneer or successional), persistent seed banks are predominant in the wet tropics whereas they are relatively less important in temperate deciduous forests (Pickett and McDonnell, Chapter 8; Garwood, Chapter 9, this volume). Almost all tropical pioneer species appear to have persistent seed banks (Garwood, Chapter 9, this volume). In contrast, many temperate deciduous forest species exhibit the ecological alternatives of yearly dispersal, and short-lived seedling or sapling banks, although some species may have persistent seed banks (e.g., *Prunus pensylvanica*). Disturbances in both forests can occur at any time during the year. In the wet tropics, where conditions for germination may be favorable throughout the year, only species with continuous dispersal would be the equal of persistent seed banks in "reaching" disturbances in a timely manner. Germination in temperate forest species is often limited to spring. Species with dispersal at any time prior to spring germination would have the same advantages as seed bank species in "reaching" the disturbance. A disturbance following the spring germination period would favor preexisting seedlings.

Seed banks can diverge within a biome due to changes in physiology or longevity associated with site quality or disturbance/stress gradients (Louda, Chapter 3, this volume). For example, in aseasonal tropics of Southeast Asia persistent seed banks are rare in contrast to the wet tropics of Australia or Central America (Garwood, Chapter 9, this volume). The aseasonal Asian tropics are found in latitudes lacking in strong winds or storms and disturbances are consequently rarer (Putz and Appanah, 1987).

IV. Seed Banks and Management

Management of vegetation has several objectives. Seed banks play a critical role in (1) maintenance, (2) restoration, (3) habitat creation, and (4) differential species management (Cavers and Benoit, Chapter 14; van der Valk and Pederson, Chapter 15; Keddy et al., Chapter 16, this volume). Vegetation management currently emphasizes maintenance of highly modified systems such as pastures or agricultural fields and maintenance of natural systems.

Research on weeds has yielded a wealth of knowledge on dormancy mechanisms, infestation rates, and treatments (Cavers and Benoit,

Chapter 14, this volume). It is necessary to determine the total size and species composition of the weed seed bank before any treatment is considered. In addition, for the major species of the seed bank, seed longevity and germination cues must be known. Only then can progress be made in practicing or devising effective crop management systems. The same base line information is necessary for maintenance, restoration, or habitat creation in natural systems (van der Valk and Pederson, Chapter 15; Keddy *et al.*, Chapter 16, this volume). Knowledge of whether seeds are transient or persistent, the nature of germination cues, and the environmental conditions suitable for establishment are fundamental to successful vegetation management (van der Valk and Pederson, Chapter 15; Keddy *et al.*, Chapter 16, this volume). With this information, management techniques have been developed to take advantage of seed bank processes. For example, in The Netherlands, seed banks are an important part of conservation management for maintaining high-diversity communities in chalk grasslands and wetlands (van der Valk and Verhoeven, 1988).

Single-species approaches are often problematic. For invasive species within natural systems, management treatments must assure maintenance of community integrity. Where the management goal is conservation of rare and endangered species, it is necessary to understand how they fit into the whole system. Rare species may rely on persistent seed banks; understanding their germination requirements allows manipulation to enhance recruitment (Keddy *et al.*, Chapter 16, this volume). Continual enhancement of rare populations, however, may degrade the community, causing their extinction.

The diversity of cues within a community requires a diversity of management treatments. Within fire-type vegetation, for example, fire per se may not be an appropriate approach; rather, fire frequency, fire intensity, season of burning, and other aspects of the overall fire regime of an area must be considered (Gill, 1975; Gill and Groves, 1981). Further, if soil seed banks are an important part of the vegetation, uniform application of prescribed fires may result in loss of species diversity (Parker and Kelly, Chapter 11, this volume).

Finally, too often it is assumed that natural systems lack weedy or invasive species. Yet just as there can be a large edge effect in agricultural fields (Cavers and Benoit, Chapter 14, this volume), there may also be significant edge effects in natural vegetation. Seed banks of such areas may contain a disproportionate representation of seeds from outside the community, as observed in tropical forest seed banks near agricultural fields (Garwood, Chapter 9, this volume). This underscores why understanding the seed bank is necessary prior to manipulation of disturbance regimes; unwanted species originating in the seed bank may become established.

V. Seed Bank Dynamics: A Perspective

Differential shifts in the relative importance of processes such as seed production, predation, dispersal, and dormancy can account for much of the differences observed in seed banks. Because these processes can significantly affect the dynamics of populations at any one site, fitness interactions among them lead to reciprocal selection and integration (Venable, Chapter 5, this volume). Each process demonstrates variability and thus divergence is possible. Variation commonly occurs in dispersal characteristics, but can also be seen in dormancy mechanisms, cycles of dormancy, and longevity (Baskin and Baskin, Chapter 4, this volume); different types and intensities of predation (Louda, Chapter 3, this volume); and fluctuations in seed production.

What seems most important is not the intensity of these processes, such as large losses to predation, but the fitness relationship in the environmental context of the population (Louda, Chapter 3, this volume). A number of environmental conditions influence seed bank dynamics via influences on production or other stages of the life cycle (Fig. 1). Within a vegetation type, variations in soil conditions, such as texture, pH, or other physical conditions, influence the development and size of seed banks (Cavers and Benoit, Chapter 14; Pickett and McDonnell, Chapter 8; Leck, Chapter 13, this volume). Likewise, competition and predation can strongly influence selection for persistence in the seed bank (Louda, Chapter 3, this volume; Venable, Chapter 5, this volume). Fitness relationships may also change in time, and selection for particular seed bank characteristics such as persistence may also change (Rice, Chapter 10, this volume). Thus considerable intraspecific variability in seed bank dynamics is possible.

As an important component of population dynamics, persistent seed banks are also the source of genetic material or evolutionary memory (Harper, 1977; Baker, Chapter 2; McGraw and Vavrek, Chapter 6, this volume). Genetic variability in the seed bank may be a source of long-term stability for a population (Epling *et al.*, 1960) or a reservoir on which selection can act differentially, i.e., year to year within populations (Baker, Chapter 2, this volume) or for differentiation among populations (Parker and Kelly, Chapter 11; Leck, Chapter 13, this volume).

Temporally, seed banks differentiate clearly into two fundamental types, transient and persistent. These can be further classified with respect to within-year or between-year patterns that are site or habitat specific. For example, the four seed bank types of Thompson and Grime (1979) reflect fall and spring germination periods and degree of persistence (see Grime, Foreword, this volume). After examining a number of communities, it becomes clear that classification of seed banks requires refinement. A diversity of alternative patterns is possible, such as the pseudopersistent seed banks found in tropical forests (Garwood, Chapter 9, this volume) and cold deserts (Kemp, Chapter 12, this volume), where continual dispersal guarantees presence of a soil seed bank.

A fundamental question in this consideration of seed banks is why persistence is selected. From the models of Cohen (1966, 1967) and a number of field studies, it appears that the delay in reproduction is adaptive and enhances long-term fitness. Most studies examining seed banks view persistence in the context of disturbance regimes, resulting in the generality that successional species rely on persistent seed banks while dominants of mature stands do not. It also seems clear, however, that disturbance selection for persistence within communities is but a subset of a more general pattern: persistence is selected when no other part of the life cycle can act as an alternative persistent stage. Within the context of disturbance, species relying on persistent seed banks are not able to maintain their populations and regenerate without the environmental changes brought on by disturbance, and thus are at risk. Dormancy, cued to disturbance, allows populations to persist. Examples of habitats lacking disturbance, but where persistence is selected because the seed is the only stage that can act as a persistent stage during certain environmental events, include prairie glacial marshes, lake shores, or monsoonal wetlands that experience water level fluctuations (Leck, Chapter 13, this volume). In these examples, temporally varying establishment environments exist due to fluctuations in climatic patterns; and accordingly, sets of species have adapted to different extremes of these fluctuations (e.g., mud flat versus submergent). In deserts, such fluctuations may not yield different establishment environments, but cause differences in favorability (Fig. 2; Kemp, Chapter 12, this volume). Thus it seems clear that seed banks are highly integrated into population and community dynamics. Characteristics of seed banks, particularly persistence, appear when species depend on seeds to maintain the population through a period of risk. Indeed, arable weed species demonstrate how rapidly selection can act on the seed bank to reduce risk (Cavers and Benoit, Chapter 14, this volume).

How do previous generalities concerning seed banks fare in view of current knowledge? Some generalities still continue to hold, but there are important exceptions. Annuals were considered more likely to have persistent seed banks than other life-forms because their yearly population turnover places them at risk in certain environments, and the only potentially persistent stage is the seed. However, not all annuals produce persistent seed banks, even in harsh environments (Kemp, Chapter 12, this volume). Moreover, in some communities perennial woody species may produce large seed banks (Parker and Kelly, Chapter 11, this volume). Another generality that often holds is that small seeds have greater longevity in the seed bank (Harper, 1977; Leck, Chapter 13, this volume). There are exceptions (Baker, Chapter 2, this volume), including large-seeded species that have aboveground seed banks.

Many other generalities must be questioned. Suggestions that tropical forests may lack persistent seed banks or that great longevity should be expected in wetlands are not true for many species and communities (Garwood, Chapter 9; Leck, Chapter 13, this volume). Suggestions that seed banks decline with elevation or latitude (Johnson, 1975; K. Thompson, 1978) or with increasing stress (K. Thompson, 1978) appear to apply only to specific circumstances. Arctic and alpine habitats, representing the extremes of latitude and altitude, have seed banks that are quite large and diverse (Miller and Cummins, 1987; McGraw and Vavrek, Chapter 6 this volume). Similarly, communities along gradients of other physical extremes such as aridity (Kemp, Chapter 12, this volume), fire (Archibold, Chapter 7; Parker and Kelly, Chapter 11, this volume), and hydrology (Leck, Chapter 13, this volume) may, on the contrary, contain quite large and diverse seed banks.

Many previous studies have indicated that great disparities between seed banks and the surface composition can exist. We find that the disparities reflect a number of circumstances rather than a fundamental dictum: distinctions were not made between persistent and transient seed banks, the timing of soil collection resulted in failure to include many species, the conditions for emergence of some species were not provided, or other persistence alternatives were present. It is important to focus on the dynamics of species in the context of the particular community. The adults of long-lived organisms are a *de facto* persistent stage; if these species also exhibit persistent seed banks, it is likely that the long-term population dynamics are at risk as a result of some predictable change in the environment. Although woody species appear not to rely on persistent seed banks, some circumstances, e.g., catastrophes such as fire or gap formation by hurricanes, do select for such seed banks in woody dominants or successional trees. In fact, the generality that community dominants do not rely on persistent seed banks does not extend to many communities, such as Mediterranean climate shrublands (Parker and Kelly, Chapter 11, this volume), arctic and alpine communities (McGraw and Vavrek, Chapter 6, this volume), and wetlands (Leck, Chapter 13, this volume).

Seed banks are important in all the vegetation types considered in this volume. It seems clear that differences in seed bank dynamics (e.g., transient versus persistent, timing, and turnover rates) are related to particular selection pressures. While selection factors may vary from community to community, responses often converge; whenever risk is high, persistent seed banks are favored. At one level climate is of overriding importance; beyond that, factors including predation, dispersal, seed longevity, and biotic interference dictate seed bank and alternative regeneration diversity within a community. Disturbance is the overarching force integrating these factors. On another scale, extreme environmental fluctuations due to rainfall, fire, or disturbance become major selection agents. Seed bank responses of species and communities may diverge over an environmental gradient.

While a general theory of seed banks is not yet at hand, a framework appears in place. Studies in the future that incorporate empirical data, experimental manipulation, and modeling offer the best approach to resolving the role seed banks play in vegetation dynamics.

Bibliography

Abbott, I., and Van Heurck, P. (1985). Comparison of insects and vertebrates as re-
movers of seed and fruit in a Western Australian forest. *Aust. J. Ecol.* **10**, 165–168.
Abramsky, Z. (1983). Experiments on seed predation by rodents and ants in the Israeli
desert. *Oecologia* **57**, 328–332.
Afanasiev, M. (1944). A study of dormancy and germination of seeds of *Cercis canaden-
sis. J. Agric. Res.* **69**, 405–420.
Ahlgren, C. E. (1960). Some effects of fire on reproduction and growth of vegetation in
northeastern Minnesota. *Ecology* **41**, 431–445.
Aikman, J. M. (1955). The ecology of balsa (*Ochroma lagopus* Swartz) in Ecuador. *Proc.
Iowa Acad. Sci.* **62**, 245–252.
Aiyar, T. V. V. (1932a). The Sholas of Palghat division. Part I. *Indian For.* **58**, 414–432.
Aiyar, T. V. V. (1932b). The Sholas of Palghat division. Part II. *Indian For.* **58**, 473–486.
Alexander, M. E., and Dubé, D. E. (1983). Fire management in wilderness areas, parks,
and other reserves. *SCOPE [Rep.]* **8**, 273–297.
Alexander, V., and Van Cleve, K. (1983). The Alaska pipeline: A success story. *Annu.
Rev. Ecol. Syst.* **14**, 443–463.
Alexandre, D. Y. (1978). Observations sur l'écologie de *Trema guineensis* en basse Côte
d'Ivoire. *Cah. ORSTOM, Ser. Biol.* **13**, 261–266.
Alexandre, D. Y. (1980). Caractère saisonnier de la fructification dans un forêt
hygrophile de Côte-d'Ivoire. *Rev. Ecol. (Terre Vie)* **34**, 335–359.
Alexandre, D. Y. (1982). Aspects de la régénération naturelle en forêt dense de Côte-
d'Ivoire. *Candollea* **37**, 579–588.
Amelin, I. S. (1947). [Securing seeds for revegetation of the deserts of central Asia.]
Dokl. Vses. Akad. S-kh. Nauk im V. I. Lenina, **12**, 28–31.
American Society of Agricultural Engineers (1981). Forest Regeneration, ASAE Pub. 10–
81. Am. Soc. Agric. Eng., St. Joseph, Missouri.
Aminuddin, bin M., and Ng, F. S. P. (1982). Influence of light on germination of *Pinus
caribaea, Gmelina arborea, Sapium baccatum,* and *Vitex pinnata. Malay. For.* **45**, 62–68.
Ammirati, J. F. (1967). The occurrence of annual and perennial plants on chaparral
burns. M.A. Thesis, San Francisco State University, San Francisco, California.
Anaya Lang, A. L., and Rovalo, M. (1976). Alelopatía en plantas superiores: Diferencias
entre el efecto de la presión osmótica y los alelopáticos sobre la germinación y cre-
cimiento de algunas especies de la vegetación secundaria de una zona cálido-
húmeda de México. *In* "Investigaciones sobre la Regeneración de Selvas Altas en
Veracruz, México" (A. Gómez-Pompa, C. Vázquez-Yanes, S. del Amo Rodríguez,
and A. Butanda Cervana, eds.), pp. 388–426. Compañia Editorial Continental,
México, D. F.
Andersen, A. N. (1987). Effects of seed predation by ants on seedling densities at a
woodland site in SE Australia. *Oikos* **48**, 171–174.
Andersen, A. N. (1988). Insect seed predators may cause far greater losses than they
appear to. *Oikos* **52**, 337–340.
Andersen, A. N., and Ashton, D. H. (1985). Rates of seed removal by ants at heath
and woodland sites in southeastern Australia. *Aust. J. Ecol.* **10**, 381–390.

Andrew, M. H. (1986). Granivory of the annual grass *Sorghum intrans* by the harvester ant *Meranoplus* sp. in tropical Australia. *Biotropica* **18**, 344–349.

Angevine, M. W., and Chabot, B. F. (1979). Seed germination syndromes in higher plants. *In* "Topics in Plant Population Biology" (O. T. Solbrig, S. Jain, G. B. Johnson, and P. H. Raven, eds.), pp. 188–206. Columbia Univ. Press, New York.

Aplet, G. H., Laven, R. D., and Smith, F. W. (1988). Patterns of community dynamics in Colorado Englemann spruce-alpine fir subalpine forests. *Ecology* **69**, 312–319.

Archibold, O. W. (1979). Buried viable propagules as a factor in postfire regeneration in northern Saskatchewan. *Can. J. Bot.* **57**, 54–58.

Archibold, O. W. (1980). Seed input into a postfire forest site in northern Saskatchewan. *Can. J. For. Res.* **10**, 129–134.

Archibold, O. W. (1981). Buried viable propagules in native prairie and adjacent agricultural sites in central Saskatchewan. *Can. J. Bot.* **59**, 701–706.

Archibold, O. W. (1984). A comparison of seed reserves in arctic, subarctic, and alpine soils. *Can. Field Nat.* **98**, 337–344.

Archibold, O. W., and Hume, L. (1983). A preliminary survey of seed input into fallow fields in Saskatchewan. *Can. J. Bot.* **61**, 1216–1221.

Arnold, R. M. (1982). Pollination, predation and seed set in *Linaria vulgaris* (Scrophulariaceae). *Am. Midl. Nat.* **107**, 360–369.

Arthur, A. E., Gale, J. S., and Lawrence, K. J. (1973). Variation in wild populations of *Papaver dubium*. VII. Germination time. *Heredity* **30**, 189–197.

Ashton, D. H. (1979). Seed harvesting by ants in forests of *Eucalyptus regrians* F. Muell. in central Victoria. *Aust. J. Ecol.* **4**, 265–277.

Ashton, P. S. (1978). Crown characteristics of tropical trees. *In* "Tropical Trees as Living Systems" (P. B. Tomlinson and M. H. Zimmermann, eds.), pp. 571–615. Cambridge Univ. Press, London and New York.

Aubréville, A. (1947). Les brousses secondaires en Afrique équatoria. l. *Bois For. Trop.* **2**, 24–49.

Auchmoody, L. R. (1979). Nitrogen fertilization stimulates germination of dormant pin cherry seed (*Prunus pensylvanica*). *Can. J. For. Res.* **9**, 514–516.

Augspurger, C. K. (1979). Irregular rain cues and the germination and seedling survival of a Panamanian shrub (*Hybanthus prunifolius*). *Oecologia* **44**, 53–59.

Augspurger, C. K. (1981). Reproductive synchrony of a tropical shrub: Experimental studies on effects of pollinators and seed predators on *Hybanthus prunifolius* (Violaceae). *Ecology* **62**, 775–788.

Augspurger, C. K. (1983). Seed dispersal of the tropical tree, *Platypodium elegans*, and the escape of its seedlings from fungal pathogens. *J. Ecol.* **71**, 759–771.

Augspurger, C. K. (1984a). Pathogen mortality of tropical tree seedlings: Experimental studies of the effects of dispersal distance, seedling density, and light conditions. *Oecologia* **61**, 211–217.

Augspurger, C. K. (1984b). Seedling survival among tropical tree species: Interactions of dispersal distance, light-gaps, and pathogens. *Ecology* **65**, 1705–1712.

Augspurger, C. K., and Franson, S. E. (1987). Wind dispersal of artificial fruits varying in mass, area, and morphology. *Ecology* **68**, 27–42.

Augspurger, C. K., and Franson, S. E. (1988). Input of wind-dispersed seeds into light-gaps and forest sites in a neotropical forest. *J. Trop. Ecol.* **4**, 239–252.

Augspurger, C. K., and Hogan, K. P. (1983). Wind dispersal of fruits with variable seed number in a tropical tree (*Lonchocarpus pentaphyllus*: Leguminosae). *Am. J. Bot.* **70**, 1031–1037.

Auld, T. D. (1983). Seed predation in native legumes of south-eastern Australia. *Aust. J. Ecol.* **8**, 367–376.

Auld, T. D. (1986a). Variation in predispersal seed predation in several Australian *Acacia* species. *Oikos* **47**, 319–326.

Auld, T. D. (1986b). Population dynamics of the shrub *Acacia suaveolens* (Sm.) Willd.: Dispersal and the dynamics of the soil seed-bank. *Aust. J. Ecol.* **11**, 235–254.

Auld, T. D. (1986c). Population dynamics of the shrub *Acacia suaveolens* (Sm.) Willd.: Fire and the transition to seedlings. *Aust. J. Ecol.* **11**, 373–385.

Auld, T. D., and Myerscough, P. J. (1986). Population dynamics of the shrub *Acacia suaveolens* (Sm.) Willd.: Seed production and predispersal seed predation. *Aust. J. Ecol.* **11**, 219–234.

Axelrod, D. I. (1973). History of the mediterranean ecosystem in California. *In* "Mediterranean-Type Ecosystems: Origin and Structure" (F. di Castri and H. A. Mooney, eds.), pp. 225–277. Springer-Verlag, Berlin.

Axelrod, D. I. (1977). Outline history of California vegetation. *In* "Terrestrial Vegetation of California" (M. G. Barbour and J. Major, eds.), pp. 140–187. Wiley, New York.

Axelrod, D. I. (1979). Desert vegetation: Its age and origin. *In* "Arid Land Plant Resources" (J. R. Goodin and D. K. Northington, eds.), pp. 1–72. Texas Tech University, Lubbock.

Axelrod, D. I. (1983). Paleobotanical history of the western deserts. *In* "Origin and Evolution of Deserts" (S. G. Wells and D. R. Haragan, eds.), pp. 113–129. Univ. of New Mexico Press, Albuquerque.

Baker, H. G. (1972). Seed weight in relation to environmental conditions in California. *Ecology* **53**, 997–1010.

Baker, H. G. (1974). The evolution of weeds. *Annu. Rev. Ecol. Syst.* **5**, 1–24.

Ballard, L. A. T. (1973). Physical barriers to germination. *Seed Sci. Technol.* **1**, 285–303.

Ballardie, R. T., and Whelan, R. J. (1986). Masting, seed dispersal and seed predation in the cycad *Macrozamia communis*. *Oecologia* **70**, 100–105.

Barbour, M. G. (1969). Age and space distribution of the desert shrub *Larrea divaricata*. *Ecology* **50**, 679–685.

Barbour, M. G. (1981). Plant-plant interactions. *In* "Arid-Land Ecosystems: Structure, Functioning and Management" (D. W. Goodall and R. A. Perry, eds.), Vol. 2, pp. 33–49. Cambridge Univ. Press, London and New York.

Barbour, M. G., and Lange, R. T. (1966). Seed populations in some natural Australian topsoils. *Ecology* **48**, 153–155.

Bard, G. E. (1952). Secondary succession on the Piedmont of New Jersey. *Ecol. Monogr.* **22**, 195–215.

Barnes, P. W., Harrison, A. T., and Heinisch, S. P. (1984). Vegetation patterns in relation to topography and edaphic variation in Nebraska sandhills prairie. *Prairie Nat.* **16**, 145–158.

Barralis, G., and Chadoeuf, R. (1980). Etude de la dynamique d'une communauté adventice. I. Evolution de la flore adventice au cours du cycle végétatif d'une culture. *Weed Res.* **20**, 231–237.

Bartholomew, B., Eaton, L. C., and Raven, P. H. (1973). *Clarkia rubicunda*: A model of plant evolution in semi-arid regions. *Evolution (Lawrence, Kans.)* **27**, 505–517.

Bartolome, J. W. (1979). Germination and seedling establishment in California annual grassland. *J. Ecol.* **67**, 273–281.

Barton, L. V. (1934). Dormancy in *Tilia* seeds. *Contrib. Boyce Thompson Inst.* **6**, 69–89.

Barton, L. V. (1944). Some seeds showing special dormancy. *Contrib. Boyce Thompson Inst.* **13**, 259–271.

Barton, L. V. (1961). "Seed Preservation and Longevity." Wiley (Interscience), New York.

Baskin, C. C., and Baskin, J. M. (1988). Studies on the germination ecophysiology of herbaceous plant species in a temperate region. *Am. J. Bot.* **75**, 286–305.

Baskin, J. M., and Baskin, C. C. (1973). Delayed germination in seeds of *Phacelia dubia* var. *dubia*. *Can. J. Bot.* **51**, 2481–2486.

Baskin, J. M., and Baskin, C. C. (1974). Some eco-physiological aspects of seed dormancy in *Geranium carolinianum* L. from central Tennessee. *Oecologia* **16**, 209–219.

Baskin, J. M., and Baskin, C. C. (1977). Predation of *Cassia marilandica* seeds by *Sennius
abbreviatus* (Coleoptera: Bruchidae). *Bull. Torrey Bot. Club* **104**, 61–64.

Baskin, J. M., and Baskin, C. C. (1980). Ecophysiology of secondary dormancy in seeds
of *Ambrosia artemisiifolia*. *Ecology* **61**, 475–480.

Baskin, J. M., and Baskin, C. C. (1981a). Seasonal changes in the germination responses of buried *Lamium amplexicaule* seeds. *Weed Res.* **21**, 299–306.

Baskin, J. M., and Baskin, C. C. (1981b). Seasonal changes in germination responses of
buried seeds of *Verbascum thapsus* and *V. blattaria* and ecological implications. *Can.
J. Bot.* **59**, 1769–1775.

Baskin, J. M., and Baskin, C. C. (1983a). Seasonal changes in the germination responses of buried seeds of *Arabidopsis thaliana* and ecological interpretation. *Bot.
Gaz. (Chicago)* **144**, 540–543.

Baskin, J. M., and Baskin, C. C. (1983b). Germination ecophysiology of eastern deciduous forest herbs: *Hydrophyllum macrophyllum*. *Am. Midl. Nat.* **109**, 63–71.

Baskin, J. M., and Baskin, C. C. (1984a). Role of temperature in regulating timing of
germination in soil seed reserves of *Lamium purpureum* L. *Weed Res.* **24**, 341–349.

Baskin, J. M., and Baskin, C. C. (1984b). Environmental conditions required for germination of Prickly sida (*Sida spinosa*). *Weed Sci.* **32**, 786–791.

Baskin, J. M., and Baskin, C. C. (1984c). Germination ecophysiology of the woodland
herb *Osmorhiza longistylis* (Umbelliferae). *Am. J. Bot.* **71**, 687–692.

Baskin, J. M., and Baskin, C. C. (1985a). The annual dormancy cycle in buried weed
seeds: A continuum. *BioScience* **35**, 492–498.

Baskin, J. M., and Baskin, C. C. (1985b). Does seed dormancy play a role in the germination ecology of *Rumex crispus*? *Weed Sci.* **33**, 340–343.

Baskin, J. M., and Baskin, C. C. (1985c). Germination ecophysiology of *Hydrophyllum
appendiculatum*, a mesic forest biennial. *Am. J. Bot.* **72**, 185–190.

Baskin, J. M., and Baskin, C. C. (1986a). Temperature requirements for after-ripening in
seeds of nine winter annuals. *Weed Res.* **26**, 375–380.

Baskin, J. M., and Baskin, C. C. (1986b). Germination ecophysiology of the mesic deciduous forest herb *Isopyrum biternatum*. *Bot. Gaz. (Chicago)* **147**, 152–155.

Baskin, J. M., and Baskin, C. C. (1987). Temperature requirements for afterripening in
buried seeds of four summer annual weeds. *Weed Res.* **27**, 385–389.

Baskin, J. M., Baskin, C. C., and McCormick, J. F. (1987). Seasonal changes in germination responses of buried seeds of *Portulaca smallii*. *Bull. Torrey Bot. Club* **114**, 169–
172.

Bate-Smith, E. C., and Ribereau-Gayon, P. (1958). Leucoanthocyanins in seeds. *Qual.
Plant. Mater. Veg.* **5**, 189–198.

Bauer, H. J. (1973). Ten year's studies of biocenological succession in the excavated
mines of the Cologne lignite district. *In* "Ecology and Reclamation of Devastated
Land" (R. J. Hutnik and G. Davis, eds.), Vol. 1, pp. 271–283. Gordon & Breach,
New York.

Baur, G. N. (1968). "The Ecological Basis of Rainforest Management" Forestry Commis-
sion of New South Wales, Sydney, Australia.

Bawa, K. S. (1974). Breeding systems of tree species of a lowland tropical community.
Evolution (Lawrence, Kans.) **28**, 85–92.

Bazzaz, F. A. (1983). Characteristics of populations in relation to disturbances. *In* "Disturbance and Ecosystems" (H. A. Mooney and M. Godron, eds.), pp. 259–273.
Springer-Verlag, Berlin.

Bazzaz, F. A. (1984). Dynamics of wet tropical forests and their species strategies. *In*
"Physiological Ecology of Plants in the Wet Tropics" (E. Medina, H. A. Mooney,
and C. Vázquez-Yanes, eds.), pp. 233–243. Junk, The Hague.

Bazzaz, F. A., and Pickett, S. T. A. (1980). Physiological ecology of tropical succession: A comparative review. *Annu. Rev. Ecol. Syst.* **11**, 287–310.

Beadle, N. C. W. (1940). Soil temperatures during forest fires and their effect on the survival of vegetation. *J. Ecol.* **28**, 180–192.

Beal, W. J. (1905). The vitality of seeds. *Bot. Gaz. (Chicago)* **40**, 140–143.

Beale, P. E. (1974). Regeneration of *Trifolium subterraneum* cv. Yarloop from seed reserves in soil on Kangaroo Island. *J. Aust. Inst. Agric. Sci.* **40**, 78–80.

Beatley, J. C. (1969). Biomass of desert winter annual plant populations in southern Nevada. *Oikos* **20**, 261–273.

Beatley, J. C. (1974). Phenological events and their environmental triggers in Mojave desert ecosystems. *Ecology* **55**, 856–863.

Beatley, J. C. (1980). Fluctuations and stability in climax shrub and woodland vegetation of the Mojave, Great Basin, and transition deserts of southern Nevada. *Isr. J. Bot.* **28**, 149–168.

Beattie, A. J. (1985). "The Evolutionary Ecology of Ant-Plant Mutualisms." Cambridge Univ. Press, London and New York.

Beattie, A. J., Breedlove, D. E., and Ehrlich, P. R. (1973). The ecology of the pollinators and predators of *Frasera speciosa*. *Ecology* **54**, 81–91.

Beauchamp, H., Lang, R., and May, M. (1975). Topsoil as a seed source for reseeding strip mine spoils. *Res. J. Wyo. Agric. Exp. Stn.* **90**, 1–8.

Becker, L., and Wong, M. (1985). Seed dispersal, seed predation and juvenile mortality of *Aglaia* sp. (Meliaceae) in a lowland dipterocarp rainforest. *Biotropica* **17**, 230–237.

Becker, P., Lee, L. L., Rothman, E. D., and Hamilton, W. D. (1985). Seed predation and the coexistence of tree species: Hubbell's models revisited. *Oikos* **44**, 382–390.

Beckwith, S. L. (1954). Ecological succession on abandoned farmland and its relationship to wildlife management. *Ecol. Monogr.* **24**, 349–375.

Becquerel, J. D. (1907). Récherches sur la vie latente des graines. *Ann. Sci. Nat., Bot. Biol. Veg.* **95**, 193–311.

Becquerel, J. D. (1934). La longevité des graines macrobiotiques. *C. R. Hebd. Seances Acad. Sci.* **199**, 1662–1664.

Bednarik, K. E. (1963). Marsh management techniques, 1960. *Game Res. Ohio* **2**, 132–144.

Begon, M., Harper, J. L., and Townsend, P. R. (1986). "Ecology: Individuals, Populations, and Communities." Sinauer, Sunderland, Massachusetts.

Bell, C. R. (1970). Seed distribution and germination experiment. *In* "A Tropical Rain Forest" (H. T. Odum and R. F. Pigeon, eds.), pp. D177–182. U.S. At. Energy Comm., Oak Ridge, Tennessee.

Bell, K. L., and Bliss, L. C. (1980). Plant reproduction in a high-arctic environment. *Arct. Alp. Res.* **12**, 1–10.

Bellrose, F. C., Jr. (1941). Duck food plants of the Illinois River Valley. *Bull. Ill. Nat. Hist. Surv.* **21**, 237–280.

Belsky, A. J. (1986). Does herbivory benefit plants? A review of the evidence. *Am. Nat.* **127**, 870–892.

Benkman, C. W., Balda, R. P., and Smith, C. C. (1984). Adaptations for seed dispersal and the compromises due to seed predation in limber pine. *Ecology* **65**, 632–642.

Benoit, D. L. (1986). Methods of sampling seed banks in arable soils with special reference to *Chenopodium* spp. Ph.D. Thesis, University of Western Ontario, London, Canada.

Bentley, S., Whittaker, J. B., and Malloch, A. J. C. (1980). Field experiments on the effects of grazing by a chrysomelid beetle (*Gastrophysa viridula*) on seed production and quality in *Rumex obtusifolius* and *Rumex crispus*. *J. Ecol.* **68**, 671–674.

Berenbaum, M. R., Zangerl, N. R., and Nitao, N. K. (1986). Constraints on chemical evolution: Wild parsnips and parsnip webworm. *Evolution* **40**, 1215–1228.

Berg, R. Y. (1966). Seed dispersal of *Dendromecon*, its ecologic, evolutionary, and taxonomic significance. *Am. J. Bot.* **53**, 61–73.

Berg, R. Y. (1975). Myrmecochorous plants in Australia and their dispersal by ants. *Aust. J. Bot.* **23**, 475–508.

Berger, A. (1985). Seed dimorphism and germination behaviour in *Salicornia patula*. *Vegetatio* **61**, 137–143.

Bergsma, B. M., Svoboda, J., and Freedman, B. (1984). Entombed plant communities released by a retreating glacier at Central Ellesmere Island, Canada. *Arctic* **37**, 49–52.

Best, K. F., and McIntyre, G. I. (1975). The biology of Canadian weeds. 9. *Thlaspi arvense* L. *Can. J. Plant Sci.* **55**, 279–292.

Best, K. F., Bowes, G. G., Thomas, A. G., and Maw, M. G. (1980). The biology of Canadian weeds. 39. *Euphorbia esula* L. *Can. J. Plant Sci.* **60**, 651–663.

Bewley, J. D., and Black, M. (1982). "Physiology and Biochemistry of Seeds in Relation to Germination." Springer-Verlag, Berlin.

Bewley, J. D., and Black, M. (1985). "Seeds: Physiology of Development and Germination." Plenum, New York.

Bhowmik, P. C., and Bandeen, J. D. (1976). The biology of Canadian weeds. 19. *Asclepias syriaca* L. *Can. J. Plant Sci.* **56**, 579–589.

Bicknell, S. H. (1979). Pattern and process of plant succession in a revegetating northern hardwood ecosystem. Ph.D. Dissertation, Yale University, New Haven, Connecticut.

Bigwood, D. W., and Inouye, D. W. (1988). Spatial pattern analysis of seed banks: An improved method and optimizing sampling. *Ecology* **69**, 497–507.

Billings, W. D. (1949). The shadscale vegetation zone of Nevada and eastern California in relation to climate and soils. *Am. Midl. Nat.* **42**, 87–109.

Billings, W. D. (1973). Arctic and alpine vegetations: similarities, differences, and susceptibility to disturbance. *BioScience* **23**, 697–704.

Billings, W. D., and Mooney, H. A. (1968). The ecology of arctic and alpine plants. *Biol. Rev. Cambridge Philos. Soc.* **43**, 481–529.

Binnie, R. C., and Clifford, P. E. (1980). Effects of some defoliation and decapitation treatments on the productivity of French beans. *Ann. Bot. (London)* [N.S.] **46**, 811–813.

Bishop, R. A., Andrews, R. D., and Bridges, R. J. (1979). Marsh management and its relationship to vegetation, waterfowl, and muskrats. *Proc. Iowa Acad. Sci.* **86**, 50–56.

Black, J. N., and Wilkinson, G. N. (1963). The role of time of emergence in determining the growth of individual plants in swards of subterranean clover (*Trifolium subterraneum*). *Aust. J. Agric. Res.* **14**, 628–638.

Black, R. A., and Bliss, L. C. (1980). Reproductive ecology of *Picea mariana* (Mill.) BSP. at tree line near Inuvik, Northwest Territories, Canada. *Ecol. Monogr.* **50**, 331–354.

Bliss, D., and Smith, H. (1985). Penetration of light into soil and its role in the control of seed germination. *Plant, Cell Environ.* **8**, 475–483.

Bliss, L. C. (1971). Arctic and alpine plant life cycles. *Annu. Rev. Ecol. Syst.* **2**, 405–538.

Blum, K. E. (1968). Contributions toward an understanding of the vegetational development in the Pacific lowlands of Panama. Ph.D. Thesis, Florida State University, Tallahassee.

Bohart, G. E., and Koerber, T. W. (1972). Insects and seed production. In "Seed Biology" (T.T. Kozlowski, ed.), Vol. 3, pp. 1–50. Academic Press, New York.

Bond, W. J. (1980). Fire and senescent fynbos in the Swartberg, southern Cape. *S. Afr. For. J.* **114**, 68–71.

Bond, W. J. (1984). Fire survival of Cape Proteaceae - influence of fire season and seed predators. *Vegetatio* **56**, 65–74.

Bond, W. J. (1985). Canopy-stored seed reserves (serotiny) in Cape Proteaceae. *S. Afr. J. Bot.* **51**, 181–186.

Bond, W. J., and Breytenbach, G. J. (1985). Ants, rodents and seed predation in Proteaceae. *S. Afr. J. Zool.* **20**, 150–154.

Bond, W. J., and Slingsby, P. (1984). Collapse of ant-plant mutualism, The Argentine ant, *Iridomyrmex humilis*, and myrmecochorous Proteaceae. *Ecology* **65**, 1031–1037.

Bond, W. J., Vlok, J., and Viviers, M. (1984). Variation in seedling recruitment of Cape Proteaceae after fire. *J. Ecol.* **72**, 209–221.

Bonde, E. K. (1969). Plant disseminules in wind-blown debris from a glacier in Colorado. *Arct. Alp. Res.* **1**, 135–140.

Borchert, M. I. (1985). Serotiny and cone-habit variation in populations of *Pinus coulteri* (Pinaceae) in the southern coast ranges of California. *Madroño* **32**, 29–48.

Borchert, M. I., and Jain, S. K. (1978). The effect of rodent seed predation on four species of California annual grasses. *Oecologia* **33**, 101–113.

Bormann, F. H., and Likens, G. E. (1979). "Pattern and Process in a Forested Ecosystem." Springer-Verlag, New York.

Borowicz, V. A., and Juliano, S. A. (1986). Inverse density-dependent parasitism of *Cornus amomom* fruit by *Rhagoletis cornivora*. *Ecology* **67**, 639–643.

Boston, H. L., and Adams, M. S. (1987). Productivity, growth and photosynthesis of two small 'isoetid' plants, *Littorella uniflora* and *Isoetes macrospora*. *J. Ecol.* **75**, 333–350.

Boucher, D. H., and Sork, V. L. (1979). Early drop of nuts in response to insect infestation. *Oikos* **33**, 440–443.

Boyd, R. S., and Brum, G. D. (1983). Postdispersal reproductive biology of a Mojave desert population of *Larrea tridentata* (Zygophyllaceae). *Am. Midl. Nat.* **110**, 25–36.

Bradford, D. F., and Smith, C. C. (1977). Seed predation and seed number in *Scheelea* palm fruits. *Ecology* **58**, 667–673.

Bradshaw, A. D., and Chadwick, M. J. (1980). "The Restoration of Land." Blackwell, Oxford.

Bradstock, R. A., and Myerscough, P. J. (1981). Fire effects on seed release and the emergence and establishment of seedlings in *Banksia ericifolia* L. f. *Aust. J. Bot.* **29**, 521–531.

Brandegee, T. S. (1891). The vegetation of "burns." *Zoe* **2**, 118–122.

Breedlove, D. E., and Ehrlich, P. R. (1968). Plant-herbivore coevolution: Lupines and lycaenids. *Science* **162**, 671–672.

Breedlove, D. E., and Ehrlich, P. R. (1972). Coevolution: Patterns of legume predation by a lycaenid butterfly. *Oecologia* **10**, 99–104.

Brenchley, W. E. (1918). Buried weed seeds. *J. Agric. Sci.* **9**, 1–31.

Brenchley, W. E., and Warington, K. (1930). The weed seed population of arable soil. I. Numerical estimation of viable seeds and observations on their natural dormancy. *J. Ecol.* **18**, 235–272.

Brenchley, W. E., and Warington, K. (1933). The weed seed population of arable soil. II. Influence of crop, soil and methods of cultivation upon the relative abundance of viable seeds. *J. Ecol.* **21**, 103–127.

Brenchley, W. E., and Warington, K. (1936). The weed seed population of arable soil. III. The re-establishment of species after reduction by fallowing. *J. Ecol.* **24**, 479–501.

Brenchley, W. E., and Warington, K. (1945). The influence of periodic fallowing on the prevalence of viable weed seeds in arable soil. *Ann. Appl. Biol.* **32**, 285–296.

Bridges, D.C., and Walker, R. H. (1985). Influence of weed management and cropping systems on sicklepod (*Cassia obtusifolia*) seed in the soil. *Weed Sci.* **33**, 800–804.

Brinkman, K. A. (1974). *Salix* L. *U.S. Dep. Agric., Agric. Handb.* **450**, 746–750.

Brinkmann, W. L. F., and Vieira, A. N. (1971). The effect of burning on germination of seeds at different soil depths of various tropical tree species. *Turrialba* **21**, 77–82.

Brokaw, N. V. L. (1985a). Treefalls, regrowth, and community structure in tropical forests. *In* "The Ecology of Natural Disturbance and Patch Dynamics" (S. T. A. Pickett and P. S. White, eds.), pp. 53–69. Academic Press, New York.

Brokaw, N. V. L. (1985b). Gap-phase regeneration in a tropical forest. *Ecology* **66**, 682–687.

Brokaw, N. V. L. (1986). Seed dispersal, gap colonization, and the case of *Cecropia insignis*. *In* "Frugivores and Seed Dispersal" (A. Estrada and T. H. Fleming, eds.), pp. 323–332. Junk, Dordrecht.

Brokaw, N. V. L. (1987). Gap-phase regeneration of three pioneer tree species in a tropical forest. *J. Ecol.* **75**, 9–19.

Brooker, R., and Corder, M. (1986). "Environmental Economy." Spon, London.

Brown, A. H. F., and Oosterhuis, L. (1981). The role of buried seed in coppicewoods. *Biol. Conserv.* **21**, 19–38.

Brown, D. E. (1982). Chihuahuan desert scrub. *Desert Plants* **4**, 169–179.

Brown, D. E., and Lowe, C. H. (1983). Biotic communities of the Southwest. *U.S., For. Serv., Rocky Mt. For. Range Exp. Stn., Gen. Tech. Rep.* **RM–78**.

Brown, J. H., and Munger, J. C. (1985). Experimental manipulation of a desert rodent community: Food addition and species removal. *Ecology* **66**, 1545–1563.

Brown, J. H., Grover, J. J., Davidson, D. W., and Leiberman, G. A. (1975). A preliminary study of seed predation in desert and montane habitats. *Ecology* **56**, 987–992.

Brown, J. H., Davidson, D. W., and Reichman, O. J. (1979a). An experimental study of competition between seed-eating desert rodents and ants. *Am. Zool.* **19**, 1129–1143.

Brown, J. H., Reichman, O. J., and Davidson, D. W. (1979b). Granivory in desert ecosystems. *Annu. Rev. Ecol. Syst.* **10**, 201–227.

Brown, J. S., and Venable, D. L. (1986). Evolutionary ecology of seed bank annuals in temporally varying environments. *Am. Nat.* **127**, 31–47.

Brown, J. S., and Venable, D. L. (1987). The evolutionary responses of seed-bank annuals to seed predation. *Am. Nat.* (in press).

Brown, M. T., and Ødum, H. T. (1985). "Studies of a Method of Wetland Reconstruction Following Phosphate Mining," Publ. No. 03–022–032. Florida Inst. Phosphate Res., Bartow.

Brown, R., and Edwards, M. (1944). The germination of the seed of *Striga lutea*. I. *Ann. Bot. (London)*, [N.S.] **8**, 131–148.

Brown, R., and Edwards, M. (1948). The germination of the seed of *Striga lutea*. II. *Ann. Bot. (London)* [N.S.] **10**, 133–142.

Brown, R., Robinson, E., and Johnson, A. W. (1949). Effect of *Striga* germination stimulant on extension growth of roots of peas. *Nature (London)* **163**, 842–843.

Brown, V. K. (1984). Secondary succession: Insect-plant relationships. *BioScience* **34**, 710–716.

Brown, V. K. (1985). Insect herbivore and succession. *Oikos* **44**, 17–22.

Brown, V. K., Gange, A. C., Evans, I. M., and Storr, A. L. (1987). The effect of insect herbivory on the growth and reproduction of two annual *Vicia* species at different stages in plant succession. *J. Ecol.* **75**, 1173–1190.

Buckley, R. C. (1982). Ant-plant interactions: A world review. *In* "Ant-Plant Interactions in Australia" (R. C. Buckley, ed.), pp. 111–162. Junk, The Hague.

Budd, A. C., Chepil, W. S., and Doughty, J. L. (1954). Germination of weed seeds. III. The influence of crops and fallow on the weed seed population of the soil. *Can. J. Agric. Sci.* **34**, 18–27.

Budowski, G. (1961). Studies on forest succession in Costa Rica and Panama. Ph.D. Thesis. Yale University, New Haven, Connecticut.

Bullock, S. H. (1978). Plant abundance and distribution in relation to types of seed dispersal in chaparral. *Madroño* **25**, 104–105.

Bullock, S. H. (1981). Life history and seed dispersal of *Dendromecon rigida*. *Gen. Tech. Rep. PSW (Pac. Southwest For. Range Exp. Stn.)* **PSW–58**, 590.

Burcham, L. T. (1957). "California Range Lands." Div. For., Dept. Nat. Resour., State of California, Sacramento.

Burrows, F. M. (1973). Calculation of the primary trajectories of plumed seeds in steady winds with variable convection. *New Phytol.* **72**, 647–664.

Byer, M. D., and Weaver, P. L. (1977). Early secondary succession in an elfin woodland in the Luquillo Mountains of Puerto Rico. *Biotropica* **9**, 35–47.

Cahoon, D. R., and Stevenson, J. C. (1986). Production, predation, and decomposition in a low-salinity *Hibiscus* marsh. *Ecology* **67**, 1341–1350.

Caldwell, M. M. (1985). Cold desert. *In* "Physiological Ecology of North American Plant Communities" (B. F. Chabot and H. A. Mooney, eds.), pp. 198–212. Chapman & Hall, New York.

Cameron, D. F. (1967). Hardseededness and seed dormancy of Townsville lucerne (*Stylosanthes humilis*). *Aust. J. Exp. Agric. Anim. Husb.* **7**, 237–240.

Cameron, E. (1935). A study of the natural control of ragwort (*Senecio jacobaea* L.). *J. Ecol.* **23**, 265–322.

Canham, C. D., and Marks, P. L. (1985). The response of woody plants to disturbance: Patterns of establishment and growth. *In* "The Ecology of Natural Disturbance and Patch Dynamics" (S. T. A. Pickett and P. S. White, eds.), pp. 197–216. Academic Press, Orlando, Florida.

Capon, B., and van Asdall, W. (1966). Heat pretreatment as a means of increasing germination of desert annual seeds. *Ecology* **48**, 305–306.

Carleton, T. J. (1982). The pattern of invasion and establishment of *Picea mariana* (Mill.) BSP. into subcanopy layers of *Pinus banksiana* Lamb. dominated stands. *Can. J. For. Res.* **12**, 973–984.

Carney, D. F., and Chabreck, R. H. (1977). Spring drawdown as a water fowl management practice in a floating fresh marsh. *Proc. Annu. Conf. Southeast. Assoc. Game Fish Comm.* **31**, 266–271.

Carpenter, F. L., and Recher, H. F. (1979). Pollination, reproduction and fire. *Am. Nat.* **113**, 871–879.

Carroll, C. R., and Janzen, D. H. (1973). Ecology of foraging ants. *Annu. Rev. Ecol. Syst.* **4**, 231–258.

Carroll, C. R., and Risch, S. J. (1984). The dynamics of seed harvesting in early successional communities by a tropical ant, *Solenopsis geminata*. *Oecologia* **61**, 388–392.

Carroll, E. J., and Ashton, D. H. (1965). Seed storage in soils of several Victorian plant communities. *Vict. Nat.* **82**, 102–110.

Castro Acuña, R., and Guevara Sada, S. (1976). Viabilidad de semillas en muestras de suelo almacenado de "Los Tuxlas", Veracruz. *In* "Investigaciones sobre la Regeneración de Selvas Altas en Veracruz, México" (A. Gómez-Pompa, C. Vázquez-Yanes, S. del Amo Rodríguez, and A. Butanda Cervana, eds.), pp. 233–249. Compañia Editorial Continental, México, D. F.

Caswell, H. (1978). Predator-mediated coexistence: A non-equilibrium model. *Am. Nat.* **112**, 127–154.

Caswell, H., and Werner, P. A. (1978). Transient behavior and life history analysis of teasel (*Dipsacus sylvestris* Huds.). *Ecology* **59**, 53–66.

Cates, R. G., and Orians, G. H. (1975). Successional status and the palatability of plants to generalist herbivores. *Ecology* **56**, 410–418.

Cavers, P. B. (1983). Seed demography. *Can. J. Bot.* **61**, 3578–3590.

Cavers, P. B., and Bough, M. A. (1986). Proso millet (*Panicum miliaceum* L.) - a crop and a weed. *In* "Studies of Plant Demography: A Festschrift for John L. Harper" (J. White, ed.), pp. 143–155. Academic Press, London.

Cavers, P. B., Heagy, M. I., and Kokron, R. F. (1979). The biology of Canadian weeds. 35. *Alliaria petiolata* (M. Bieb.) Cavara and Grande. *Can. J. Plant Sci.* **59**, 217–229.

Chabot, B. F., and Hicks, D. J. (1982). The ecology of leaf life spans. *Annu. Rev. Ecol. Syst.* **13**, 229–259.

Chadoeuf, R., Barralis, G., and Longchamp, J. P. (1984). Evolution du potentiel semencier de mauvaises herbes annuelles dans un sol cultivé. *Colloq. Int. Ecol., Biol. Syst. Mauvaises Herbes, COLUMA, 7th* pp. 63–70.

Chadoeuf-Hannel, R. (1985). La dormance chez les semences de mauvaises herbes. *Agronomie* **5**, 761–769.

Champness, S. S., and Morris, K. (1948). The population of buried viable seeds in relation to contrasting pasture and soil types. *J. Ecol.* **36**, 149–173.

Chancellor, R. J. (1980). New weeds for old in annual crops. *In* "Opportunities for Increasing Crop Yields" (R. G. Hurd, P. V. Biscoe, and C. Dennis, eds.), pp. 313–322. Pitman, London.

Chandler, C., Cheney, P., Thomas, P., Trabaud, L., and Williams, D. (1983). "Fire in Forestry," Vol. 1. Wiley, New York.

Chapin, F. S., III, and Chapin, M. C. (1980). Revegetation of an arctic disturbed site by native tundra species. *J. Appl. Ecol.* **17**, 449–456.

Cheke, A. S., Nanakorn, W., and Yankoses, C. (1979). Dormancy and dispersal of seeds of secondary forest species under the canopy of a primary tropical rain forest in northern Thailand. *Biotropica* **11**, 88–95.

Chepil, W. S. (1946). Germination of weed seeds. II. The influence of tillage treatments on germination. *Sci. Agric.* **26**, 347–357.

Chesson, P. L. (1986). Environmental variation and the coexistence of species. *In* "Community Ecology" (J. Diamond and T. J. Case, eds.), pp. 240–256. Harper & Row, New York.

Chester, A. L., and Shaver, G. R. (1982a). Reproductive effort in cotton grass tussock tundra. *Holarctic Ecol.* **5**, 200–206.

Chester, A. L., and Shaver, G. R. (1982b). Seedling dynamics of some cotton grass tussock tundra species during the natural revegetation of small disturbed areas. *Holarctic Ecol.* **5**, 207–211.

Chew, R. M. (1974). Consumers as regulators of ecosystems: An alternative to energetics. *Ohio J. Sci.* **74**, 359–370.

Chew, R. M., and Chew, A. E. (1970). Energy relationships of the mammals of a desert shrub (*Larrea tridentata*) community. *Ecol. Monogr.* **40**, 1–21.

Childs, S., and Goodall, D. W. (1973). Seed reserves of desert soils. *US/IBP Desert Biome Res. Mem.* pp. 73–75.

Chin, H. F., and Roberts, E. H. (1980). "Recalcitrant Crop Seeds." Tropical Press, Kuala Lumpur, Malaysia.

Chippendale, H. G., and Milton, W. E. J. (1934). On the viable seeds present in the soil beneath pastures. *J. Ecol.* **22**, 508–531.

Christensen, N. L. (1985). Shrubland fire regimes and their evolutionary consequences. *In* "The Ecology of Natural Disturbance and Patch Dynamics" (S. T. A. Pickett and P. S. White, eds.), pp. 86–100. Academic Press, Orlando, Florida.

Christensen, N. L., and Muller, C. H. (1975a). Relative importance of factors controlling germination and seedling survival in chaparral. *Am. Midl. Nat.* **93**, 71–78.

Christensen, N. L., and Muller, C. H. (1975b). Effects of fire on factors controlling plant growth in *Adenostoma* chaparral. *Ecol. Monogr.* **45**, 29–55.

Christiansen, J. E., and Low, J. B. (1970). Water requirements of waterfowl marshlands in northern Utah. *Utah Div. Fish Game Publ.* **69–12**.

Christiansen, M. N., and Moore, R. P. (1959). Seed coat structural differences that influence water uptake and seed quality in hard seed cotton. *Agron. J.* **51**, 582–584.

Clark, D. A., and Clark, D. B. (1987). Análisis de la regeneración deárboles del dosel en bosque muy húmedo tropical: Aspectos teóricos y práticos. *Rev. Biol. Trop.* **35**, Suppl. 1, 41–54.

Clark, N. (1953). The biology of *Hypericum perforatum* L. var. *angustifolium* DC (St. John's Wort) in the Ovens Valley, Victoria with particular reference to entomological control. *Aust. J. Bot.* **1**, 98–120.

Clegg, M. T., Kahler, A. L., and Allard, R. W. (1978). Genetic demography of plant populations. *In* "Ecological Genetics: The Interface" (P. F. Brussard, ed.), pp. 173–188. Springer-Verlag, Berlin and New York.

Clemens, J., Jones, P. G., and Gilbert, N. H. (1977). Effect of seed treatments on germination in *Acacia. Aust. J. Bot.* **25**, 269–276.

Cody, M. L. (1978). Distribution ecology of *Haplopappus* and *Chrysothamnus* in the Mohave Desert. I. Niche position and nich shifts on north-facing granitic slopes. *Am. J. Bot.* **65**, 1107–1116.

Cody, M. L., and Mooney, H. A. (1978). Convergence versus nonconvergence in mediterranean-climate ecosystems. *Annu. Rev. Ecol. Syst.* **9**, 265–321.

Coetzee, J. H., and Giliomee, J. H. (1987). Seed predation and survival in the infructescences of *Protea repens* (Proteaceae). *S. Afr. J. Bot.* **53**, 61–64.

Cohen, D. (1966). Optimizing reproduction in a randomly varying environment. *J. Theor. Biol.* **12**, 119–129.

Cohen, D. (1967). Optimizing reproduction in a randomly varying environment when a correlation may exist between the conditions at the time a choice has to be made and the subsequent outcome. *J. Theor. Biol.* **16**, 1–14.

Cohen, D. (1987). Community ESS for the germination and establishment of seeds. *Int. Bot. Congr. 14th* Abstr., p. 377.

Coker, W. C. (1909). Vitality of pine seeds and the delayed opening of cones. *Am. Nat.* **43**, 677–681.

Cole, L. (1954). Population consequences of life history phenomena. *Q. Rev. Biol.* **29**, 103–137.

Cole, N. H. A. (1977). Effect of light, temperature, and flooding on seed germination of the Neotropical *Panicum laxum* Sw. *Biotropica* **9**, 191–194.

Coley, P. D., Bryant, J. P., and Chapin, F. S., III (1985). Resource availability and plant anti-herbivore defense. *Science* **230**, 895–899.

Collinge, S. K., and Louda, S. M. (1987). Patterns of resource use by a drosophilid (Diptera) leaf miner on a native crucifer. *Ann. Entomol. Soc. Am.* (in press).

Collinge, S. K., and Louda, S. M. (1988a). Herbivory by leaf miners in response to experimental shading of a native crucifer. *Oecologia* (in press).

Collinge, S. K., and Louda, S. M. (1988b). Influence of plant phenology on the herbivore/bittercress interaction. *Oikos* **75**, 559–566.

Collins, B. S., and Pickett, S. T. A. (1987). Influence of canopy opening on the environment and herb layer in a Northern Hardwoods forest. *Vegetatio* **70**, 3–10.

Collins, B. S., Dunne, K. P., and Pickett, S. T. A. (1985). Responses of forest herbs to canopy gaps. *In* "The Ecology of Natural Disturbance and Patch Dynamics" (S. T. A. Pickett and P. S. White, eds.), pp. 217–234. Academic Press, Orlando, Florida.

Collins, S. L., and Uno, G. E. (1985). Seed predation, seed dispersal, and disturbance in grasslands: A comment. *Am. Nat.* **125**, 866–872.

Colosi, J. C., Cavers, P. B., and Bough, M. (1988). Dormancy and survival in buried seeds of proso millet (*Panicum miliaceum* L.). *Can. J. Bot.* **66**, 161–168.

Comins, H. N., Hamilton, W. D., and May, R. (1980). Evolutionarily stable dispersal strategies. *J. Theor. Biol.* **82**, 205–230.

Conn, E. C. (1979). Cyanide and cyanogenic glycosides. *In* "Herbivores" (G. A. Rosenthal and D. H. Janzen, eds.), pp. 387–412. Academic Press, New York.

Conn, J. S., Cochrane, C. L., and DeLapp, J. A. (1984). Soil seed bank changes after forest clearing and agricultural use in Alaska. *Weed Sci.* **32**, 343–347.

Connell, J. H. (1971). On the role of natural enemies in preventing competitive exclusion in some marine animals and in rain forest trees. *In* "Dynamics of Populations" (P. J. den Boer and G. R. Gradwell, eds.), pp. 298–312. Centre for Agricultural Publishing and Documentation, Wageningen.

Conrad, C. E., and Oechel, W. C., eds. (1982). "Dynamics and Management of Mediterranean-type Ecosystems," U.S. For. Serv., Gen. Tech. Rep., PSW–58. USDA, Washington, D. C.

Conti, R. S., and Gunther, P. P. (1984). Relations of phenology and seed germination to the distribution of dominant plants in Okefenokee Swamp. *In* "The Okefenokee Swamp: Its Natural History, Geology, and Geochemistry" (A. D. Cohen, D. J. Casagrande, M. J. Andrejko, and G. R. Best, eds.), pp. 144–167. Wetland Surveys, Los Alamos, New Mexico.

Cook, C. E., Wichard, L. P., and Wall, M. E. (1972). Germination stimulants. II. The structure of strigol—a potent seed germination stimulant for witchweed (*Striga lutea*). *J. Am. Chem. Soc.* **94**, 6198–6199.

Cook, P., Atsatt, P. R., and Simon, C. (1971). Doves and dove weed: Multiple defenses against avian predation. *BioScience* **21**, 277–281.

Cook, R. E. (1979). Patterns of juvenile mortality and recruitment in plants. *In* "Topics in Plant Population Biology" (O. T. Solbrig, S. Jain, G. B. Johnson, and P. R. Raven, eds.), pp. 207–231. Columbia Univ. Press, New York.

Cook, R. E. (1980). The biology of seeds in the soil. *In* "Demography and Evolution in Plant Populations" (O. T. Solbrig, ed.), pp. 107–129. Univ. of California Press, Berkeley.

Cottam, D. A. (1986). The effects of slug-grazing on *Trifolium repens* and *Dactylis glomerata* in monoculture and mixed sward. *Oikos* **47**, 275–279.

Cottam, D. A., Whittaker, J. B., and Malloch, A. J. C. (1986). The effects of chrysomelid beetle grazing and plant competition on the growth of *Rumex obtusifolius*. *Oecologia* **70**, 452–456.

Courtney, N. D. (1968). Seed dormancy and field emergence in *Polygonum aviculare*. *J. Appl. Ecol.* **5**, 675–684.

Coutinho, L. M., and Struffaldi, Y. (1971). Observações sôbre a germinacção das sementes e o crescimento das plântulas de uma leguminosa da mata amazônica de igapó (*Parkia auriculata* Spruce Mss.). *Phyton* **28**, 149–159.

Cowardin, L. M., Carter, V., Golet, F. C., and LaRoe, E. T. (1979). "Classification of Wetlands and Deepwater Habitats of the United States," FWS/OBS–79/31. Off. Biol. Serv., Fish Wildl. Serv., U.S. Dept. Int., Washington, D.C.

Cowling, R. M., and Lamont, B. B. (1985a). Seed release in *Banksia*: The role of wet-dry cycles. *Aust. J. Ecol.* **10**, 169–171.

Cowling, R. M., and Lamont, B. B. (1985b). Variation in serotiny of three *Banksia* species along a climatic gradient. *Aust. J. Ecol.* **10**, 345–350.

Cowling, R. M., Pierce, S. M., and Moll, E. J. (1986). Conservation and utilisation of south coast renosterveld, an endangered South African vegetation type. *Biol. Conserv.* **37**, 363–377.

Cowling, R. M., Lamont, B. B., and Pierce, S. M. (1987). Seed bank dynamics of four co-occurring *Banksia* species. *J. Ecol.* **75**, 289–302.

Cox, C. S., and McEvoy, P. B. (1983). Effect of summer moisture stress on the capacity of tansy ragwort (*Senecio jacobaea*) to compensate for defoliation by cinnabar moth (*Tyria jacobaeae*). *J. Appl. Ecol.* **20**, 225–234.

Crawley, M. J. (1983). "Herbivory, The Dynamics of Animal-Plant Interactions." Univ. of California Press, Berkeley.

Crawley, M. J. (1985). Reduction of oak fecundity by low-density herbivore populations. *Nature (London)* **314**, 163–164.

Crawley, M. J. (1988a). Herbivores and plant population dynamics. *In* "Plant Population Biology" (A. J. Davy, M. J. Hutchings, and A. R. Watkinson, eds.). Blackwell, Oxford (in press).

Crawley, M. J. (1988b). The relative importance of vertebrate and invertebrate herbivores in plant population dynamics. *In* "Plant/Animal Interactions" (E. A. Bernays, ed.). CRC Press, Boca Raton, Florida (in press).

Crawley, M. J., and Nachapong, M. (1985). The establishment of seedlings from primary and regrowth seeds of ragwort (*Senecio jacobaea*). *J. Ecol.* **73**, 255–261.

Critchfield, W. B. (1971). Profiles of California vegetation. *U.S., For. Serv., Res. Pap. PSW* **76**.

Currie, J. A. (1973). The seed-soil system. *In* "Seed Ecology" (W. Heydecker, ed.), pp. 463–480. Butterworth, London.

Cussans, G. W. (1987). Weed management in cropping systems. *Proc. Aust. Weeds Conf., 8th, 1987* pp. 337–347.

Cuthbertson, E. G. (1970). *Chondrilla juncea* in Australia. 3. Seed maturity and other factors affecting germination and establishment. *Aust. J. Exp. Agric. Anim. Husb.* **10**, 62–66.

Dale, H. M. (1964). Influence of soil on weed vegetation on a drained river millpond. *Can. J. Bot.* **42**, 823–830.

Dale, H. M. (1965). Influence of soil on weed vegetation on a drained river millpond. II. *Can. J. Bot.* **43**, 557–561.

Daniel, T. W., and Schmidt, J. (1972). Lethal and nonlethal effects of the organic horizons of forested soils on the germination of seeds from several associated conifer species of the Rocky Mountains. *Can. J. For. Res.* **2**, 179–184.

Darlington, H. T. (1922). Dr. W. J. Beal's seed viability experiment. *Am. J. Bot.* **9**, 266–269.

Darlington, H. T. (1931). The fifty-year period of Dr. W. J. Beal's viability experiment. *Am. J. Bot.* **18**, 262–265.

Darlington, H. T., and Steinbaur, G. P. (1961). The eighty year period for Dr. W. J. Beal's seed viability experiment. *Am. J. Bot.* **48**, 321–325.

Darwin, C. (1857). On the action of sea-water on the germination of seeds. *J. Linn. Soc. London* **1**, 130–140.

Darwin, C. (1859). "The Origin of the Species by Means of Natural Selection or the Preservation of Favoured Races in the Struggle for Life." Murray, London.

Darwin, C. (1897). "The Formation of Vegetable Mould Through the Action of Worms." Murray, London.

Datta, S. C., and Sen, S. (1982). Environment-controlled seed dormancy in *Cassia sophera* var. *purpurea*. *Geobios* **9**, 86–88.

Daubenmire, R. (1968). Ecology of fire in grasslands. *Adv. Ecol. Res.* **5**, 209–266.

Daubenmire, R. F. (1974). "Plants and Environment. A Textbook of Autecology," 3rd ed. Wiley, New York.

Davey, J. R. (1982). Stand replacement in *Ceanothus crassifolius*. M.S. Thesis, California State Polytechnic University, Pomona.

Davidson, D. W. (1977). Foraging ecology and community organization in desert seed-eating ants. *Ecology* **58**, 725–737.

Davidson, D. W. (1980). Some consequences of diffuse competition in a desert ant community. *Am. Nat.* **116**, 92–105.

Davidson, D. W., and Morton, S. R. (1981). Myrmecochory in some plants (F. Chenopodiaceae) of the Australian arid zone. *Oecologia* **50**, 357–366.

Davidson, D. W., Samson, D. A., and Inouye, R. S. (1985). Granivory in the Chihuahuan Desert: Interactions within and between trophic levels. *Ecology* **66**, 486–502.

Davis, M. A. (1981). The effect of pollinators, predators, and energy constraints on the floral ecology and evolution of *Trillium erectum*. *Oecologia* **48**, 400–406.

de Candolle, A. C. P. (1887). The latent vitality of seeds. *Pop. Sci. Mon.* **51**, 106–111.

de Foresta, H., and Prévost, M. F. (1986). Végétation pionnière et graines du sol en Forêt Guyanaise. *Biotropica* **18**, 279–286.

de Foresta, H., Charles-Dominique, P., Erard, C., and Prévost, M. F. (1984). Zoochorie et premiers stades de la régénération naturelle après coupe en forêt guyanaise. *Rev. Ecol. (Terre Vie)* **39**, 369–400.

de la Mensbruge, G. (1966). "La Germination et les Plantules des Essences Arborées de la Forêt Dense Humide de la Côte d'Ivoire," Trop. Publ. No. 26. Centre Technique Forestier, Nogent-sur-Marne, France.

Dell, B. (1980). Structure and function of the strophiolar plug in seeds of *Albizia lophantha*. *Am. J. Bot.* **67**, 556–563.

Denslow, J. S. (1980). Gap partitioning among tropical rainforest trees. *Biotropica* **12**, Suppl. , 47–95.

Denslow, J. S. (1987). Tropical rainforest gaps and tree species diversity. *Annu. Rev. Ecol. Syst.* **18**, 431–451.

Denslow, J. S., and Moermond, T. C. (1982). The effect of accessibility on rates of fruit removal from tropical shrubs: An experimental study. *Oecologia* **54**, 170–176.

De Steven, D. (1981). Predispersal seed predation in a tropical shrub (*Mabea occidentalis*, Euphorbiaceae). *Biotropica* **13**, 146–150.

De Steven, D. (1983). Reproductive consequences of insect seed predation in *Hamamelis virginiana*. *Ecology* **64**, 89–98.

De Steven, D., and Putz, F. E. (1984). Impact of mammals on early recruitment of a tropical canopy tree, *Dipteryx panamensis*, in Panama. *Oikos* **43**, 207–216.

deVlaming, V., and Proctor, V. W. (1968). Dispersal of aquatic organisms: Viability of seeds recovered from the droppings of captive killdeer and mallard ducks. *Am. J. Bot.* **55**, 20–26.

de Wet, J. M. J. (1975). Evolutionary dynamics of cereal domestication. *Bull. Torrey Bot. Club* **102**, 307–312.

di Castri, F., and Mooney, H. A., eds. (1973). "Mediterranean-type Ecosystems: Origin and Structure." Springer-Verlag, Berlin.

Dirzo, R. (1984). Herbivory: A phytocentric overview. *In* "Perspectives on Plant Population Ecology" (R. Dirzo and J. Sarukhán, eds.), pp. 141–165. Sinauer, Sunderland, Massachusetts.

Dirzo, R. (1986). The role of the grazing animal. *In* "Studies of Plant Demography: A Festschrift for John L. Harper" (J. White, ed.), pp. 343–355. Academic Press, London.

Dolinger, P. M., Ehrlich, P. R., Fitch, E. L., and Breedlove, D. E. (1973). Alkaloid and predation patterns in Colorado lupine populations. *Oecologia* **13**, 191–204.

Donald, W. W. (1985). AC-94,377 for breaking dormancy of wild mustard seed in soil. *N. D. Farm Res.* **43**, 28–31.

Donelan, M., and Thompson, K. (1980). Distribution of buried viable seeds along a successional series. *Biol. Conserv.* **17**, 297–311.

Dore, W. G., and Raymond, L. C. (1942). Pasture studies. XXIV. Viable seeds in pasture soil and manure. *Sci. Agric.* **23**, 69–79.

Dospekhov, B. A., and Chekryzhov, A. D. (1972). Counting weed seeds in the soil by the method of small samples (in Russian). *Izv. Timiryazev sk. S-kh. Akad.* **2**, 213–215.

Douglas, G. (1965). The weed flora of chemically-renewed lowland swards. *J. Br. Grassl. Soc.* **20**, 91–100.

Duggan, A. E. (1985). Predispersal seed predation by *Anthocharis cardamines* (Pieridae) in the population dynamics of the perennial *Cardamine pratensis* (Brassicaceae). *Oikos* **44**, 99–106.

Dunn, W. J., and Best, G. R. (1984). Enhancing ecological succession. 5. Seed bank survey of some Florida marshes and role of seed banks in marsh reclamation. In "Surface Mining, Hydrology, Sedimentology, and Reclamation" (D. H. Graves, ed.), pp. 365–370. Off. Eng. Serv., University of Kentucky, Lexington.

DuRant, J. A. (1982). Influence of southern corn billbug (Coleoptera: Curculionidae) population density and plant growth stage infested on injury to corn. *J. Econ. Entomol.* **75**, 892–894.

Duvel, J. W. T. (1902). Seeds buried in soil. *Science* **17**, 872–873.

Duvel, J. W. T. (1904). The vitality and germination of seeds. *U.S. Dep. Agric. Bur. Plant Ind. Bull.* **58**, 1–96.

Dvořák, J., and Krejčíř, J. (1980). [Reserve of weed seed in topsoil under different crop rotations and herbicide application.] *Acta Univ. Agric. Fac. Agron. (Brno)* **28**, 9–23.

Dye, A. J. (1969). Germination potentials and accumulation of native plant seeds from southern New Mexico. M.S. Dissertation, New Mexico State University, Las Cruces.

Edmisten, J. (1970). Some autecological studies of *Ormosia krugii*. In "A Tropical Rain Forest" (H. T. Odum and R. F. Pigeon, eds.), pp. B291–298. U.S. At. Energy Comm., Oak Ridge, Tennessee.

Edwards, M. (1980). Aspects of the population ecology of charlock. *J. Appl. Ecol.* **17**, 151–171.

Edwards, P. J., and Wratten, S. D. (1980). "Ecology of Insect-Plant Interactions." Edward Arnold, London.

Egler, F. E. (1954). Vegetation science concepts. I. Initial floristic composition—a factor in old-field vegetation development. *Vegetatio* **4**, 412–417.

Egley, G. H. (1986). Stimulation of weed seed germination in soil. *Rev. Weed Sci.* **2**, 67–89.

Egley, G. H., and Elmore, C. D. (1987). Germination and the potential persistence of weedy and domestic okra (*Abelmoschus esculentus*) seeds. *Weed Sci.* **35**, 45–51.

Egley, G. N., Paul, R. N., Jr., and Lax, A. R. (1986). Seed coat imposed dormancy: Histochemistry of the region controlling onset of water entry into *Sida spinosa* seeds. *Physiol. Plant.* **67**, 320–327.

Ehleringer, J. (1985). Annuals and perennials of warm deserts. In "Physiological Ecology of North American Plant Communities" (B. F. Chabot and H. A. Mooney, eds.), pp. 162–180. Chapman & Hall, New York.

Ehrlich, P. R., and Raven, P. H. (1964). Butterflies and plants: A study in coevolution. *Evolution (Lawrence, Kans.)* **18**, 586–608.

Elliott, D. L. (1979). The current regenerative capacity of the northern Canadian trees, Keewatin, N.W.T., Canada: Some preliminary observations. *Arct. Alp. Res.* **11**, 243–251.

Ellner, S. (1985a). ESS germination strategies in randomly varying environments. I. Logistic type models. *Theor. Popul. Biol.* **28**, 50–79.

Ellner, S. (1985b). ESS germination strategies in randomly varying environments. II. Reciprocal yield law models. *Theor. Popul. Biol.* **28**, 80–116.

Ellner, S. (1987a). Competition and dormancy: A reanalysis and review. *Am. Nat.* **130**, 798–803.

Ellner, S. (1987b). Alternate plant life-history strategies and coexistence in randomly varying environments. *Vegetatio* **69**, 199–208.

Ellner, S., and Schmida, A. (1981). Why are adaptations for long-range seed dispersal rare in desert plants? *Oecologia* **51**, 133–144.

Elven, R., and Ryvarden, L. (1975). Dispersal and primary establishment of vegetation. *In* "Fennoscandian Tundra Ecosystems. Part 1. Plants and Microorganisms" (F. E. Wielgolaski, ed.), p. 82–85. Springer-Verlag, Berlin.

Engel, J. L. (1983). The ecology of seed banks of the upland-salt marsh transition zone. M.S. Thesis, Rutgers University, Piscataway, New Jersey.

Enright, N. (1985). Evidence of a soil seed bank under rain forest in New Guinea. *Aust. J. Ecol.* **10**, 67–71.

Epling, C., Lewis, H., and Ball, F. (1960). The breeding group and seed storage: A study in population dynamics. *Evolution (Lawrence, Kans.)* **14**, 238–55.

Epp, G. A. (1987). The seed bank of *Eupatorium odoratum* along a successional gradient in a tropical rain forest in Ghana. *J. Trop. Ecol.* **3**, 139–149.

Erlandson, C. S. (1987). The potential role of seed banks in the restoration of drained prairie wetlands. M.S. Thesis, Iowa State University, Ames.

Ernst, W. H. O. (1985). Some considerations of and perspectives in coastal ecology. *Vegetatio* **62**, 533–545.

Estrada, A., and Coates-Estrada, R. (1986). Frugivory in howling monkeys (*Alouatta palliata*) at Los Tuxtlas, Mexico: Dispersal and fate of seeds. *In* "Frugivores and Seed Dispersal" (A. Estrada and T. H. Fleming, eds.), pp. 93–104. Junk, Dordrecht.

Evans, E. W., and Smith, C. C. (1988). Role of phenology of flower presentation in the seed predation by weevils on *Baptisia*. *Ann. Entomol. Soc. Am.* (in review).

Evanari, M. (1980/81). The history of germination research and the lesson it contains for today. *Israel J. Bot.* **29**, 4–21.

Evenari, M., Shanan, L., and Tadmor, N. (1982). "The Negev, the Challenge of a Desert," 2nd ed. Harvard Univ. Press, Boston, Massachusetts.

Ewart, A. J. (1908). On the longevity of seeds. *Proc. R. Soc. Victoria* [N.S.] **21**, 1–120.

Ewel, J., and Conde, L. (1979). Seeds in soils of former everglades farmlands. *In* "Proceedings of the First Conference on Scientific Research in the National Parks" (R. M. Linn, ed.), Vol. 1, Trans. Proc. Ser. No. 5, pp. 225–234. Natl. Park Serv., U.S. Dept. Int., Washington, D.C.

Ewel, J., Ojima, D., and DeBusk, W. (1979). Seed fluxes among everglades ecosystems. *In* "Proceedings of the Second Conference on Scientific Research in the National Parks" (R. M. Linn, ed.), Vol. 2, Terrestrial Biology: Botany, Coastal Biology," pp. 96–98. Natl. Park Serv., U.S. Dept. Int., Washington, D.C.

Ewel, J., Berish, C., Brown, B., Price, N., and Raich, J. (1981). Slash and burn impacts on a Costa Rican wet forest site. *Ecology* **62**, 816–829.

Fancourt, E. (1856). Vitality of seeds. *Gard. Chron.* **16**, 39.

Farmer, R. E., Jr., Cunningham, M., and Barnhill, M. A. (1982). First-year development of plant communities originating from forest topsoils placed on southern Appalachain minesoils. *J. Appl. Ecol.* **19**, 283–294.

Farris, M. (1985). Demographic variation and natural selection in an annual plant growing along soil moisture gradients. Ph.D. Thesis, University of Colorado, Boulder.

Fatubarin, A. (1987). Observations on the natural regeneration of the woody plants in a savanna ecosystem in Nigeria. *Trop. Ecol.* **28**, 1–8.

Fekete, R. (1975). Comparative weed-investigations in traditionally-cultivated and chemically-treated wheat and maize crops. IV. Study of the weed-seed contents of the soils of maize crops. *Acta Biol. (Szeged)* **21**, 9–20.

Fenner, M. (1978). A comparison of the abilities of colonizers and closed-turf species to establish from seed in artificial swards. *J. Ecol.* **66**, 953–963.

Fenner, M. (1980a). The inhibition of germination of *Bidens pilosa* seeds by leaf canopy shade in some natural vegetation types. *New Phytol.* **84**, 95–101.

Fenner, M. (1980b). Germination tests on thirty-two East African weed species. *Weed Res.* **20**, 135–138.

Fenner, M. (1985). "Seed Ecology." Chapman & Hall. London.

Fenner, M. (1987). Seedlings. *New Phytol.* **106**, Suppl., 35–47.

Fernald, M. L. (1921). The Gray Herbarium expedition to Nova Scotia 1920. *Rhodora* **23**, 89–111, 130–152, 153–171, 184–195, 223–245, 257–278, 284–300.

Fernald, M. L. (1922). Notes on the flora of western Nova Scotia 1921. *Rhodora* **24**, 157–181, 201–208.

Ferren, W. R., Jr., and Good, R. E. (1977). Habitat, morphology and phenology of southern wild rice (*Zizania aquatica* L.) from the Wading River in New Jersey. *Bull. Torrey Bot. Club* **104**, 392–396.

Ferren, W. R., Jr., and Schuyler, A. E. (1980). Intertidal vascular plants of river systems near Philadelphia. *Proc. Acad. Nat. Sci. Philadelphia* **132**, 86–120.

Fiedler, P. L. (1987). Life history and population dynamics of rare and common mariposa lilies (*Calochortus* Pursh: Liliaceae). *J. Ecol.* **75**, 977–995.

Fleming, T. H., and Heithaus, E. R. (1981). Frugivorous bats, seed shadows, and the structure of tropical forests. *Biotropica* **13**, Suppl., 45–53.

Fleming, T. H., Williams, C. F., Bonaccorso, F. J., and Herbst, L. H. (1985). Phenology, seed dispersal, and colonization in *Muntingia calabura*, a neotropical tree. *Am. J. Bot.* **72**, 383–391.

Flinn, M. A., and Wein, R. W. (1977). Depth of underground plant organs and theoretical survival during fire. *Can. J. Bot.* **55**, 2550–2554.

Floyd, A. G. (1976). Effect of burning on regeneration from seeds in wet sclerophyll forest. *Aust. For.* **39**, 210–220.

Foley, M. E. (1987). The effect of wounding on primary dormancy in wild oat (*Avena fatua*) caryopses. *Weed Sci.* **35**, 180–184.

Fonteyn, P. J., and Mahall, B. E. (1981). An experimental analysis of structure in a desert plant community. *J. Ecol.* **69**, 883–896.

Forcella, F. (1984). A species-area curve for buried viable seeds. *Aust. J. Agric. Res.* **35**, 645–652.

Foster, J. (1964). Studies on the population dynamics of the daisy, *Bellis perennis*. Ph.D. Thesis, University of Wales.

Foster, R. B. (1980). Heterogeneity and disturbance in tropical vegetation. *In* "Conservation Biology" (M. E. Soulé and B. A. Wilcox, eds.), pp. 75–92. Sinauer, Sunderland, Massachusetts.

Foster, R. B., Arce, B. J., and Wachter, T. S. (1986). Dispersal and the sequential plant communities in Amazonian Peru floodplain. *In* "Frugivores and Seed Dispersal" (A. Estrada and T. H. Fleming, eds.), pp. 357–370. Junk, Dordrecht.

Foster, S. A. (1986). On the adaptive value of large seeds for tropical moist forest trees: A review and synthesis. *Bot. Rev.* **52**, 260–299.

Foster, S. A., and Janson, C. H. (1985). The relationship between seed size and establishment conditions in tropical woody plants. *Ecology* **66**, 773–780.

Fowells, H. A. (1965). "The Sylvics of Forest Trees of the United States," Agric. Handb. No. 271. USDA For. Serv., Washington, D.C.

Fowler, H. G. (1979). Seed predator responses. *Oecologia* **41**, 361–363.

Fowler, H. G., and Whitford, W. G. (1983). Predispersal seed feeding insects of a desert legume, *Astragalus nutallianus*. *Southwest. Nat.* **28**, 466–469.

Fox, J. F. (1983). Germinable seed banks of interior Alaskan U.S.A. tundra. *Arct. Alp. Res.* **15**, 405–412.

Fox, L. R., and Morrow, P. A. (1986). On comparing herbivore damage in Australian and north temperate systems. *Aust. J. Ecol.* **11**, 387–393.

Fox, M. D., and Fox, B. J. (1986). The effect of fire frequency on the structure and floristic composition of a woodland understorey. *Aust. J. Ecol.* **11**, 77–85.

Frank, R. M., and Safford, L. O. (1970). Lack of viable seeds in the forest floor after clearcutting. *J. For.* **68**, 776–778.

Franklin, B. (1980). Phytochrome and seed germination. *What's New Plant Physiol.* **11**(8), 28–32.

Franklin, J. F., and Dyrness, C. T. (1973). Natural vegetation of Oregon and Washington. *U.S., For. Serv., Pac. Northwest For. Range Exp. Stn., Gen. Tech. Rep.* **PNW–8**.

Fraser, W. (1976). Viability of black spruce seed in or on a boreal forest seedbed. *For. Chron.* **52**, 229–231.

Freas, K. E., and Kemp, P. R. (1983). Some relationships between environmental reliability and seed dormancy in desert annual plants. *J. Ecol.* **71**, 211–217.

Fredrickson, L. H., and Taylor, T. S. (1982). "Management of Seasonally Flooded Impoundments for Wildlife," Resour. Publ. 148. U.S. Fish Wildl. Serv., Washington, D.C.

Freedman, B., Hill, N., Svoboda, J., and Henry, G. (1982). Seed banks and seedling occurrence in a high arctic oasis at Alexandra Fjord, Ellesmere Island, Canada. *Can. J. Bot.* **60**, 2112–2118.

Froud-Williams, R. J., Chancellor, R. J., and Drennan, D. S. H. (1983). Influence of cultivation regime upon buried weed seeds in arable cropping systems. *J. Appl. Ecol.* **20**, 199–208.

Furniss, R. L., and Carolin, V. M. (1977). Western forest insects. *USDA For. Serv., Misc. Publ.* **1339**.

Futuyma, D. J., and Slatkin, M., eds. (1983). "Coevolution." Sinauer, Sunderland, Masschusetts.

Galinato, M. I., and van der Valk, A. G. (1986). Seed germination traits of annuals and emergents recruited during drawdowns in the Delta Marsh, Manitoba, Canada. *Aquat. Bot.* **26**, 89–102.

Gardner, J. L. (1951). Vegetation of the creosotebush areas of the Rio Grande river valley in New Mexico. *Ecol. Monogr.* **21**, 379–403.

Gartner, B. L., Chapin, F. S., III, and Shaver, G. R. (1983). Demographic patterns of seedling establishment and growth of native graminoids in an Alaskan tundra disturbance. *J. Appl. Ecol.* **20**, 965–980.

Gartner, B. L., Chapin, F. S., III, and Shaver, G. R. (1986). Reproduction of *Eriophorum vaginatum* by seed in Alaskan tussock tundra. *J. Ecol.* **74**, 1–18.

Garwood, N. C. (1982). Seasonal rhythm of seed germination in a semideciduous tropical forest. *In* "The Ecology of a Tropical Forest: Seasonal Rhythms and Long-term Changes" (E. G. Leigh, A. S. Rand, and D. M. Windsor, eds.), pp. 173–185. Smithsonian Institution Press, Washington, D.C.

Garwood, N. C. (1983). Seed germination in a seasonal tropical forest in Panama: A community study. *Ecol. Monogr.* **53**, 159–181.

Garwood, N. C. (1985). Earthquake-caused landslides in Panama: Recovery of the vegetation. *Natl. Geogr. Soc. Res. Rep.* **21**, 181–184.

Garwood, N. C., Janos, D. P., and Brokaw, N. (1979). Earthquake-caused landslides: A major disturbance to tropical forests. *Science* **205**, 997–999.

Gashwiler, J. S. (1970). Further study of conifer seed survival in a western Oregon clearcut. *Ecology* **51**, 849–854.

Gaudet, J. J. (1977). Natural drawdown on Lake Naivasha, Kenya, and the formation of papyrus swamps. *Aquat. Bot.* **3**, 1–47.

Gilbert, L. E., and Raven, P. H. (1975). "Coevolution of Animals and Plants." Univ. of Texas Press, Austin.

Gill, A. M. (1975). Fire and the Australian flora: A review. *Aust. For.* **38**, 4–25.

Gill, A. M. (1981a). Adaptive responses of Australian vascular plant species to fires. *In* "Fire and the Australian Biota" (A. M. Gill, R. H. Groves, and I. R. Noble, eds.), pp. 243–272. Australian Academy of Sciences, Canberra.

Gill, A. M. (1981b). Coping with fire. *In* "The Biology of Australian Plants" (J. S. Pate and A. J. McComb, eds.), pp. 65–87. Univ. of Western Australia Press, Nedlands.

Gill, A. M., and Groves, R. H. (1981). Fire regimes in heathlands and their plant-ecological effects. *In* "Ecosystems of the World" (R. L. Specht, ed.), Vol. 9B, pp. 61–84. Elsevier, Amsterdam.

Gill, A. M., and Ingwersen, F. (1976). Growth of *Xanthorrhea australis* R.Br. in relation to fire. *J. Appl. Ecol.* **13**, 195–203.

Gill, A. M., and McMahon, A. (1986). A post-fire chronosequence of cone, follicle, and seed production in *Banksia ornata*. *Aust. J. Bot.* **34**, 425–433.

Gillon, D. (1983). The fire problem in tropical savannas. *In* "Ecosystems of the World" (F. Bourlière, ed.), Vol. 13, pp. 617–638. Elsevier, Amsterdam.

Givens, L. S., and Atkeson, T. Z. (1957). The use of dewatered land in southeastern waterfowl management. *J. Wildl. Manage.* **21**, 465–467.

Godwin, H. (1968). Evidence for the longevity of seeds. *Nature (London)* **220**, 708–709.

Godwin, H., and Willis, E. H. (1964). The viability of lotus seeds (*Nelumbium nucifera* Gaertn.). *New Phytol.* **63**, 410–412.

Gogue, G. J., and Emino, E. R. (1979). Seed coat scarification of *Albizia julibrissin* Durazz. by natural mechanisms. *J. Am. Soc. Hortic. Sci.* **104**, 421–423.

Goldberg, A. D., and Kigel, J. (1986). Dynamics of the weed community in coffee plantations grown under shade trees: Effect of clearing. *Isr. J. Bot.* **35**, 121–131.

Goldberg, D. E. (1985). Effects of soil pH, competition, and seed predation on the distribution of two tree species. *Ecology* **66**, 503–511.

Gómez Lepe, B., and Jiménez Avila, E. (1972). Germinación de las semillas de dos especies de plantas secundarias tropicales: *Heliocarpus donnell-smithii* Rose y *Piper auritum* H.B.K. *An. Inst. Biol., Univ. Nac. Auton. Mex., Ser. Biol. Exp.* **43**, 17–34.

Gómez-Pompa, A. (1967). Some problems of tropical forest plant ecology. *J. Arnold Arbor., Harv. Univ.* **48**, 105–121.

Gómez-Pompa, A., and Vázquez-Yanes, C. (1974). Studies of the secondary succession of tropical lowlands: The life cycle of secondary species. *Proc. Int. Congr. Ecol., 1st* pp. 336–342.

Gopal, B. (1986). Vegetation dynamics in temporary and shallow freshwater habitats. *Aquat. Bot.* **23**, 391–396.

Górski, T., Górska, K., and Nowicki, J. (1977). Germination of seeds of various herbaceous species under leaf canopy. *Flora (Jena)* **166**, 249–259.

Gottsberger, G. (1978). Seed dispersal by fish in the inundated regions of Humaitá, Amazonia. *Biotropica* **10**, 170–183.

Goyeau, H., and Fablet, G. (1982). Etude du stock de semences de mauvaises herbs dans le sol: Le problème de l'echantillonnage. *Agronomie (Paris)* **2**, 545–552.

Graaff, J. L., and van Staden, J. (1983). The effect of different chemical and physical treatments on seed coat structure and seed germination of *Sesbania* species. *Z. Pflanzenphysiol.* **112**, 221–230.

Grace, J. B. (1983). Autotoxic inhibition of seed germination by *Typha latifolia* - an evaluation. *Oecologia* **59**, 366–369.

Grace, J. B. (1984). Effects of tubificid worms on the germination and establishment of *Typha latifolia*. *Ecology* **65**, 1689–1693.

Granström, A. (1987). Seed viability of fourteen species during five years of storage in a forest soil. *J. Ecol.* **75**, 321–331.

Granström, A., and Fries, C. (1985). Depletion of viable seeds of *Betula pubescens* and

Betula verrucosa sown onto some north Swedish forest soils. *Can. J. For. Res.* **15**, 1176–1180.

Grant Lipp, A. E., and Ballard, L. A. T. (1963). Germination patterns shown by the light-sensitive seed of *Anagallis arvensis*. *Aust. J. Biol. Sci.* **16**, 572–584.

Gratkowski, H. J. (1962). Heat as a factor in germination of seeds of *Ceanothus velutinus* var. *laevigatus* T. & G. Ph.D. Thesis, Oregon State University, Corvallis.

Gratkowski, H. J. (1973). Pregermination treatments for redstem *Ceanothus* seeds. *U.S., For. Serv., Res. Pap.* **PNW 156**.

Gray, P. N., and Bolen, E. G. (1987). Seed reserves in the tailwater pits of playa lakes in relation to waterfowl management. *Wetlands* **7**, 11–23.

Green, T. W., and Palmbald, I. G. (1975). Effects of insect seed predators on *Astragalus cibarius* and *Astragalus utahensis* (Leguminosae). *Ecology* **56**, 1435–1440.

Greenwood, J. D. (1985). Frequency-dependent selection by seed-predators. *Oikos* **44**, 195–210.

Griffin, J. R. (1971). Oak regeneration in the Upper Carmel Valley, California. *Ecology* **52**, 862–868.

Griffin, J. R. (1978). Maritime chaparral and endemic shrubs of the Monterey Bay region, California. *Madroño* **25**, 65–81.

Griffith, B. G. (1931). The natural regeneration of spruce in central B.C. *For. Chron.* **7**, 204–219.

Griggs, R. F. (1956). Competition and succession on a Rocky Mountain fellfield. *Ecology* **37**, 8–20.

Griggs T. (1981). Life histories of vernal pool annual grasses. *Fremontia* **9**, 14–17.

Grime, J. P. (1977). Evidence for the existence of three primary strategies in plants and its relevance to ecological and evolutionary theory. *Am. Nat.* **111**, 1169–1194.

Grime, J. P. (1979). "Plant Strategies and Vegetation Process." Wiley, New York.

Grime, J. P. (1980). An ecological approach to management. *In* "Amenity Grassland: An Ecological Perspective" (I. H. Rorlson and R. Hunt, eds.), pp. 13–35. Wiley, New York.

Grime, J. P. (1981). The role of seed dormancy in vegetation dynamics. *Ann. Appl. Biol.* **98**, 555–558.

Grime, J. P. (1987). Dominant and subordinate components of plant communities: Implications for succession, stability and diversity. *In* "Colonization, Succession and Stability" (A. Gray, P. Edwards, and M. Crawley, eds.), pp. 413–428. Blackwell, Oxford.

Grime, J. P., and Hillier, S. H. (1981). Predictions based upon the laboratory characteristics of seeds. *In* "Annual Report 1981," p. 6. Unit of Comparative Plant Ecology (NERC), University of Sheffield, Sheffield, England.

Grime, J. P., and Hunt, R. (1975). Relative growth rate: Its range and adaptive significance in a local flora. *J. Ecol.* **63**, 393–422.

Grime, J. P., and Jarvis, B. C. (1975). Shade avoidance and shade tolerance in flowering plants. II. Effects of light on the germination of species of contrasted ecology. *In* "Light as an Ecological Factor: II" (G. C. Evans, R. Bainbridge, and O. Rackham, eds.), pp. 525–532. Blackwell, Oxford.

Grime, J. P., Mason, G., Curtis, A. V., Rodman, J., Band, S. R., Mowforth, M. A. G., Neal, A. M., and Shaw, S. (1981). A comparative study of germination characteristics in a local flora. *J. Ecol.* **69**, 1017–1059.

Grime, J. P., Hodgson, J. G., and Hunt, R. (1988). "Comparative Plant Ecology: A Functional Approach to Common British Species." Unwin Hyman, London.

Gross, K. L. (1984). Effects of seed size and growth form on seedling establishment of six monocarpic perennial plants. *J. Ecol.* **72**, 369–387.

Grubb, P. J. (1977a). Control of forest growth and distribution on wet tropical mountains: With special reference to mineral nutrition. *Annu. Rev. Ecol. Syst.* **8**, 83–107.

Grubb, P. J. (1977b). The maintenance of species-richness in plant communities: The importance of the regeneration niche. *Biol. Rev. Cambridge Philos. Soc.* **52**, 107–145.

Grulke, N. E., and Bliss, L. C. (1983). A note on winter seed rain in the high arctic. *Arct. Alp. Res.* **15**, 261–265.

Grushvitzky, I. V. (1967). After-ripening of seeds of primitive tribes of angiosperms, conditions and peculiarities. *Physiol. Öekol. Biochem. Keimung, Mater. Int. Symp., 1963* Vol. 1, pp. 329–336.

Guariguata, M. R., and Azocar, A. (1988). Seed bank dynamics and germination ecology in *Espeletia timotensis* (Compositae), an Andean giant rosette. *Biotropica* **20**, 54–59.

Guevara Sada, S. (1986). Plant species availability and regeneration in Mexican tropical rain forest. Ph.D. Thesis, Uppsala University, Sweden.

Guevara Sada, S., and Gómez-Pompa, A. (1972). Seeds from surface soils in a tropical region of Veracruz, Mexico. *J. Arnold Arbor., Harv. Univ.* **53**, 312–335. [Reprinted in Spanish. (1976). Determinación del contendido de semillas en muestras de suelo superficial de un selva tropical de Veracruz, México. *In* "Investigaciones sobre la Regeneración de Selvas Altas en Veracruz, México" (A. Gómez-Pompa, C. Vázquez-Yanes, S. del Amo Rodríguez, and A. Butanda Cervana, eds.), pp. 202–323. Compañia Editorial Continental, México D. F.]

Gunn, C. R., and Dennis, J. V. (1976). "World Guide to Tropical Drift Seeds and Fruits." Quadrangle/New York Times Book Co., New York.

Gunther, P. P., Casagrande, D. J., and Cherney, R. R. (1984). The viability and fate of seeds as a function of depth in the peats of Okefenokee Swamp. *In* "The Okefenokee Swamp: Its Natural History, Geology, and Geochemistry " (A. D. Cohen, D. J. Casagrande, M. J., Andrejko, and G. R. Best, eds.), pp. 168–179. Wetland Surveys, Los Alamos, New Mexico.

Gutterman, Y. (1980–1981). Influences on seed germinability: Phenotypic maternal effects during seed maturation. *Isr. J. Bot.* **29**, 105–117.

Gutterman, Y. (1982). Survival mechanisms of desert winter annual plants in the Negev highlands of Israel. *In* "Scientific Reviews on Arid Zone Research" (H. S. Mann, ed.), Vol. 1, pp. 249–283. Scientific Publishers, Jodhpur, India.

Gutterman, Y. (1983). Mass germination of plants under desert conditions. Effects of environmental factors during seed maturation, dispersal, germination and establishment of desert annual and perennial plants in the Negev highlands, Israel. *In* "Developments in Ecology and Environmental Quality" (H. I. Shuval, ed.), pp. 1–10. Balaban ISS, Philadelphia, Pennsylvania.

Haag, R. W. (1983). Emergence of seedlings of aquatic macrophytes from lake sediments. *Can. J. Bot.* **61**, 148–156.

Hacker, J. B. (1984). Genetic variation in seed dormancy in *Digitaria milanjiana* in relation to rainfall at the collection site. *J. Appl. Ecol.* **21**, 947–959.

Haddock, R. C., and Chaplin, S. J. (1982). Pollination and seed production in two phenotypically divergent prairie legumes (*Baptisia leucophaea* and *B. leucantha*). *Am. Midl. Nat.* **108**, 175–186.

Hadley, E. B. (1961). Influence of temperature and other factors on *Ceanothus megacarpus* seed germination. *Madroño* **16**, 132–138.

Hagon, M. W., and Ballard, L. A. T. (1970). Reversibility of strophiolar permeability to water in seeds of subterranean clover (*Trifolium subterraneum* L.). *Aust. J. Biol. Sci.* **23**, 519–528.

Hainsworth, F. R., Wolf, L. L., and Mercier, T. (1984). Pollination and predispersal seed predation: Net effects on reproduction and inflorescence characteristics in *Ipomopsis aggregata*. *Oecologia* **63**, 405–409.

Hall, A. V., and Veldhuis, H. A. (1985). "South African Red Data Book: Plants - Fynbos and Karoo Biome," S. Afr. Natl. Sci. Programmes Rep. No. 11.7. CSIR, Pretoria, South Africa.

Hall, I. V. (1955). Floristic changes following the cutting and burning of a woodlot for blueberry production. *Can. J. Agric. Sci.* **35**, 143–152.

Hall, J. B., and Swaine, M. D. (1980). Seed stocks in Ghanaian forest soils. *Biotropica* **12**, 256–263.

Hallman, G., Teetes, G., and Johnson, J. (1984). Weight compensation of undamaged kernels in response to damage by sorghum midge (Diptera: Cecidomyiidae). *J. Econ. Entomol.* **77**, 1033–1036.

Hallwachs, W. (1986). Agoutis (*Dasyprocta punctata*), the inheritors of guapinol (*Hymenaea courbaril*: Leguminosae). *In* "Frugivores and Seed Dispersal" (A. Estrada and T. H. Fleming, eds.), pp. 285–304. Junk, Dordrecht.

Hamilton, W. D., and May, R. (1977). Dispersal in stable habitats. *Nature (London)* **269**, 578–581.

Hamley, D. H. (1932). Softening of the seeds of *Melilotus alba*. *Bot. Gaz. (Chicago)* **93**, 345–375.

Hamrick, J. L., Linhart, Y. B., and Mitton, J. B. (1979). Relationships between life history characteristics and electrophoretically detectable genetic variation. *Annu. Rev. Ecol. Syst.* **10**, 173–200.

Handel, S. N., Fisch, S. B., and Schatz, G. E. (1981). Ants disperse a majority of herbs in a mesic forest community in New York State. *Bull. Torrey Bot. Club* **108**, 430–437.

Hanes, T. L. (1971). Succession after fire in the chaparral of southern California. *Ecol. Monogr.* **41**, 27–52.

Hanes, T. L. (1977). California chaparral. *In* "Terrestrial Vegetation of California" (M. G. Barbour and J. Major, eds.), pp. 417–469. Wiley, New York.

Hanes, T. L., and Jones, H. (1967). Postfire chaparral succession in southern California. *Ecology* **48**, 259–264.

Hanna, P. J. H. (1984). Anatomical features of the seed coat of *Acacia kempeana* (Mueller) which relate to increased germination rate induced by heat treatment. *New Phytol.* **96**, 23–29.

Hare, J. D. (1980). Variation in fruit size and susceptibility to seed predation among and within populations of the cocklebur, *Xanthium strumarium* L. *Oecologia* **46**, 217–222.

Harlan, J. R. (1982). Relationships between weeds and crops. *In* "Biology and Ecology of Weeds" (W. Holzner and M. Numata, eds.), pp. 91–96. Junk, The Hague.

Harmon, G. W., and Keim, F. D. (1934). The percentage and viability of weed seeds recovered in the faeces of farm animals and their longevity when buried in manure. *J. Am. Soc. Agron.* **26**, 762–767.

Harper, J. L. (1969). The role of predation in vegetational diversity. *Brookhaven Symp. Biol.* **22**, 48–62.

Harper, J. L. (1977). "Population Biology of Plants." Academic Press, New York.

Harper, J. L. (1982). After description. *In* "The Plant Community as a Working Mechanism" (E. I. Newman, ed.), pp. 11–25. Blackwell, Oxford.

Harper, J. L., and White, J. (1974). The demography of plants. *Annu. Rev. Ecol. Syst.* **5**, 419–463.

Harper, J. L., Williams, J. T., and Sagar, G. R. (1965). The behaviour of seeds in soil. I. The heterogeneity of soil surfaces and its role in determining the establishment of plants from seed. *J. Ecol.* **53**, 273–286.

Harper, J. L., Lovell, P. H., and Moore, K. G. (1970). The shapes and sizes of seeds. *Annu. Rev. Ecol. Syst.* **1**, 327–356.

Harrington, J. F. (1972). Seed storage and longevity. *In* "Seed Biology" (T. T. Kozlowski, ed.), Vol. 3, pp. 145–245. Academic Press, New York.

Harris, P. (1973). Insects in the population dynamics of plants. *In* "Insect/Plant Relationships" (H. F. van Emden, ed.), pp. 201–209. Blackwell, Oxford.

Harris, P. (1980). Effects of *Urophora affinis* Frfld. and *U. quadrifasciata* (Meig.) (Diptera: Tephritidae) on *Centaurea diffusa* Lam. and *C. maculosa* Lam. (Compositae). *Z. Angew. Entomol.* **90**, 190–201.

Harris, S. W., and Marshall, W. H. (1963). Ecology of water level manipulations on a northern marsh. *Ecology* **44**, 331–343.

Hartesveldt, R. J., Harvey, H. T., Shellhammer, H. S., and Sticker, R. E. (1969). Sequoia's dependence on fire. *Science* **166**, 552–553.

Hartshorn, G. S. (1978). Tree falls and tropical forest dynamics. *In* "Tropical Trees as Living Systems" (P. B. Tomlinson and M. H. Zimmermann, eds.), pp. 617–638. Cambridge Univ. Press, London and New York.

Hartshorn, G. S. (1980). Neotropical forest dynamics. *Biotropica* **12**, Suppl., 23–30.

Harvey, R. G. (1979). Serious new weed threat: Wild proso. *Crops Soils* **31**, 10–13.

Hassan, M. A., and West, N. E. (1986). Dynamics of soil seed pools in burned and unburned sagebrush semi-deserts. *Ecology* **67**, 269–272.

Hauptli, H., and Jain, S. K. (1978). Biosystematics and agronomic potential of some weedy and cultivated amaranths. *Theor. Appl. Genet.* **52**, 177–185.

Hawthorne, W. R., and Hayne, P. D. (1978). Seed production and predispersal seed predation in the biennial composite species, *Arctium minus* (Hill) Bernh. and *A. lappa* L. *Oecologia* **34**, 283–295.

Hayashi, I., and Numata, M. (1975). Viable buried seed population in grasslands in Japan. *JIBP Synth.* **13**, 58–69.

Hedlin, A. F. (1974). "Cone and Seed Insects of British Columbia," Can. For. Serv. BC-X–90. Environ. Can., Ottawa, Ontario.

Heinselman, M. L. (1973). Fire in the virgin forests of the Boundary Waters Canoe Area, Minnesota. *Quat. Res. (N.Y.)* **3**, 329–382.

Heithaus, E. R. (1981). Seed predation by rodents on three ant-dispersed plants. *Ecology* **62**, 136–145.

Heithaus, E. R., Culver, D.C., and Beattie, A. J. (1980). Models of some ant-plant mutualisms. *Am. Nat.* **116**, 347–361.

Heithaus, E. R., Stashko, E., and Anderson, P. K. (1982). Cumulative effects of plant-animal interactions on seed production by *Bauhinia ungulata*, a neotropical legume. *Ecology* **63**, 1294–1302.

Hemsley, W. B. (1895). Vitality of seeds. *Nature (London)* **27**, 5–6.

Hendrix, S. D. (1979). Compensatory reproduction in a biennial herb following insect defloration. *Oecologia* **42**, 107–118.

Hendrix, S. D. (1984). Reactions of *Heracleum lanatum* to floral herbivory by *Depressaria pastinacella*. *Ecology* **65**, 191–197.

Hendrix, S. D. (1988). Seed predation. *In* "Plant Reproductive Ecology" (J. Lovett-Doust, ed.). CRC Press, Boca Raton, Florida (in press).

Hendry, G. W. (1931). The adobe brick as a historical source. *Agric. Hist.* **5**, 110–127.

Herrera, D. M. (1984a). Selective pressures on fruit seediness: Differential predation of fly larvae on the fruits of *Berberis hispanica*. *Oikos* **42**, 166–170.

Herrera, D. M. (1984b). Avian interference with insect frugivory: An exploration into the plant-bird-fruit pest evolutionary triad. *Oikos* **42**, 203–210.

Herrera, D. M. (1984c). Seed dispersal and fitness determinants in wild rose: Combined effects of hawthorne, birds, mice, and browsing ungulates. *Oecologia* **63**, 386–393.

Herwitz, S. R. (1981). Regeneration of selected tropical tree species in Corcovardo National Park, Costa Rica. *Univ. Calif., Berkeley, Publ. Geogr.* **24**, 1–78.

Hey, D. L. (1987). The Des Plaines River wetlands demonstration project: Creating wetlands hydrology. *Natl. Wetlands Newsl.* **9**, 12–14.

Heydecker, W., ed. (1973). "Seed Ecology." Pennsylvania State Univ. Press, University Park and London.

Hickman, J. E. (1979). The basic biology of plant numbers. In "Topics in Plant Population Biology" (O. T. Solbrig, S. Jain, G. B. Johnson, and P. H. Raven, eds.), pp. 232–263. Columbia Univ. Press, New York.

Higgens, M. L. (1979). Intensity of seed predation on Brosimum utile by mealy parrots. Biotropica 11, 80.

Hilbert, D. W. (1988). A model of life history strategies of chaparral shrubs in relation to fire frequency. In "Plant Response to Stress-Functional Analysis in Mediterranean Ecosystems" (J. D. Tenhunen, F. M. Catarino, O. L. Lange, and W. C. Oechel, eds.). Springer-Verlag, Berlin (in press).

Hilbig, W. (1982). Preservation of agrestal weeds. In "Biology and Ecology of Weeds" (W. Holzner and M. Numata, eds.), pp. 57–69. Junk, The Hague.

Hill, A. W. (1917). The flora of the Somme battlefield. Kew Bull. pp. 297–300.

Hill, M. O., and Stevens, P. A. (1981). The density of viable seed in soils of forest plantations in upland Britain. J. Ecol. 69, 693–709.

Hobbs, R. J. (1985). Harvestor ant foraging and plant species distribution in annual grassland. Oecologia 67, 519–523.

Hodgkinson, K. C., Harrington, G. H., and Miles, G. E. (1980). Composition, spatial and temporal variability of the soil seed pool in a Eucalyptus populnea shrub woodland in central New South Wales. Aust. J. Ecol. 5, 23–29.

Hodkinson, I. D., and Hughes, M. K. (1982). "Insect Herbivory." Chapman & Hall, New York.

Hofmann, J. V. (1917). Natural reproduction from seed stored in the forest floor. J. Agric. Res. 11, 1–26.

Holm, R. E. (1972). Volatile metabolites controlling germination in buried weed seeds. Plant Physiol. 50, 293–297.

Holthuijzen, A. M. A., and Boerboom, J. H. A. (1982). The Cecropia seedbank in the Surinam lowland rain forest. Biotropica 14, 62–68.

Holthuijzen, A. M. A., and Sharik, T. L. (1985). The red cedar (Juniperus virginiana L.) seed shadow along a fencerow. Am. Midl. Nat. 113, 200–202.

Holzner, W. (1982). Concepts, categories and characteristics of weeds. In "Biology and Ecology of Weeds" (W. Holzner and M. Numata, eds.), pp. 3–20. Junk, The Hague.

Hopkins, D. M., and Sigafoos, R. S. (1951). Frost action and vegetation patterns on Seward Peninsula, Alaska. Geol. Surv. Bull. (U.S.) 974-C.

Hopkins, D. R., and Parker, V. T. (1984). A study of the seed bank of a salt marsh in northern San Francisco Bay. Am. J. Bot. 71, 348–355.

Hopkins, M. S., and Graham, A. W. (1983). The species composition of soil seed banks beneath lowland tropical rainforests in North Queensland, Australia. Biotropica 15, 90–99.

Hopkins, M. S., and Graham, A. W. (1984a). Viable soil seed banks in disturbed lowland tropical rainforest sites in North Queensland, Australia. Aust. J. Ecol. 9, 71–79.

Hopkins, M. S., and Graham, A. W. (1984b). The role of soil seed banks in regeneration in canopy gaps in Australian tropical lowland rainforest-preliminary field experiments. Malay. For. 47, 146–158.

Hopkins, M. S., and Graham, A. W. (1987). The viability of seeds of rain forest species after experimental soil burials under tropical wet lowland forest in north-eastern Australia. Aust. J. Ecol. 12, 97–108.

Horton, J. S., and Kraebel, C. J. (1955). Development of vegetation after fire in the chamise chaparral of southern California. Ecology 36, 244–262.

Horvitz, C. C., and Schemske, D. W. (1984). Effects of ants and an ant-tended herbivore on seed production of a neotropical herb. Ecology 65, 1369–1378.

Horvitz, C. C., and Schemske, D. W. (1986). Seed dispersal and environmental hetero-geneity in a neotropical herb: A model of population and patch dynamics. *In* "Fru-givores and Seed Dispersal" (A. Estrada and T. H. Fleming, eds.), pp. 169–186. Junk, Dordrecht.

Howard, T. M. (1974). *Nothofagus cunninghamii* ecotonal stages: Buried viable seed in North West Tasmania. *Proc. R. Soc. Victoria* **86**, 137–142.

Howard, T. M., and Ashton, D. H. (1967). Studies of soil seed in snow gum woodland. *Vict. Nat.* **84**, 331–335.

Howe, H. F. (1986). Seed dispersal by fruit-eating bats and mammals. *In* "Seed Disper-sal" (D. R. Murray, ed.), pp. 123–189. Academic Press, Sydney.

Howe, H. F., and Smallwood, J. (1982). Ecology of seed dispersal. *Annu. Rev. Ecol. Syst.* **13**, 201–228.

Howe, H. F., and Westley, L. C. (1986). Ecology of pollination and seed dispersal. *In* "Plant Ecology" (M. J. Crawley, ed.), pp. 185–215. Blackwell, Oxford.

Howell, J. T. (1946). Carbonated landscape. *Sierra Club Bull.* **37**(7), 18–23.

Howell, J. T. (1947). Marin County miscellany. IV. *Leafl. West. Bot.* **5**, 41–45.

Howell, J. T. (1970). "Marin Flora and Supplement." Univ. of California Press, Berkeley.

Hubbell, S. P. (1980). Seed predation and the coexistence of tree species in tropical for-ests. *Oikos* **35**, 214–229.

Hubbell, S. P., and Foster, R. B. (1986a). Canopy gaps and the dynamics of a neo-tropical forest. *In* "Plant Ecology" (M. J. Crawley, ed.), pp. 77–96. Blackwell, Oxford.

Hubbell, S. P., and Foster, R. B. (1986b). Commonness and rarity in neotropical forests: Implications for tropical tree conservation. *In* "Conservation Biology: Science of Scarcity and Diversity" (M. E. Soulé, ed.), pp. 201–231. Sinauer, Sunderland, Massachusetts.

Hubbell, S. P., and Werner, P. A. (1979). On measuring the intrinsic rate of increase of populations with heterogeneous life histories. *Am. Nat.* **113**, 277–293.

Huenneke, L. F., and Sharitz, R. R. (1986). Microsite abundance and distribution of woody seedlings in a South Carolina cypress-tupelo swamp. *Am. Midl. Nat.* **115**, 328–335.

Hull, A. C. (1973). Germination of range plant seeds after long periods of uncontrolled storage. *J. Range Manage.* **26**, 198–200.

Hultén, E. (1968). "Flora of Alaska and Neighboring Territories. Stanford Univ. Press, Stanford, California.

Hume, L., and Archibold, O. W. (1986). The influence of a weedy habitat on the seed bank of an adjacent cultivated field. *Can. J. Bot.* **64**, 1879–1883.

Hume, L., Martinez, J., and Best, K. (1983). The biology of Canadian weeds. 60. *Poly-gonum convolvulus* L. *Can. J. Plant Sci.* **63**, 959–971.

Hunt, C. B. (1966). Plant Ecology of Death Valley, California. *Geol. Surv. Prof. Pap. (U.S.)* **509**.

Hurle, K. (1974). Effect of long-term weed control measures on viable weed seeds in the soil. *Proc. Br. Weed Control Conf.* **12**, 1145–1152.

Hutchings, M. L. (1986). Plant population biology. *In* "Methods in Plant Ecology" (P. D. Moore and S. B. Chapman, eds.), pp. 377–435. Blackwell, Oxford.

Hutchinson, I., Colosi, J., and Lewin, R. A. (1984). The biology of Canadian weeds. 63. *Sonchus asper* (L.) Hill and *S. oleraceus* L. *Can. J. Plant Sci.* **64**, 731–744.

Inouye, D. W. (1982). The consequences of herbivory: A mixed blessing for *Jurinea mol-lis* (Asteraceae). *Oikos* **39**, 269–272.

Inouye, D. W., and Taylor, O. R. (1979). A temperate region plant-ant-seed predator system: Consequences of extra-floral nectar secretion by *Helianthella quinquenervis*. *Ecology* **60**, 1–7.

Inouye, R. S. (1980). Density-dependent germination response by seeds of desert annuals. *Oecologia* **46**, 235–238.

Inouye, R. S., Byers, G. S., and Brown, J. H. (1980). Effects of predation and competition on survivorship, fecundity, and community structure of desert annuals. *Ecology* **6**, 1344–1351.

Isaac, L. A. (1935). Life of Douglas fir seed in the forest floor. *J. For.* **33**, 61–66.

Ivens, G. (1978). Some aspects of the seed ecology of gorse. *Proc. N.Z. Weed Pest Control Conf.* **31**, 53–57.

Ives, J. D. (1970). Arctic tundra: How fragile? A geomorphologist's point of view. *In* "The Tundra Environment" (F. K. Hare, ed.), pp. 39–42. Univ. of Toronto Press, Toronto.

Iwata, E. (1966). Germination behaviour of shrubby Lespedeza (*Lespedeza cyrtobotrya* Miq.) seeds with special reference to burning. *Ecol. Rev.* **16**, 217–227.

Jacobsen, J. V., and Pressman, E. (1979). A structural study of germination in celery (*Apium graveolens* L.) seed with emphasis on endosperm breakdown. *Planta* **144**, 241–248.

Jaeger, E. C. (1940). "Desert Wild Flowers" (rev. ed.). Stanford Univ. Press, Stanford, California.

Jain, S. K., ed. (1976). "Vernal Pools: Their Ecology and Conservation," Inst. Ecol. Publ. No. 9. University of California, Davis.

Jain, S. K. (1982). Variation and adaptive role of seed dormancy in some annual grassland species. *Bot. Gaz. (Chicago)* **143**, 101–106.

Jalloq, M. C. (1975). The invasion of molehills by weeds as a possible factor in the degeneration of reseeded pasture. I. The buried viable seed population of molehills from four reseeded pastures in West Wales. *J. Appl. Ecol.* **12**, 643–657.

Janzen, D. H. (1969). Seed-eaters versus seed size, number, toxicity and dispersal. *Evolution (Lawrence, Kans.)* **23**, 1–27.

Janzen, D. H. (1970). Herbivores and the number of tree species in tropical forests. *Am. Nat.* **104**, 501–528.

Janzen, D. H. (1971a). Seed predation by animals. *Annu. Rev. Ecol. Syst.* **2**, 465–492.

Janzen, D. H. (1971b). Escape of *Cassia grandis* L. beans from predators in time and space. *Ecology* **52**, 964–979.

Janzen, D. H. (1971c). Escape of juvenile *Dioclea megacarpa* (Leguminosae) vines from predators in a deciduous tropical forest. *Am. Nat.* **105**, 97–112.

Janzen, D. H. (1972a). Association of a rain forest palm and seed-eating beetles in Puerto Rico. *Ecology* **53**, 258–261.

Janzen, D. H. (1972b). Escape in space by *Sterculia apetala* seeds from the bug *Dysdercus fasciatus* in a Costa Rican deciduous forest. *Ecology* **53**, 350–361.

Janzen, D. H. (1974a). The role of the seed predator guild in a tropical deciduous forest, with some reflections on tropical biological control. *In* "Biology in Pest and Disease Control" (D. Rice Jones and M. E. Solomon, eds.), pp. 3–14. Blackwell, Oxford.

Janzen, D. H. (1974b). Tropical blackwater rivers, animals, and mast fruiting by the Dipterocarpaceae. *Biotropica* **6**, 69–103.

Janzen, D. H. (1975a). Intra- and interhabitat variations in *Guazuma ulmifolia* (Sterculiaceae) seed predation by *Amblycerus cistelinus* (Bruchidae) in Costa Rica. *Ecology* **56**, 1009–1013.

Janzen, D. H. (1975b). Behavior of *Hymenaea courbaril* when its predispersal seed predator is absent. *Science* **189**, 145–147.

Janzen, D. H. (1975c). Interactions of seeds and their insect predators/parasitoids in a tropical deciduous forest. *In* "Evolutionary Strategies of Parasitic Insects and Mites" (P. W. Price, ed.), pp. 154–186. Plenum, New York.

Janzen, D. H. (1976). Why bamboos wait so long to flower. *Annu. Rev. Ecol. Syst.* **7**, 347–392.

Janzen, D. H. (1977). The interaction of seed predators and seed chemistry. *Colloq. Int. C. N. R. S.* **265**, 415–428.

Janzen, D. H. (1978a). The ecology and evolutionary biology of seed chemistry as related to seed predation. *In* "Biochemical Aspects of Plant and Animal Coevolution" (J. B. Harborne, ed.), pp. 163–206. Academic Press, New York.

Janzen, D. H. (1978b). Reduction in seed predation on *Bauhinia pauletia* (Leguminosae) through habitat destruction in a Costa Rican deciduous forest. *Brenesia* **14/15**, 325–335.

Janzen, D. H. (1980). Specificity of seed-attacking beetles in a Costa Rican deciduous forest. *J. Ecol.* **68**, 929–952.

Janzen, D. H. (1982a). Neotropical anachronisms: The fruits the Gomphotheres ate. *Science* **215**, 19–27.

Janzen, D. H. (1982b). Seeds in tapir dung in Santa Rosa National Park, Costa Rica. *Brenesia* **19/20**, 129–135.

Janzen, D. H. (1983a). Dispersal of seeds by vertebrate guts. *In* "Coevolution" (D. J. Futuyma and M. Slatkin, eds.), pp. 232–262. Sinauer, Sunderland, Massachusetts.

Janzen, D. H. (1983b). No park is an island: Increase in interference from outside as park size decreases. *Oikos* **41**, 402–410.

Jefferies, R. L., Davy, A. J., and Rudmik, T. (1981). Population biology of the salt marsh annual *Salicornia europaea* agg. *J. Ecol.* **69**, 17–31.

Jefferies, R. L., Jensen, A., and Bazely, D. (1983). The biology of the annual *Salicornia europaea* agg. at the limits of its range in Hudson Bay. *Can. J. Bot.* **61**, 762–773.

Jensen, H. A. (1969). Content of buried seed in arable soil in Denmark and its relation to the weed population. *Dan. Bot. Ark.* **27**, 1–55.

Jerling, L. (1985). Population dynamics of *Plantago maritima* along a distributional gradient on a Baltic seashore meadow. *Vegetatio* **61**, 155–161.

Joenje, W. (1979). Plant succession and nature conservation of newly embanked tidal flats in Lauwerszeepolder. *In* "Ecological Processes in Coastal Environments" (R. L. Jefferies and A. J. Davy, eds.), pp. 617–634. Blackwell, Oxford.

Joern, A. (1987). Insect herbivory in the transition to California annual grasslands: Did grasshoppers deliver the coup de grass? *In* "California Grasslands" (H. A. Mooney and L. Hunnecke, eds.) (in press).

Johnson, C. D. (1981). Interactions between bruchid (Coleoptera) feeding and behavioral patterns of pods of the Leguminosae. *Environ. Entomol.* **10**, 249–253.

Johnson, E. A. (1975). Buried seed populations in the subarctic forest east of Great Slave Lake, Northwest Territories. *Can. J. Bot.* **53**, 2933–2941.

Johnson, M. S., and Bradshaw, A. D. (1979). Ecological principles for the restoration of disturbed and degraded land. *Appl. Biol.* **4**, 141–200.

Johnson, W. C., Sharpe, D. M., DeAngelis, D. L., Fields, D. E., and Olson, R. J. (1981). Modeling seed dispersal and forest island dynamics. *In* "Forest Island Dynamics in Man-Dominated Landscapes" (R. L. Burgess and D. M. Sharpe, eds.), pp. 215–239. Springer-Verlag, New York.

Johnston, A., Smoliak, S., and Stringer, P. W. (1969). Viable seed populations in Alberta prairie topsoils. *Can. J. Plant Sci.* **49**, 75–82.

Johnston, M. C. (1977). Brief resume of botanical, including vegetational, features of the Chihuahuan desert region with special emphasis on their uniqueness. *In* "Transactions of the Symposium on the Biological Resources of the Chihuahuan Desert Region, U.S. and Mexico" (R. H. Wauer and D. H. Riskind, eds.), Trans. Proc., Ser. 3, pp. 335–362. U.S. Dept. Int., Natl. Park Serv., Washington, D.C.

Johnstone, I. M. (1986). Plant invasion windows: A time based classification of invasion potential. *Biol. Rev. Cambridge Philos. Soc.* **61**, 369–394.

Jolivet, P. (1985). "Insects and Plants," Flora and Fauna Publ. 2. Gainesville, Florida.

Jolivet, P. (1986). "Les Fourmis et les Plantes: Un Exemple de Coévolution." Société Nouvelle des Editions, Boubee, Paris.

Jones, C. S., and Schlesinger, W. H. (1980). *Emmenanthe penduliflora*: Further consideration of germination response. *Madroño* **27**, 122–127.

Jones, R. M., and Evans, T. R. (1977). Soil seed levels of *Lotonis bainesii, Desmodium intortum*, and *Trifolium repens* in subtropical pastures. *J. Aust. Inst. Agric. Sci.* **43**, 164–166.

Jordano, P. (1983). Fig seed predation and dispersal by birds. *Biotropica* **15**, 38–41.

Juhren, M., Went, F. W., and Phillips, E. (1956). Ecology of desert plants. IV. Combined field and laboratory work on germination of annuals in the Joshua Tree National Monument, California. *Ecology* **37**, 318–330.

Juliano, J. B. (1940). Viability of some Philippine weed seeds. *Philipp. Agric.* **29**, 313–326.

Junk, W. (1970). Investigations on the ecology and production-biology of the "floating meadows" (*Paspalo-Echinochloetum*) on the middle Amazon. *Amazoniana* **2**, 449–495.

Kadlec, J. A. (1962). Effects of a drawdown on a waterfowl impoundment. *Ecology* **43**, 267–281.

Kadlec, J. A., and Smith, L. M. (1984). Marsh plant establishment on newly flooded salt flats. *Wildl. Soc. Bull.* **12**, 388–394.

Kadlec, J. A., and Wentz, W. A. (1974). "State-of-the-Art Survey and Evaluation of Marsh Plant Establishment Technique: Induced and Natural," Vols. 1 and 2. Natl. Tech. Inf. Serv., Springfield, Virginia.

Kamm, J. A. (1979). Plant bugs: Effects of feeding on grass seed development, and cultural control. *Environ. Entomol.* **8**, 73–76.

Karssen, C. M. (1980–1981). Environmental conditions and endogenous mechanisms involved in secondary dormancy of seeds. *Isr. J. Bot.* **29**, 45–64.

Karssen, C. M. (1982). Seasonal patterns of dormancy in weed seeds. *In* "The Physiology and Biochemistry of Seed Development, Dormancy and Germination" (A. A. Khan, ed.), pp. 243–270. Elsevier, Amsterdam.

Kartawinata, K., Riswan, S., and Soedjito, H. (1980). The floristic change after disturbances in lowland dipertocarp forest in East Kalimantan, Indonesia. *In* "Tropical Ecology and Development" (J. I. Furtado, ed.), pp. 47–54. International Society of Tropical Ecology, Kuala Lumpur, Malaysia.

Kassas, M., and Batanouny, K. H. (1984). Plant ecology. *In* "Sahara Desert" (J. L. Cloudsley-Thompson, ed.), pp. 77–90. Pergamon Press, New York.

Keay, R. W. (1960). Seeds in forest soil. *Niger. For. Inf. Bull.* **4**, 1–4.

Keddy, P. A. (1981a). Vegetation with Atlantic coastal plain affinities in Axe Lake, near Georgian Bay, Ontario. *Can. Field Nat.* **95**, 241–248.

Keddy, P. A. (1981b). Matchedash Lake: A rare ecosystem threatened. *Packet* (Orillia, Ontario) Nov. 3, p.5.

Keddy, P. A. (1985a). Wave disturbance on lakeshores and the within-lake distribution of Ontario's Atlantic coastal plain flora. *Can. J. Bot.* **63**, 656–660.

Keddy, P. A. (1985b). Lakeshores in the Tusket River Valley, Nova Scotia: Distribution and status of some rare species, including *Coreopsis rosea* Nutt. and *Sabatia kennedyana* Fern. *Rhodora* **87**, 309–320.

Keddy, P. A., and Constabel, P. (1986). Germination of ten shoreline plants in relation to seed size, particle size and water level: An experimental study. *J. Ecol.* **74**, 133–141.

Keddy, P. A., and Ellis, T. H. (1985). Seedling recruitment of 11 wetland plant species along a water level gradient: Shared or distinct responses? *Can. J. Bot.* **63**, 1876–1879.

Keddy, P. A., and Reznicek, A. A. (1982). The role of seed banks in the persistence of Ontario's coastal plain flora. *Am. J. Bot.* **69**, 13–22.

Keddy, P. A., and Reznicek, A. A. (1986). Great Lakes vegetation dynamics: The role of fluctuating water levels and buried seeds. *J. Great Lakes Res.* **12**, 25–36.

Keeler, K. H. (1988). Ant-plant interactions. *In* "Plant-Animal Interactions" (W. G. Abrahamson, ed.). Macmillan, New York (in press).

Keeler, K. H., Harrison, A. T., and Vescio, L. S. (1980). The flora and sandhills prairie communities of Arapaho Prairie, Arthur County, Nebraska. *Prairie Nat.* **12**, 65–78.

Keeley, J. E. (1977a). Seed production, seed populations in the soil, and seedling production after fire for two congeneric pairs of sprouting and non-sprouting chaparral shrubs. *Ecology* **58**, 820–829.

Keeley, J. E. (1977b). Fire-dependent reproduction strategies in *Arctostaphylos* and *Ceanothus. Gen. Tech. Rep. WO—U.S., For. Serv. [Wash. off.]* **GTR—WO-3**, 391–396.

Keeley, J. E. (1986a). Resilience of mediterranean shrub communities to fires. *In* "Resilience in Mediterranean-type Ecosystems" (B. Dell, A. J. M. Hopkins, and B. B. Lamont, eds.), pp. 95–112. Junk, Dordrecht.

Keeley, J. E. (1986b). Seed germination patterns of *Salvia mellifera* in fire-prone environments. *Oecologia* **71**, 1–5.

Keeley, J. E. (1987a). Ten years of change in seed banks of the chaparral shrubs, *Arctostaphylos glauca* and *A. glandulosa. Am. Midl. Nat.* **117**, 446–448.

Keeley, J. E. (1987b). Role of fire in seed germination of woody taxa in California chaparral. *Ecology* **68**, 434–443.

Keeley, J. E., and Hays, R. L. (1976). Differential seed predation on two species of *Arctostaphylos* (Ericaceae). *Oecologia* **24**, 71–81.

Keeley, J. E., and Keeley, S. C. (1981). Post-fire regeneration of southern California chaparral. *Am. J. Bot.* **68**, 524–530.

Keeley, J. E., and Keeley, S. C. (1987). Role of fire in the germination of chaparral herbs and suffrutescents. *Madroño* **34**, 240–249.

Keeley, J. E., and Nitzberg, M. E. (1984). Role of charred wood in the germination of the chaparral herbs *Emmenanthe penduliflora* (Hydrophyllaceae) and *Eriophyllum confertiflorum* (Asteraceae). *Madroño* **31**, 208–218.

Keeley, J. E., and Zedler, P. H. (1978). Reproduction of chaparral shrubs after fire, comparison of sprouting and seedling strategies. *Am. Midl. Nat.* **99**, 146–161.

Keeley, J. E., Keeley, S. C., Swift, C. C., and Lee, J. (1984). Seed predation due to the *Yucca*-moth symbiosis. *Am. Midl. Nat.* **112**, 187–191.

Keeley, J. E., Morton, B. A., Pedrosa, A., and Trotter, P. (1985). Role of allelopathy, heat and charred wood in the germination of chaparral herbs and suffrutescents. *J. Ecol.* **73**, 445–458.

Keeley, J. E., Keeley, S. C., and Ikeda, D. A. (1986). Seed predation by yucca moths on semelparous, iteroparous and vegetatively reproducing sub species of *Yucca whipplei* (Agavaceae). *Am. Midl. Nat.* **115**, 1–9.

Keeley, S. C., and Johnson, A. W. (1977). A comparison of the pattern of herb and shrub growth in comparable sites in Chile and California. *Am. Midl. Nat.* **97**, 20–32.

Keeley, S. C., and Pizzorno, M. (1986). Charred wood stimulated germination of two fire-following herbs of the California chaparral and the role of hemicellulose. *Am. J. Bot.* **73**, 1289–1297.

Keeley, S. C., Keeley, J. E, Hutchinson, S. M., and Johnson, A. W. (1981). Postfire succession of the herbaceous flora in southern California chaparral. *Ecology* **62**, 1608–1621.

Keen, F. P. (1958). Cone and seed insects of western forest trees. *U.S., Dep. Agric., Tech. Bull.* **1169**.

Kellman, M. (1970a). "Secondary Plant Succession in Tropical Montane Mindanao," Publ. BG/2. Dept. of Biogeography and Geomorphology, Australian National University, Canberra.

Kellman, M. (1970b). The viable seed content of some forest soil in coastal British Columbia. *Can. J. Bot.* **48**, 1383–1385.

Kellman, M. (1974a). Preliminary seed budgets for two plant communities in coastal British Columbia. *J. Biogeogr.* **1**, 123–133.

Kellman, M. (1974b). The viable weed seed content of some tropical agricultural soils. *J. Appl. Ecol.* **11**, 669–677.

Kellman, M. (1978). Microdistribution of viable weed seed in two tropical soils. *J. Biogeogr.* **5**, 291–300.

Kelly, V. R. (1986). Seed banks and reproductive life histories in sprouting and non-sprouting *Arctostaphylos* species. M.A. Thesis, San Francisco State University, San Francisco, California.

Kelly, V. R., and Parker, V. T. (1989). Seed banks and reproductive life histories in sprouting and nonsprouting species of *Arctostaphylos* (Ericaceae) (submitted).

Kelsall, J. P., Telfer, E. S., and Wright, T. D. (1977). The effects of fire on the ecology of the forest with particular reference to the Canadian North: Review and selected bibliography. *Can. Wildl. Serv. Rep.* **69**, 1–32.

Kemp, P. R. (1983). Phenological patterns of Chihuahuan desert plants in relation to the timing of water availability. *J. Ecol.* **71**, 427–436.

Khan, M. A., and Ungar, I. A. (1986). Life history and population dynamics of *Atriplex triangularis*. *Vegetatio* **66**, 17–25.

Kiew, R. (1982). Germination and seedling survival in kemenyan, *Styrax benzoin*. *Malay. For.* **45**, 69–80.

Kigel, J., Ofir, M., and Koller, D. (1977). Control of the germination of *Amaranthus retroflexus* L. seeds by their parental photothermal environment. *J. Exp. Bot.* **28**, 1125–1136.

King, L. J. (1966). "Weeds of the World. Biology and Control." Wiley (Interscience), New York.

King, T. J. (1976). The viable seed contents of ant-hill and pasture soil. *New Phytol.* **77**, 143–147.

Kinsman, S., and Platt, W. J. (1984). The impact of herbivores (*Heliodines nyctaginella*: Lepidoptera) upon *Mirabilis hirsuta*, a fugitive prairie plant. *Oecologia* **65**, 2–6.

Kivilaan, A., and Bandurski, R. S. (1973). The ninety year period for Dr. W. J. Beal's seed viability experiment. *Am. J. Bot.* **60**, 140–145.

Kivilaan, A., and Bandurski, R. S. (1981). The one hundred year period for Dr. W. J. Beal's seed viability experiment. *Am. J. Bot.* **68**, 1290–1292.

Kjellsson, G. (1985). Seed fate in a population of *Carex pilulifera* L. II. Seed predation and its consequences for dispersal and the seed bank. *Oecologia* **67**, 424–429.

Kjoller, A., and Ødum, S. (1971). Evidence for longevity of seeds and microorganisms in permafrost. *Arctic* **24**, 230–232.

Knab, W., and Hurle, K. (1986). [Influence of primary tillage on weeds—a contribution to the prediction of infestations.] *Proc. EWRS Symp. 1986, Econ. Weed Control* pp. 309–316. [De].

Koebernik, J. (1971). Germination of palm seed. *Principes* **15**, 134–137.

Koller, D., and Cohen, D. (1959). germination-regulating mechanisms in some desert seeds. VI. *Convolvulus lanatus* Vah., *Convolvulus negevensis* Zoh. and *Convolvulus secundus* Desr. *Bull. Res. Counc. Isr. Sect. D* **7**, 175-180.

Kozlowski, T. T., and Ahlgren, C. E., eds. (1974). "Fire and Ecosystems." Academic Press, New York.

Krajina, V. J. (1969). Ecology of forest trees in British Columbia. *In* "Ecology of Western North America," Vol. 2, No. 1, pp. 1–146. Dept. of Bot., University of British Columbia, Vancouver.

Kramer, I. F. (1933). De natuurlijke verjonging in het Goenoeng-Gedehcomplex. *Tectona* **26**, 155–185.

Kramer, N. B., and Johnson, F. D. (1987). Mature forest seed banks of three habitat types in central Idaho. *Can. J. Bot.* **65**, 1961–1966.

Kremer, R. J. (1986). Antimicrobial activity of Velvetleaf (*Abutilon theophrasti*) seeds. *Weed Sci.* **34**, 617–622.

Kropáč, Z. (1966). Estimation of weed seeds in arable soil. *Pedobiologia* **6**, 105–128.

Kruger, F. J. (1977). Ecology of Cape fynbos in relation to fire. *Gen. Tech. Rep. WO— U.S., For. Serv. [Wash. off.]* **GTP–WO–3**, 230–244.

Kruger, F. J. (1983). Plant community diversity and dynamics in relation to fire. *In* "Mediterranean-type Ecosystems, the Role of Nutrients" (F. J. Kruger, D. T. Mitchell, and J. U. M. Jarvis, eds.), pp. 446–472. Springer-Verlag, Berlin.

Kruger, F. J., Mitchell, D. T., and Jarvis, J. U. M., eds. (1983). "Mediterranean-type Ecosystems, the Role of Nutrients." Springer-Verlag, Berlin.

Krusi, B. O., and Debussche, M. (1988). The fate of flowers and fruits of *Cornus sanguinea* L. in three contrasting Mediterranean habitats. *Oecologia* **74**, 592–599.

Kuchler, A. W. (1970). Potential natural vegetation (map at scale 1: 750,000). *In* "The National Atlas of the U.S.A.," U.S. Geological Survey, U.S. Gov. Printing Office, Washington, D.C.

LaCroix, L. J., and Staniforth, D. W. (1964). Seed dormancy in Velvetleaf. *Weeds* **12**, 171–174.

Lal, R. (1987). "Tropical Ecology and Physical Edaphology." Wiley, Chichester, England.

Lamont, B. B. (1985). The comparative reproductive biology of three *Leucospermum* species (Proteaceae) in relation to fire responses and breeding system. *Aust. J. Bot.* **33**, 139–145.

Lamp, W. O., and McCarty, M. K. (1979). A preliminary study of seed predators of Platte thistle. *Trans. Nebr. Acad. Sci. Affil. Soc.* **7**, 71–74.

Lamp, W. O., and McCarty, M. K. (1981). Biology and ecology of the Platte Thistle (*Cirsium canescens*). *Weed Sci.* **29**, 686–692.

Lamp, W. O., and McCarty, M. K. (1982). Predispersal seed predation of a native thistle, *Circium canescens*. *Environ. Entomol.* **11**, 847–851.

Lamprey, H. F., Halevy, G., and Makacha, S. (1974). Interactions between *Acacia*, bruchid seed beetles and large herbivores. *East Afr. Wildl. J.* **12**, 81–85.

Lawton, R. O., and Putz, F. E. (1988). Natural disturbance and gap-phase regeneration in a wind-exposed tropical cloud forest. *Ecology* **69**, 764–777.

Lebrón, M. L. (1979). An autecological study of *Palicourea riparia* Bentham as related to rain forest disturbance in Puerto Rico. *Oecologia* **42**, 31–46.

Leck, M. A. (1979). Germination behavior of *Impatiens capensis* Meerb. (Balsaminaceae). *Bartonia* **46**, 1–14.

Leck, M. A. (1980). Germination in Barrow, Alaska tundra soil cores. *Arct. Alp. Res.* **12**, 343–349.

Leck, M. A., and Graveline, K. J. (1979). The seed bank of a freshwater tidal marsh. *Am. J. Bot.* **66**, 1006–1015.

Leck, M. A., and Simpson, R. L. (1987a). Seed bank of a freshwater tidal wetland: Turnover and relationship to vegetation change. *Am. J. Bot.* **74**, 360–370.

Leck, M. A., and Simpson, R. L. (1987b). Spore bank of a Delaware River freshwater tidal wetland. *Bull. Torrey Bot. Club* **114**, 1–7.

Leck, M. A., Simpson, R. L., Whigham, D. F., and Leck, C. F. (1988). Plants of the Hamilton Marshes: A Delaware River freshwater tidal wetland. *Bartonia* **54**, 1–17.

Leck, M. A., Simpson, R. L., and Parker, V. T. (1989). The seed bank of a freshwater tidal wetland and its relationship to vegetation dynamics. *In* "Freshwater Wetlands and Wildlife" (R. R. Sharitz and J. W. Gibbons, eds.) DOE-CONS 860326, Off. Sci. Tech. Info. USDOE. (in press).

Lee, J. A., and Ignaciuk, R. (1985). The physiological ecology of strandline plants. *Vegetatio* **62**, 319–326.

Lefkovitch, L. P. (1971). Some comments on the invariants of population growth. *In* "Statistical Ecology" (E. C. Pielou and W. E. Waters, eds.), Vol. 2, pp. 337–360. Pennsylvania State Univ. Press, University Park.

Leighton, M., and Wirawan, N. (1986). Catatrophic drought and fire in Borneo tropical rain forest associated with the 1982–1983 El Niño Southern Oscillation Event. *In* "Tropical Rain Forests and the World Atmosphere" (G. T. Prance, ed.), pp. 75–102. Westbury Press, Boulder, Colorado.

Leite, A. M. C., and Rankin, J. M. (1981). Ecologia de sementes de *Pithecolobium racemosum* Ducke. *Acta Amazonica* **11**, 309–318.

Leon, J. A. (1985). Germination strategies. *In* "Evolution: Essays in Honour of John Maynard Smith" (P. J. Greenwood, P. H. Harvey, and M. Slatkin, eds.), pp. 129–143. Cambridge Univ. Press, London.

Lerman, J. C., and Cigliano, E. (1971). New carbon–14 evidence for six hundred years old *Canna compacta* seed. *Nature (London)* **232**, 568–570.

Leslie, P. H. (1945). On the use of matrices in certain population mathematics. *Biometrika* **35**, 185–212.

Levin, D. A. (1978). Some genetic consequences of being a plant. *In* "Ecological Genetics: The Interface" (P. F. Brussard, ed.), pp. 189–212. Springer-Verlag, Berlin and New York.

Levin, D. A., and Turner, B. L. (1977). Clutch size in Compositae. *In* "Evolutionary Ecology" (B. Stonehouse and C. M. Perrine, eds.), pp. 215–222. Macmillan, New York.

Levin, S., Cohen, D., and Hastings, A. (1984). Dispersal strategies in patchy environments. *Theor. Popul. Biol.* **26**, 165–191.

Levins, R. (1969). Dormancy as an adaptive strategy. *In* "Dormancy and Survival" (H. W. Woolhouse, ed.), pp. 1–10. Academic Press, New York.

Levyns, M. R. (1966). *Haemanthus canaliculatus*, a new fire-lily from the western Cape Province. *J. Bot. S. Afr.* **32**, 73–75.

Lewandowska, A., and Skapski, H. (1979). [Evaluation of viable weed seeds in the soil following onion production in six regions of Poland.] *Biul. Warzywniczy* **23**, 285–305.

Lewis, H., and Epling, C. (1959). *Delphinium gypsophilum*, a diploid species of hybrid origin. *Evolution (Lawrence, Kans.)* **13**, 511–525.

Lewis, J. (1973). Longevity of crop and weed seeds: Survival after 20 years in soil. *Weed Res.* **13**, 179–191.

Lewis, R. R., Jr. (1982). "Creation and Restoration of Coastal Plant Communities." CRC Press, Boca Raton, Florida.

Lewontin, R. C. (1974). "The Genetic Basis of Evolutionary Change." Columbia Univ. Press, New York.

Lieberman, M., and Lieberman, D. (1986). An experimental study of seed ingestion and germination in a plant-animal assemblage in Ghana. *J. Trop. Ecol.* **2**, 113–126.

Lieffers, V. J., and Shay, J. M. (1981). The effects of water level on growth and reproduction of *Scirpus maritimus* var. *paludosus* on the Canadian prairies. *Can. J. Bot.* **59**, 118–121.

Liew, T. C. (1973). Occurrence of seeds in virgin forest top soil with particular reference to secondary species in Sabah. *Malay. For.* **36**, 185–193.

Liew, T. C., and Wong, F. O. (1973). Density, recruitment, mortality and growth of Dipterocarp seedlings in virgin and logged-over forests in Sabah. *Malay. For.* **36**, 3–15.

Lippert, R. D., and Hopkins, H. H. (1950). Study of viable seeds in various habitats in mixed prairies. *Trans. Kans. Acad. Sci.* **53**, 355–364.

Livingston, R. B., and Allessio, M. L. (1968). Buried viable seed in successional fields and forest stands, Harvard Forest, Massachusetts. *Bull. Torrey Bot. Club* **95**, 58–69.

Longman, K. A. (1969). The dormancy and survival of plants in the humid tropics. *Symp. Soc. Exp. Biol.* **23**, 471–488.

López-Quiles, M. M., and Vázquez-Yanes, C. (1976). Estudio sobre germinación de semillas en condiciones naturales controladas. *In* "Investigaciones sobre la Regeneración de Selvas Altas en Veracruz, México" (A. Gómez-Pompa, C. Vázquez-Yanes, S. del Amo Rodríguez, and A. Butanda Cervana, eds.), pp. 250–262. Compañia Editorial Continental, México, D. F.

Louda, S. M. (1978). A test of predispersal seed predation in the population dynamics of *Haplopappus* (Asteraceae). Ph.D. Thesis, University of California, Riverside, and San Diego State University, San Diego, California.

Louda, S. M. (1982a). Limitation of the recruitment of the shrub *Haplopappus squarrosus* (Asteraceae) by flower- and seed-feeding insects. *J. Ecol.* **70**, 43–53.

Louda, S. M. (1982b). Distribution ecology: Variation in plant recruitment over a gradient in relation to insect seed predation. *Ecol. Monogr.* **52**, 25–41.

Louda, S. M. (1982c). Inflorescence spiders: A cost/benefit analysis for the host plant, *Haplopappus venetus* Blake (Asteraceae). *Oecologia* **55**, 185–191.

Louda, S. M. (1983). Seed predation and seedling mortality in the recruitment of a shrub, *Haplopappus venetus* (Asteraceae), along a climatic gradient. *Ecology* **64**, 511–521.

Louda, S. M. (1984). Herbivore effect on stature, fruiting and leaf dynamics of a native crucifer. *Ecology* **65**, 1379–1386.

Louda, S. M. (1986). Insect herbivory in response to root-cutting and flooding stress on a native crucifer under field conditions. *Acta Oecol. [Ser.]: Oecol. Gen.* **7**, 37–53.

Louda, S. M. (1988). Insect pests and plant stress as considerations for revegetation of disturbed ecosystems. *In* "Rehabilitating Damaged Ecosystems" (J. Cairns, ed.), Vol. 2, pp. 51–67. CRC Press, Boca Raton, Florida.

Louda, S. M., and Rodman, J. E. (1983a). Ecological patterns in the glucosinolate content of a native mustard, *Cardamine cordifolia*, in the Rocky Mountains. *J. Chem. Ecol.* **9**, 397–422.

Louda, S. M., and Rodman, J. E. (1983b). Concentration of glucosinolates in relation to habitat and insect herbivory for the native crucifer *Cardamine cordifolia*. *Biochem. Syst. Ecol.* **11**, 199–207.

Louda, S. M., and Zedler, P. H. (1985). Predation in insular plant dynamics: An experimental assessment of postdispersal fruit and seed survival, Enewetak Atoll, Marshall Islands. *Am. J. Bot.* **72**, 438–445.

Louda, S. M., Dixon, P., and Huntly, N. J. (1987a). Herbivory in sun versus shade at a natural meadow-woodland ecotone in the Rocky Mountains. *Vegetatio* **72**, 87–93.

Louda, S. M., Farris, M. A., and Blua, M. J. (1987b). Variation in methylglucosinolate and insect damage to *Cleome serrulata* (Capparaceae) along a natural soil moisture gradient. *J. Chem. Ecol.* **13**, 569–581.

Louda, S. M., Handel, S. N., and Mischken, J. (1987c). Habitat related predispersal seed predation on a native plant, *Aster divaricatus* (Asteraceae). *Oikos* (in review).

Louda, S. M., Huntly, N. J., and Dixon, P. M. (1987d). Insect herbivory across a sun/shade gradient: Response to experimentally-induced in situ plant stress. *Acta Oecol. [Ser.] Oecol. Gen.* **8**, 357–361.

Louda, S. M., Keeler, K. H., and Holt, R. D. (1989a). Herbivore influences on plant performance and competitive interactions. *In* "Perspectives in Plant Competition" (J. B. Grace and D. Tilman, eds.). Academic Press, New York (in press).

Louda, S. M., Potvin, M. A., and Collinge, S. K. (1989b). Biological factors in the establishment of seedlings of Platte Thistle (*Circium canescens*) in sandhills prairie. *Proc. Prairie Conf., 10th* (in press).

Lubchenco, J. (1979). Consumer terms and concepts. *Am. Nat.* **113**, 315–317.

Lutz, H. J. (1956). The ecological effects of forest fires in the interior of Alaska. U.S., *Dep. Agric.*, Tech. Bull. **1133**.

Lyle, E. S., Jr. (1987). "Surface Mine Reclamation Manual." Am. Elsevier, New York.

Lyon, J. L., and Stickney, P. F. (1976). Early vegetational succession following large northern Rocky Mountain wildfires. *Proc. Tall Timbers Fire Ecol. Conf.* **14**, 355–375.

MacArthur, R. H. (1972). "Geographical Ecology: Patterns in the Distribution of Species." Harper & Row, New York.

Macedo, M. (1977). Dispersão de plantas lenhosas de uma Campina Amazônica. *Acta Amazonica* **7**, Suppl., 1–69.

Mack, R. N., and Pyke, D. A. (1983). The demography of *Bromus tectorum*: Variation in time and space. *J. Ecol.* **71**, 69–93.

MacMahon, J. A. (1979). North American deserts: Their floral and faunal components. *In* "Arid-land Ecosystems: Structure, Functioning and Management" (D. W. Goodall and R. A. Perry, eds.), Vol. 1, pp. 21–82. Cambridge Univ. Press, London and New York.

MacMahon, J. A., and Wagner, F. H. (1985). The Mojave, Sonoran and Chihuahuan deserts of North America. *In* "Ecosystems of the World" (M. Evenari, I. Noy-Meir, and D. W. Goodall, eds.), Vol. 12, pp. 105–202. Am. Elsevier, New York.

Madsen, C. (1986). Wetland restoration: A pilot project. *J. Soil Water Conserv.* **41**, 159–160.

Maher, R. V., White, D. J., Argus, G. W., and Keddy, P. A. (1978). "The Rare Vascular Plants of Nova Scotia," Syllogeus No. 18. National Museums of Canada, Ottawa, Ontario.

Major, J., and Pyott, W. T. (1966). Buried, viable seeds in two California bunchgrass sites and their bearing on the definition of a flora. *Vegetatio* **13**, 253–282.

Malanson, G. P., and O'Leary, J. F. (1985). Effects of fire and habitat on post-fire regeneration in mediterranean-type ecosystems: *Ceanothus spinosus* chaparral and Californian coastal sage scrub. *Acta Oecol. [Ser.] Oecol. Plant.* **6**, 169–181.

Mallick, A. U., and Gimingham, C. H. (1985). Ecological effects of heather burning. II. Effects on seed germination and vegetative regeneration. *J. Ecol.* **73**, 633–644.

Manders, P. T. (1986). Seed dispersal and seedling recruitment in *Protea laurifolia*. *S. Afr. J. Bot.* **52**, 421–424.

Mann, R. K., Rieck, C. E., and Witt, W. W. (1981). Germination and emergence of Burcucumber (*Sicyos angulatus*). *Weed Sci.* **29**, 83–86.

Manzur, M. I., and Courtney, S. P. (1984). Influence of insect damage in fruits of hawthorn on bird foraging and seed dispersal. *Oikos* **43**, 265–270.

Mares, M. A., and Rosenzweig, M. L. (1978). Granivory in North and South American deserts: Rodents, birds and ants. *Ecology* **59**, 235–241.

Marín, D., and Medina, E. (1981). Duración foliar, contenido de nutrientes y esclerofilia en árboles de un bosque muy seco tropical. *Acta Cient. Venez.* **32**, 508–514.

Marks, M. K. (1983a). Timing of seedling emergence and reproduction in some tropical dicotyledonous weeds. *Weed Res.* **23**, 325–332.

Marks, M. K. (1983b). Periodicity of seedling emergence in six monocotyledonous weeds of southeastern Nigeria. *Acta Oecol. [Ser.]: Oecol. Appl.* **4**, 75–85.

Marks, M. K., and Nwachuku, A. C. (1986). Seed bank characteristics in a group of tropical weeds. *Weed Res.* **26**, 151–157.

Marks, P. L. (1974). The role of pin cherry (*Prunus pensylvanica* L.) in the maintenance of stability in northern hardwood ecosystems. *Ecol. Monogr.* **44**, 73–88.

Marks, P. L. (1983). On the origin of field plants in the northeastern United States. *Am. Nat.* **122**, 210–228.

Marks, P. L., and Bormann, F. H. (1972). Revegetation following forest cutting: Mechanisms for return to steady-state nutrient cycling. *Science* **176**, 914–915.

Marks, P. L., and Mohler, C. L. (1985). Succession after elimination of buried seeds from a recently plowed field. *Bull. Torrey Bot. Club* **112**, 376–382.

Marks, T. C., and Truscott, A. J. (1985). Variation in seed production of *Spartina anglica* within a zoned saltmarsh. *J. Ecol.* **73**, 695–705.

Marquis, D. A. (1975). Seed storage and germination under northern hardwood forests. *Can. J. For. Res.* **5**, 478–484.

Marquis, D. A., Eckert, P. L., and Roach, B. A. (1976). Acorn weevils, rodents and deer all contribute to oak-regeneration difficulties in Pennsylvannia. *USDA For. Serv. Res. Pap. NE* **NE–356**.

Marquis, R. J., and Braker, H. E. (1988). Plant/herbivore interactions at La Selva: Diversity, specialization and impact on plant populations. *In* "La Selva: Ecology and Natural History of a Neotropical Rainforest" (K. S. Bawa, G. S. Hartshorn, H. H. Hespenheide, and L. DeLade, eds.). Organization for Tropical Studies, Durham, North Carolina (in press).

Marrero, J. (1949). Tree seed data from Puerto Rico. *Caribb. For.* **10**, 11–30.

Marshall, D. L., Levin, D. A., and Fowler, N. L. (1985). Plasticity in yield components in response to fruit predation and date of fruit initiation in three species of *Sesbania* (Leguminosae). *J. Ecol.* **73**, 71–81.

Marshall, D. R., and Jain, S. K. (1970). Seed predation and dormancy in the population dynamics of *Avena fatua* and *A. barbata. Ecology* **51**, 886–891

Martell, A. M. (1979). Selection of conifer seeds by deer mice and red backed voles. *Can. J. For. Res.* **9**, 201–204.

Martin, A. C., and Barkley, W. D. (1961). "Seed Identification Manual." Univ. of California Press, Berkeley.

Martin, A. R. H. (1966). The plant ecology of the Grahamstown Nature Reserve. II. Some effects of burning. *J. Bot. S. Afr.* **32**, 1–39.

Martin, J. L. (1955). Observations on the origin and early development of a plant community following a forest fire. *For. Chron.* **31**, 154–161.

Martineau, R. (1984). "Insects Harmful to Forest Trees." Multiscience Publications Ltd. in co-operation with Can. For. Serv., Environment Canada and the Government Publishing Centre, Supply and Services Canada, Ottawa, Ontario.

Martínez-Ramos, M., and Alvarez-Buylla, E. (1986). Seed dispersal, gap dynamics and tree recruitment: The case of *Cecropia obtusifolia* at Los Tuxtlas, Mexico. *In* "Frugivores and Seed Dispersal" (A. Estrada and T. H. Fleming, eds.), pp. 333–346. Junk, Dordrecht.

Matthews, J. D. (1963). Factors affecting the production of seed by forest trees. *For. Abstr.* **24**, 1–13.

Maun, M. A., and Cavers, P. B. (1971a). Seed production and dormancy in *Rumex crispus*. I. The effects of removal of cauline leaves at anthesis. *Can. J. Bot.* **49**, 1123–1130.

Maun, M. A., and Cavers, P. B. (1971b). Seed production and dormancy in *Rumex crispus*. II. The effects of removal of various proportions of flowers at anthesis. *Can. J. Bot.* **49**, 1841–1848.

Mayer, A. M. (1980–1981). Germination research—the state of the art. *Isr. J. Bot.* **29**, 1–3.

Mayer, A. M., and Poljakoff-Mayber, A. (1975). "The germination of Seeds." Macmillan, New York.

McCanny, S. J. (1986). The analysis and prediction of seed dispersal patterns. Ph.D. Thesis, Univ. of Western Ontario, London, Canada.

McCanny, S. J., and Cavers, P. B. (1988). The spread of proso millet (*Panicum miliaceum*) in Ontario, Canada. II. Dispersal by combines. *Weed Res.* **28**, 67–72.

McCarthy, K. A. (1987). Spatial and temporal distributions of species in two intermittant ponds in Atlantic County, New Jersey. M.S. Thesis, Rutgers University, Piscataway, New Jersey.

McCarty, M. K., and Lamp, W. O. (1982). Effect of a weevil, *Rhinocyllus conicus* on Musk thistle (*Carduus thoermeri*) seed production. *Weed Sci.* **30**, 136–140.

McDade, L. A., and Kinsman, S. (1980). The impact of floral parasitism in two neo-tropical hummingbird-pollinated plant species. *Evolution (Lawrence, Kans.)* **34**, 944–958.

McDonnell, M. J. (1986). Old field vegetation height and the dispersal pattern of bird-disseminated woody plants. *Bull. Torrey Bot. Club* **113**, 6–11.

McDonnell, M. J., and Stiles, E. W. (1983). The structural complexity of old field vegetation and the recruitment of bird-dispersed plant species. *Oecologia* **56**, 109–116.

McGraw, J. B. (1980). Seed bank size and distribution of seeds in cottongrass tussock tundra, Eagle Creek, Alaska. *Can. J. Bot.* **58**, 1607–1611.

McGraw, J. B. (1987). Seed bank properties of an Appalachian sphagnum bog and a model of the depth distribution of viable seeds. *Can. J. Bot.* **65**, 2028–2035.

McGraw, J. B., and Shaver, G. R. (1982). Seedling density and seedling survival in Alaskan cotton grass tussock tundra. *Holarct. Ecol.* **5**, 212–217.

McKeon, G. M., and Mott, J. J. (1982). The effect of temperature on the field softening of hard seed of *Stylosanthes humilis* and *S. hanata* in a dry monsoonal climate. *Aust. J. Agric. Res.* **33**, 75–85.

McKeon, G. M., and Mott, J. J. (1984). Seed biology of *Stylosanthes*. In "The Biology and Agronomy of *Stylosanthes*" (H. M. Stace and L. A. Edye, eds.), pp. 311–332. Academic Press, Sydney.

McKey, D. (1975). The ecology of coevolved seed dispersal systems. In "Coevolution of Animals and Plants" (L. E. Gilbert and P. H. Raven, eds.), pp. 159–191. Univ. of Texas Press, Austin.

McKey, D. (1978). Soils, vegetation and seed eating by black Colobus monkeys. In "The Ecology of Arboreal Folivores" (G. G. Montgomery, ed.), pp. 423–437. Smithsonian Institution Press, Washington, D.C.

McKey, D. (1979). The distribution of secondary compounds within plants. In "Herbivores" (G. A. Rosenthal and D. H. Janzen, eds.), pp. 56–133. Academic Press, New York.

McLaughlin, S. P. (1986). Floristic analysis of the Southwestern United States. *Great Basin Nat.* **46**, 46–65.

McLean, A. (1969). Fire resistance of forest species as influenced by root systems. *J. Range Manage.* **22**, 120–122.

McMaster, G. S., and Zedler, P. H. (1981). Delayed seed dispersal in *Pinus torreyana* (Torrey pine). *Oecologia* **51**, 62–66.

McMillan, C. (1981). Seed reserves and seed germination for two seagrasses, *Halodule wrightii* and *Syringodium filiforme*, from the western Atlantic. *Aquat. Bot.* **11**, 279–296.

McMillan, C. (1983). Seed germination in *Halodule wrightii* and *Syringodium filiforme* from Texas and the U.S. Virgin Islands. *Aquat. Bot.* **15**, 217–220.

McNaughton, S. J. (1983). Physiological and ecological implications of herbivory. In "Physiological Plant Ecology" (O. T. Lange, P. S. Nobel, C. B. Osmond, and H. Ziegler, eds.), Vol. 3, pp. 657–677. Springer-Verlag, Berlin.

McPherson, J. K., and Muller, C. H. (1969). Allelopathic effect of *Adenostoma fasciculatum*, 'chamise,' in the California chaparral. *Ecol. Monogr.* **39**, 177–198.

McRill, M., and Sagar, G. R. (1973). Earthworms and seeds. *Nature (London)* **243**, 482.

Medellin-Leal, F. (1982). The Chihuahuan Desert. In "Reference Handbook on the Deserts of North America" (G. L. Bender, ed.), pp. 321–371. Greenwood Press, Westport, Connecticut.

Meeks, R. L. (1969). Effect of drawdown date on wetland plant succession. *J. Wildl. Manage.* **38**, 817–821.

Meeuse, B. J. D. (1974). Seed and fruit. *In* "Encyclopedia Britannica" (Macropedia), Vol. 16, pp. 480–488. Encyclopedia Britannica, Chicago, Illinois.

Mehlhop, P. (1981). Forager size and polymorphism in ants of the genus *Veromessor* and their relation to foraging ecology. Ph.D. Dissertation, University of New Mexico, Albuquerque.

Mellanby, K. (1968). The effects of some mammals and birds on regeneration of oak. *J. Appl. Ecol.* **5**, 359–366.

Menges, R. M. (1987). Weed seed population dynamics during six years of weed management systems in crop rotations on irrigated soil. *Weed Sci.* **35**, 328–332.

Meredith, T. C. (1985). Factors affecting recruitment from the seed bank of sedge (*Cladius mariscus*) dominated communities at Wicken Fen, Cambridgeshire, England. *J. Biogeogr.* **12**, 463–472.

Mertz, D. B. (1971). Life history phenomena in increasing and decreasing populations. *In* "Statistical Ecology" (E. C. Pielou and W. E. Waters, eds.), Vol. 2, pp. 361–399. Pennsylvania State Univ. Press, University Park.

Midgley, A. R. (1926). Effects of alternate freezing and thawing on the impermeability of alfalfa and dodder seeds. *J. Am. Soc. Agron.* **18**, 1087-1098.

Miège, J., and Tchoumé, M. (1963). Influence d'arrosages reégulièrementrépétés sur la germination des graines en saison sècheè Dakar (Sénégal). *Ann. Fac. Sci., Univ. Dakar* **9**, 81–109.

Mika, V. (1978). Der Vorrat an keimfahigen Samen in Sudbohmischen Niedermoorboden. *Z. Acker-Pflanzen bau* **146**, 222–234.

Miller, G. E., Hedin, A. F., and Ruth, D. S. (1984). Damage by two Douglas Fir cone and seed insects: Correlation with cone crop size. *J. Entomol. Soc. B. C.* **81**, 46–50.

Miller, G. R., and Cummins, R. P. (1987). Role of buried viable seeds in the recolonization of disturbed ground by heather (*Calluna vulgaris* [L.] Hull) in the Cairgorm Mountains, Scotland, U. K. *Arct. Alp. Res.* **19**, 396–401.

Miller, P. C. (1981). "Resource Use by Chaparral and Matorral: A Comparison of Vegetation Function in Two Mediterranean-type Ecosystems." Springer-Verlag, Berlin.

Mills, J. N. (1983). Herbivory and seedling establishment in post-fire southern California charparral. *Oecologia* **60**, 267–270.

Milton, W. E. J. (1939). The occurrence of buried viable seeds in soils at different elevations and in a salt marsh. *J. Ecol.* **27**, 149–159.

Mitchell, R. (1977). Bruchid beetles and seed packaging by palo verde. *Ecology* **58**, 644–651.

Mitsch, W. J., and Gosselink, J. G. (1986). "Wetlands." Van Nostrand-Reinhold, New York.

Mittelbach, G. G., and Gross, K. L. (1984). Experimental studies of seed predation in old-fields. *Oecologia* **65**, 7–13.

Mladenoff, D. J. (1985). Dynamics of soil seed banks, vegetation, and nitrogen availability in treefall gaps. Ph.D. Dissertation, University of Wisconsin, Madison.

Molofsky, J., and Augspurger, C. K. (1988). The effect of litter on seed germination and seedling establishment in a tropical seasonal forest. *Bull. Ecol. Soc. Am. (Suppl.)* **69**, 236.

Montgomery, F. H. (1977). "Seeds and Fruits of Plants of Eastern Canada and Northeastern United States." Univ. of Toronto Press, Toronto.

Mooney, H. A., and Conrad, C. E., eds. (1977). "Proceedings of the Symposium on the Environmental Consequences of Fire and Fuel Management in Mediterranean Ecosystems," USDA For. Serv. Gen. Tech. Rep. WO-3. USDA, Washington, D.C.

Mooney, H. A., and Godron, M. (1983). "Disturbance and Ecosystems: Components of Response." Springer-Verlag, Berlin.

Moore, D. R. J., and Keddy, P. A. (1988). Effects of a water-depth gradient on the germination of lakeshore plants. *Can. J. Bot.* **66**, 548–552.

Moore, J. M., and Wein, R. W. (1977). Viable seed populations by soil depth and potential site recolonization after disturbance. *Can. J. Bot.* **55**, 2408–2412.

Moore, L. A., and Willson, M. F. (1982). The effect of micro-habitat, spatial distribution, and display size on dispersal of *Lindera benzoin* by avian frugivores. *Can. J. Bot.* **60**, 557–560.

Moore, L. R. (1978a). Seed predation in the legume *Crotalaria*. I. Intensity and variability of seed predation in native and introduced populations of *C. pallida* Ait. *Oecologia* **34**, 185–202.

Moore, L. R. (1978b). Seed predation in the legume *Crotalaria*. II. Correlates of interplant variability in predation intensity. *Oecologia* **34**, 203–223.

Moore, P. (1982). How to reproduce in bogs and fens. *New Sci.* **111**, 369–371.

Morash, R., and Freedman, B. (1983). Seedbanks in several recently clear-cut and mature hardwood forests in Nova Scotia. *Proc. N. S. Inst. Sci.* **33**, 85–94.

Moreno Casasola, P. (1976a). Viabilidad de semillas de árboles tropicales y templados: Una revisión bibliográfica. *In* "Investigaciones sobre la Regeneración de Selvas Altas en Veracruz, México" (A. Gómez-Pompa, C. Vázquez-Yanes, S. del Amo Rodríguez, and A. Butanda Cervana, eds.), pp. 471–526. Compañia Editorial Continental, México, D. F.

Moreno Casasola, P. (1976b). Latencia y viabilidad de semillas de vegetación primaria. *In* "Investigaciones sobre la Regeneración de Selvas Altas en Veracruz, México" (A. Gómez-Pompa, C. Vázquez-Yanes, S. del Amo Rodríguez, and A. Butanda Cervana, eds.), pp. 527–547. Compañia Editorial Continental, México, D. F.

Moreno Cassola, P. (1977). Latencia y viabilidad de semillas de árboles tropicales. *Interciencia* **2**, 298–301.

Morrow, P. A., and LaMarche, V. C. (1977). Tree ring evidence for chronic insect suppression of productivity in subalpine *Eucalyptus*. *Science* **201**, 1244–1245.

Mortimer, A. M. (1983). On weed demography. *In* "Recent Advances in Weed Research" (W. W. Fletcher, ed.), pp. 3–41. Commonwealth Agricultural Bureaux, Farnham Royal, United Kingdom.

Morton, S. R. (1985). Granivory in arid regions: Comparisons of Australia with North and South America. *Ecology* **66**, 1859–1866.

Mott, J. J., and Groves, R. H. (1981). Germination strategies. *In* "The Biology of Australian Plants" (J. S. Pate and A. J. McComb, eds.), pp. 307–341. Univ. of Western Australia Press, Nedlands.

Muller, C. H. (1947). Vegetation and climate of Coahuila, Mexico. *Madroño* **9**, 33–57.

Muller, C. H. (1953). The association of desert annuals with shrubs. *Am. J. Bot.* **40**, 53–60.

Muller, C. H., Hanawalt, R. B., and McPherson, J. K. (1968). Alleopathic control of herb growth in the fire cycle of California chaparral. *Bull. Torrey Bot. Club* **95**, 225–231.

Muller, R. N., and Bormann, F. H. (1976). Role of *Erythronium americanum* Ker. in energy flow and nutrient dynamics of a northern hardwood forest ecosystem. *Science* **193**, 1126–1128.

Mulroy, T. W., and Rundel, P. W. (1977). Annual plants: Adaptations to desert environments. *BioScience* **27**, 109–114.

Munz, P. A., and Keck, D.C. (1970). "A California Flora." Univ. of California Press, Berkeley.

Murray, D. R., ed. (1986a). "Seed Dispersal." Academic Press, Sydney.

Murray, D. R. (1986b). Seed dispersal by water. *In* "Seed Dispersal" (D. R. Murray, ed.), pp. 49–85. Academic Press, Sydney.

Murray, K. G. (1986). Consequences of seed dispersal for gap-dependent plants: Relationships between seed shadows, germination requirements, and forest dynamic processes. *In* "Frugivores and Seed Dispersal" (A. Estrada and T. H. Fleming, eds.), pp. 187–198. Junk, Dordrecht.

Murray, K. G. (1988). Avian seed dispersal of three neotropical gap-dependent plants. *Ecol. Monogr.* **58**, 271–298.

Naim, P. A. (1987). Wetland seed banks: Implications in vegetation management. Ph.D. Dissertation, Iowa State University, Ames.

Naka, K., and Yoda, K. (1984). Community dynamics of evergreen broadleaf forest in southwestern Japan. II. Species composition and density of seeds buried in the soil of a climax evergreen oak forest. *Bot. Mag.* **97**, 61–79.

Nakagoshi, N. (1981). Notes on the buried viable seeds in soils of forest communities on Mt. Futatabi, Kobe. *In* "Studies on Vegetation and Soil of the Permanent Natural Reserve Area in Mt. Futatabi, Kobi," Rep. 2nd Invest., pp. 69–94. Park Dept., Public Works Bureau, Kobe City Govt., Kobe, Japan.

Nakagoshi, N. (1984a). Buried viable seed populations in forest communities on the Hiba Mountains, southwestern Japan. *J. Sci. Hiroshima Univ., Ser. B, Div. 2* **19**, 1–56.

Nakagoshi, N. (1984b). Ecological studies on the buried viable seed population in soil of the forest communities of Miyajima Island, southwestern Japan. II. *Hikobia* **9**, 109–122.

Nakagoshi, N. (1985a). Buried viable seeds in temperate forests. *In* "The Population Structure of Vegetation" (J. White, ed.), pp. 551–570. Junk, Dordrecht.

Nakagoshi, N. (1985b). Notes on the buried viable seeds in soils of forest communities on Mt. Futatabi, Kobe. II. *In* "Studies on Vegetation and Soil of the Permanent Natural Reserve Area in Mt. Futatabi, Kobe," Rep. 3rd Invest. Park Dept., Public Works Bureau, Kobe City Govt., Kobe, Japan.

Narang, A. K., and Bhardwaj, N. (1974). Seed coat regulated germination in *Tephrosia appolina* DC. and its significance. *Int. J. Ecol. Environ. Sci.* **1**, 47–51.

Naveh, Z. (1973). The ecology of fire in Israel. *Proc. Tall Timbers Fire Ecol. Conf.* **13**, 131–170.

Naveh, Z. (1974). Effects of fire in the Mediterranean region. *In* "Fire and Ecosystems" (T. T. Kozlowski and C. E. Ahlgren, eds.), pp. 401–434. Academic Press, New York.

Naylor, R. E. L. (1972). Aspects of the population dynamics of the weed *Alopecurus myosuroides* Huds. in winter cereal crops. *J. Appl. Ecol.* **9**, 127–139.

Naylor, R. E. L. (1984). Seed ecology. *Adv. Res. Technol. Seeds* **9**, 61–91.

Naylor, R. E. L. (1986). Establishment and peri-establishment mortality. *In* "Studies of Plant Demography: A Festschrift for John L. Harper" (J. White, ed.), pp. 95–109. Academic Press, London.

Nelson, D. M., and Johnson, C. D. (1983). Stabilizing selection on seed size in *Astragalus* (Leguminosae) due to differential predation and differential germination. *J. Kans. Entomol. Soc.* **56**, 169–174.

Nelson, J. F., and Chew, R. M. (1977). Factors affecting seed reserves in the soil of a Mojave desert ecosystem, Rock Valley, Nye County, Nevada. *Am. Midl. Nat.* **97**, 300–321.

Nelson, J. R., and Wilson, A. M. (1969). Influence of age and awn removal on dormancy of medusahead seeds. *J. Range Manage.* **22**, 289–290.

New, T. R. (1983). Seed predation of some Australian Acacias by weevils (Coleoptera: Curculionidae). *Aust. J. Zool.* **31**, 345–352.

Ng, F. S. P. (1978). Strategies of establishment in Malayan forest trees. *In* "Tropical Trees as Living Systems" (P. B. Tomlinson and M. H. Zimmermann, eds.), pp. 129–162. Cambridge Univ. Press, London and New York.

Ng, F. S. P. (1980). Germination ecology of Malaysian woody plants. *Malay. For.* **43**, 406–437.

Ng, F. S. P. (1983). Ecological principles of tropical lowland rain forest conservation. *In* "Tropical Rain Forest: Ecology and Management" (S. L. Sutton, T. C. Whitmore, and A. C. Chadwick, eds.), pp. 359–375. Blackwell, Oxford.

Nichols, H. (1976). Historical aspects of the northern Canadian treeline. *Arctic* **29**, 38–47.

Nicholson, A., and Keddy, P. A. (1983). The depth profile of a shoreline seed bank in Matchedash Lake, Ontario. *Can. J. Bot.* **61**, 3293–3296.

Nikolaeva, M. G. (1977). Factors controlling the seed dormancy pattern. *In* "The Physiology and Biochemistry of Seed Dormancy and Germination" (A. A. Khan, ed.), pp. 51–74. North-Holland Publ., Amsterdam.

Nilsen, E., and Schlesinger, W. H. (1981). Phenology, productivity and nutrient accumulation in the post-fire shrub, *Lotus scoparius*. *Oecologia* **50**, 217–224.

Noble, I. R., and Slatyer, R. O. (1977). Post-fire succession of plants in Mediterranean ecosystems. *Gen. Tech. Rep. WO—U.S., For. Serv. [Wash. Off.]* **GTR-WO-3**, 27–36.

Noble, I. R., and Slatyer, R. O. (1980). The use of vital attributes to predict successional changes in plant communities subject to recurrent disturbances. *Vegetatio* **43**, 5–21.

Nobs, M. (1963). Experimental studies on species relationships in *Ceanothus*. *Carnegie Inst. Washington Publ.* **623**.

Numata, M., Hayashi, K., Komura, T., and Oki, K. (1964). Ecological studies on the buried-seed population in the soil as related to plant succession. 1. *Jpn. Ecol.* **14**, 207–215.

O'Dowd, D. J., and Gill, A. M. (1984). Predator satiation and site alteration following fire: Mass reproduction of alpine ash (*Eucalyptus delegatensis*) in southeastern Australia. *Ecology* **65**, 1052–1066.

O'Dowd, D. J., and Hay, M. E. (1980). Mutualism between harvester ants and a desert ephemeral: Seed escape from rodents. *Ecology* **61**, 531–540.

Ødum, S. (1965). Germination of ancient seeds. Floristical observations and experiments with archaeologically dated soil samples. *Dan. Bot. Ark.* **24**, 1–70.

Ohga, I. (1923). On the longevity of fruit of *Nelumbo nucifera*. *Bot. Mag.* **37**, 87–95.

Ohga, I. (1926a). The germination of century old and recently harvested Indian lotus fruits, with special reference to the effect of oxygen supply. *Am. J. Bot.* **13**, 754–759.

Ohga, I. (1926b). On the structure of some ancient, but still viable fruits of Indian lotus, with special reference to their prolonged dormancy. *Jpn. J. Bot.* **3**, 1–20.

Ohga, I. (1926c). A report on the longevity of the fruit of *Nelumbium*. *J. Bot., Br. For.* **64**, 154–157.

Ohmart, C. P. (1982). Destructive insects of native and planted *Pinus radiata* in California and their relevance to Australian forestry. *Aust. For. Res.* **12**, 151–161.

Oka, H. (1984). Secondary succession of weed communities in lowland habitats of Taiwan in relation to the introduction of wild rice (*Oryza perennis*) populations. *Vegetatio* **56**, 177–187.

Oldeman, R. A. A. (1978). Architecture and energy exchange of dicotyledonous trees in the forest. *In* "Tropical Trees as Living Systems" (P. B. Tomlinson and M. H. Zimmermann, eds.), pp. 535–560. Cambridge Univ. Press, London and New York.

Oliver, C. D. (1981). Forest development in North America following major disturbances. *For. Ecol. Manage.* **3**, 153–168.

Olmstead, N. W., and Curtis, J. D. (1947). Seeds of the forest floor. *Ecology* **28**, 49–52.

Olson, D. F. (1974). *Quercus* L. *U.S., Dep. Agric. Agric. Handb.* **450**, 692–743.

Oosting, H. F., and Humphreys, M. E. (1940). Buried viable seeds in successional series of old fields and forest soils. *Bull. Torrey Bot. Club* **67**, 253–273.

Orians, G. H. (1982). The influence of tree-falls in tropical forests in tree species richness. *Trop. Ecol.* **23**, 255–279.

Orians, G. H., and Janzen, D. H. (1974). Why are embryos so tasty? *Am. Nat.* **108**, 581–592.

Orme, M. L., and Leege, T. A. (1976). Emergence and survival of redstem (*Ceanothus sanquineus*) following prescribed burning. *Proc. Tall Timbers Fire Ecol. Conf.* **14**, 391–420.

Orozco Segovia, A. (1986). Fisiología ecológica del fotoblastismo en semillas de cuatro especies del género *Piper* L. Ph.D. Thesis, Universidad Nacional Autónoma de México, México, D. F.

O'Toole, J. J. (1982). Seed banks of *Panicum miliaceum* in three crops. M.Sc. Thesis, University of Western Ontario, London, Canada.

O'Toole, J. J., and Cavers, P. B. (1983). Input to seed banks of proso millet (*Panicum miliaceum*) in Southern Ontario. *Can. J. Plant Sci.* **63**, 1023–1030.

Owen, D. F., and Wiegert, R. G. (1981). Mutualism between grasses and grazers: An evolutionary hypothesis. *Oikos* **36**, 376–378.

Owen, O. S. (1980). "Natural Resource Conservation: an Ecological Approach." Macmillan, New York.

Paatela, J., and Erviö, L.-A. (1971). Weed seeds in cultivated soils in Finland. *Ann. Agric. Fenn.* **10**, 141 152.

Paige, K. N., and Whitham, T. G. (1987). Overcompensations in response to mammalian herbivory: The advantage of being eaten. *Am. Nat.* **129**, 407–416.

Pandya, S. M., and Baghela, N. (1973). Ecological studies of *Celosia argentea* Linn., a weed. I. Seed germination, seedling emergence and growth performance in different soil types. *Trop. Ecol.* **14**, 39–51.

Panetta, F. D. (1985). Population studies on pennyroyal mint (*Mentha pulegium* L.). II. Seed banks. *Weed Res.* **25**, 311–315.

Pareja, M. R., and Staniforth, D. W. (1985). Seed-soil microsite characteristics in relation to weed seed germination. *Weed Sci.* **33**, 190–195.

Pareja, M. R., Staniforth, D. W., and Pareja, G. P. (1985). Distribution of weed seed among soil structural units. *Weed Sci.* **33**, 182–189.

Parker, M. A., and Root, R. B. (1981). Insect herbivores limit habitat distribution of a native composite, *Machaeranthera canescens*. *Ecology* **62**, 1390–1392.

Parker, V. T. (1984). Correlation of physiological divergence with reproductive mode in chaparral shrubs. *Madroño* **31**, 231–242.

Parker, V. T. (1986). Effect of wet-season management burns on chaparral regeneration. *Program Issue Int. Congr. Ecol., 7th, 1986* Abstr., p. 266.

Parker, V. T. (1987a). Can native flora survive prescribed burns? *Fremontia* **15**(2), 3–6.

Parker, V. T. (1987b). Effect of wet-season management burns on chaparral regeneration: Implications for rare species. *In* "Rare and Endangered Plants: A Conference on Their Conservation and Management" (T. S. Elias, ed.), pp. 233–237. California Native Plant Soc. Publ., Sacramento.

Parker, V. T., and Leck, M. A. (1985). Relationships of seed banks to plant distribution patterns in a freshwater tidal wetland. *Am. J. Bot.* **72**, 161–174.

Parmenter, R. R., and MacMahon, J. A. (1983). Factors determining the abundance and distribution of rodents in a shrub-steppe ecosystem: The role of shrubs. *Oecologia* **59**, 145–156.

Patric, J. H., and Hanes, T. L. (1964). Chaparral succession in a San Gabriel Mountain area of California. *Ecology* **45**, 353–360.

Patten, D. T. (1978). Productivity and production efficiency of an upper Sonoran Desert ephemeral community. *Am. J. Bot.* **65**, 891–895.

Pawlowski, F. (1963). [Content and specific composition of weed seeds in the more important soils of the Lublin district.] *Ann. Univ. Mariae Curie-Sklodowska, Sect. E* **18**, 125–154.

Pawlowski, F., and Malicki, L. (1968). [Effects of different methods of ploughing on the vertical distribution of weed seeds in soil formed from loess.] *Ann. Univ. Mariae Curie-Sklodowska, Sect. E* **23**, 161–174.

Payette, S, Deshaye, J., and Gilbert, H. (1982). Tree seed populations at the treeline in Riviere aux Feuilles area, northern Quebec, Canada. *Arct. Alp. Res.* **14**, 215–221.

Payne, W. W. (1970). Preliminary reports on the flora of Wisconsin No. 62. Compositae. VI. Composite family. VI. The genus *Ambrosia*-The ragweeds. *Wis. Acad. Sci. Arts Lett.* **58**, 353–371.

Pederson, R. L. (1981). Seed bank characteristics of the Delta Marsh, Manatoba: Applications for wetland management. *In* "Selected Proceedings of the Midwest Conference on Wetland Values and Management" (B. Richardson, ed.), pp. 61–69. Freshwater Soc., Navarre, Minnesota.

Pederson, R. L. (1983). Abundance, distribution, and diversity of buried seed in the Delta Marsh, Manitoba, Canada. Ph.D. Dissertation, Iowa State University, Ames.

Pederson, R. L., and van der Valk, A. G. (1984). Vegetation change and seed banks in marshes: Ecological and management implications. *Trans. North Am. Wildl. Nat. Resour. Conf.* **49**, 271–280.

Pérez-Nasser, N. (1985). Viabilidad en el suelo de las semillas de la vegetación de Los Tuxtlas, Veracruz. Tesis Profesional, Universidad Nacional Autónoma de México, México, D. F.

Petal, J. (1978). The role of ants in ecosystems. *In* "Production Ecology of Ants and Termites" (M. V. Brian, ed.), pp. 293–325. Cambridge Univ. Press, London and New York.

Petrov, V. V., and Belyaeva, Y. E. (1981). Store of plant seeds in soil under forest canopy. *Lesovedenie*, pp. 19–38.

Pfadenhauer, J., and Maas, D. (1987). Samenpotential in Niedermoorböden des Alpenvorlandes bei Grünlandnutzung unterschiedlicher Intensität. *Flora (Jena)* **179**, 85–97.

Pfister, R. D., Kovalchik, B. L., Arno, S. F., and Presby, R. C. (1977). Forest habitat types of Montana. *USDA For. Serv. Gen. Tech. Rep. Int* **INT-34**.

Philipupillai, J., and Ungar, I. A. (1984). The effect of seed dimorphism on the germination and survival of *Salicornia europaea* L. populations. *Am. J. Bot.* **71**, 542–549.

Pickett, S. T. A., and Baskin, J. M. (1973). The role of temperature and light in the germination behavior of *Ambrosia artemisiifolia*. *Bull. Torrey Bot. Club* **100**, 165–170.

Pickett, S. T. A., and White, P. S. (1985a). "The Ecology of Natural Disturbance and Patch Dynamics." Academic Press, Orlando, Florida.

Pickett, S. T. A., and White, P. S. (1985b). Patch dynamics: A synthesis. *In* "The Ecology of Natural Disturbance and Patch Dynamics (S. T. A. Pickett and P. S. White, eds.), pp. 371–384. Academic Press, Orlando, Florida.

Pickett, S. T. A., Collins, S. L., and Armento, J. J. (1987). Models, mechanisms and pathways of succession. *Bot. Rev.* **53**, 335–371.

Pieterse, P. J., and Cairns, A. L. P. (1986). The effect of fire on an *Acacia longifolia* seed bank in the south-western Cape. *S. Afr. J. Bot.* **52**, 233–236.

Piroznikow, E. (1983). Seed bank in the soil of stabilized ecosystem of a deciduous forest (*Tilio-Carpinetum*) in the Bialowieza National Park. *Ekol. Pol.* **31**, 145–172.

Pitty, A., Staniforth, D. W., and Tiffany, L. H. (1987). Fungi associated with caryopses of *Setaria* species from field-harvested seeds and from soil under two tillage systems. *Weed Sci.* **35**, 319–323.

Platt, W. J. (1975). The colonization and formation of equilibrium plant species associations on badger disturbances in a tallgrass prairie. *Ecol. Monogr.* **45**, 285–305.

Platt, W. J. (1976). The natural history of a fugitive prairie plant (*Mirabilis hirsuta* (Pursh.) MacM.). *Oecologia* **22**, 399–409.

Platt, W. J., and Weiss, I. M. (1977). Resource partitioning and competition within a guild of fugitive prairie plants. *Am. Nat.* **111**, 479–513.

Platt, W. J., and Weiss, I. M. (1985). An experimental study of competition among fugitive prairie plants. *Ecology* **66**, 708–720.

Platt, W. J., Hill, G. R., and Clark, S. (1974). Seed production in a prairie legume (*Astragalus canadensis* L.): Interactions between pollination, predispersal seed predation and plant density. *Oecologia* **17**, 55–63.

Poiani, K. A. (1987). The effect of hydroperiod on seed banks in semi-permanent prairie wetlands. M.S. Thesis, Virginia Polytechnic Institute and State University, Blacksburg, Virginia.

Ponce de León G, L. (1982). L'écophysiologie de la germination d'espèces forestières et de savane den rapport avec la dynamique de la végétation den Côte d'Ivoire. *Bull. Liaison Cher. Lamto*, [N.S.] **1**, 1–144.

Pons, T. L. (1983). Significance of inhibition of seed germination under the leaf canopy in ash coppice. *Plant, Cell Environ.* **6**, 385–392.

Porsild, A. E., Harington, C. R., and Mulligan, G. A. (1967). *Lupinus arcticus* Wats. grown from seeds of Pleistocene age. *Science* **158**, 113–114.

Pratt, D. W., Black, R. A., and Zamora, B. A. (1984). Buried viable seed in a ponderosa pine community. *Can. J. Bot.* **62**, 44–52.

Prévost, M. F. (1981). Mise en évidence de graines d'espèces pionnières dans le sol de forêt primaire en Guyane. *Turrialba* **31**, 121–127.

Price, M. V. (1978a). The role of microhabitat in structuring desert rodent communities. *Ecology* **59**, 910–921.

Price, M. V. (1978b). Seed dispersion preferences of coexisting desert rodent species. *J. Mammal.* **59**, 624–626.

Price, M. V., and Heinz, K. M. (1984). Effects of body size, seed density, soil characteristics on rates of seed harvest by heteromyid rodents. *Oecologia* **61**, 420–425.

Price, M. V., and Jenkins, S. H. (1987). Rodents as seed consumers and dispersers. *In* "Seed Dispersal" (D. R. Murray, ed.), pp. 191–235. Academic Press, Sydney.

Price, M. V., and Reichman, O. J. (1987). Distribution of seeds in Sonoran desert soils: Implications for heteromyid rodent foraging. *Ecology* **68**, 1797–1811.

Priestley, D. A. (1986). "Seed Aging: Implications for Seed Storage and Persistence in the Soil." Cornell Univ. Press, Ithaca, New York.

Pulliam, H. R., and Brand, M. R. (1975). The production and utilization of seeds in plains grassland of southeastern Arizona. *Ecology* **56**, 1156–1166.

Purata, S. E. (1986). Floristic and structural changes during old-field succession in the Mexican tropics in relation to site history and species availability. *J. Trop. Ecol.* **2**, 257–276.

Purdie, R. W. (1977). Early stages of regeneration after burning in dry sclerophyll vegetation. I. Regeneration by seed germination. *Aust. J. Bot.* **25**, 35–46.

Purdie, R. W., and Slatyer, R. O. (1976). Vegetation succession after fire in sclerophyll woodland communities in south-eastern Australia. *Aust. J. Ecol.* **1**, 223–296.

Putz, F. E. (1983). Treefall pits and mounds, buried seeds, and the importance of soil disturbance to pioneer trees on Barro Colorado Island, Panama. *Ecology* **64**, 1069–1074.

Putz, F. E., and Appanah, B. (1987). Buried seeds, newly dispersed seeds, and the dynamics of a lowland forest in Malaysia. *Biotropica* **19**, 326–339.

Putz, F. E., and Brokaw, N. V. L. (in review). Resprouting of broken trees in a tropical moist forest. Unpublished manuscript.

Quick, C. R. (1935). Notes on the germination of *Ceanothus* seeds. *Madroño* **3**, 23–30.

Quick, C. R. (1956). Viable seeds from the duff and soil of sugar pine forests. *For. Sci.* **2**, 36–42.

Quick, C. R. (1959). *Ceanothus* seeds and seedlings on burns. *Madroño* **15**, 79–81.

Quick, C. R. (1975). Seed longevity of the Sierra gooseberry. *Madroño* **23**, 236.

Quick, C. R., and Quick, A. S. (1961). Germination of *Ceanothus* seeds. *Madroño* **16**, 23–30.

Quinlivan, B. J. (1961). The effect of constant and fluctuating temperatures on the permeability of hard seeds of some legume species. *Aust. J. Agric. Res.* **12**, 1009–1022.

Quinlivan, B. J. (1966). The relationship between temperature fluctuations and the softening of hard seeds of some legume species. *Aust. J. Agric. Res.* **17**, 625–631.

Quinlivan, B. J. (1968). Seed coat impermeability in the common annual legume pasture species of Western Australia. *Aust. J. Exp. Agric. Anim. Husb.* **8**, 695–701.

Quinlivan, B. J., and Nicol, H. I. (1971). Embryo dormancy in subterranean clover seeds. I. Environmental control. *Aust. J. Agric. Res.* **22**, 599–606.

Rabinowitz, D. (1978). Early growth of mangrove seedlings in Panama, and an hypothesis concerning the relationship of dispersal and zonation. *J. Biogeogr.* **5**, 113–133.

Rabinowitz, D. (1981). Buried viable seed in a North American tall-grass prairie: The resemblance of their abundance and composition to dispersing seeds. *Oikos* **36**, 191–195.

Rabinowitz, D., and Rapp, J. K. (1980). Seed rain in a North American tall grass prairie. *J. Appl. Ecol.* **17**, 793–802.

Radwan, M. A., and Crouch, G. L. (1977). Seed germination and seedling establishment of redstem *Ceanothus*. *J. Wildl. Manage.* **41**, 760–766.

Ramsbottom, J. (1942a). Duration of viability in seeds. *Gard. Chron.* **111**, 234.

Ramsbottom, J. (1942b). Recent work on germination. *Nature (London)* **149**, 658–659.

Ratcliffe, D. (1961). Adaptation to habitat in a group of annual plants. *J. Ecol.* **49**, 187–203.

Rathke, B., and Lacey, E. P. (1985). Phenological patterns of terrestrial plants. *Annu. Rev. Ecol. Syst.* **16**, 179–214.

Rauh, W. (1985). The Peruvian-Chilean deserts. *In* "Ecosystems of the World" (M. Evenari, I. Noy-Meir, and D. W. Goodall, eds.), Vol. 12, pp. 239–267. Am. Elsevier, New York.

Raunkiaer, C. (1934). "The Life Forms of Plants and Statistical Plant Geography." Oxford Univ. Press (Clarendon), London and New York.

Raup, H. M. (1957). Vegetational adjustment to the instability of the site. *Proc. Pap. Tech. Meet., Int. Union Conserv. Nat. Nat. Resour., 6th.*

Raup, H. M. (1975). Species versatility in shore habitats. *J. Arnold Arbor., Harv. Univ.* **56**, 126–163.

Raven, P. H. (1960). Notes on plants of the 1959 Refugio Pass burn, Santa Barbara County, California. *Leafl. West. Bot.* **9**, 87–91.

Reader, R. J., and Buck, J. (1986). Topographic variation in the abundance of *Hieracium floribundum*: Relative importance of differential seed dispersal, seedling establishment, plant survival and reproduction. *J. Ecol.* **74**, 815–822.

Reichman, O. J. (1975). Relation of desert rodent diets to available resources. *J. Mammal.* **56**, 731–749.

Reichman, O. J. (1977). Optimization of diets through food preferences by heteromyid rodents. *Ecology* **58**, 454–457.

Reichman, O. J. (1979). Desert granivore foraging and its impact on seed densities and distributions. *Ecology* **60**, 1085–1092.

Reichman, O. J. (1984). Spatial and temporal variation of seed distributions in Sonoran Desert soils. *J. Biogeogr.* **11**, 1–11.

Reichman, O. J., and Oberstein, D. (1977). Selection of seed distribution types by *Dipodomys merriami* and *Perognathus amplus*. *Ecology* **58**, 636–643.

Reid, R. (1987). A little bit of Atlantic. *Seasons* **27**(4), 20–23.

Reukema, D. L. (1982). Seedfall in a young growth Douglas-fir stand: 1950–1978. *Can. J. For. Res.* **12**, 249–254.

Reynolds, D. (1984). Populational dynamics of three annual species of alpine plants in the Rocky Mountains. *Oecologia* **62**, 250–255.

Reynolds, H. G. (1958). The ecology of Merriam kangaroo rat (*Dipodomys merriami* Mearns) on the grazing lands of southern Arizona. *Ecol. Monogr.* **28**, 111–127.

Rice, B., and Westoby, M. (1981). Myrmecochory in sclerophyll vegetation of the West Head, New South Wales. *Aust. J. Ecol.* **6**, 291–298.

Rice, K. J. (1985). Responses of *Erodium* to varying microsites: The role of germination cueing. *Ecology* **66**, 1651–1657.

Rice, K. J. (1986). Interaction of disturbance patch size and herbivory in *Erodium* colonization. *Ecology* **68**, 1113–1115.

Richards, P. W. (1952). "The Tropical Rain Forest." Cambridge Univ. Press, London and New York.

Ricklefs, R. E. (1977). Environmental heterogeneity and plant species diversity: A hypothesis. *Am. Nat.* **111**, 376–381.

Ridley, H. N. (1930). "The Dispersal of Plants Throughout the World." Reeve, Ashford.

Riera, B. (1985). Importance des buttes de déracinement dans la régénération forestière en Guyane française. *Rev. Ecol. (Terre Vie)* **40**, 321–330.

Riggan, P. J., Franklin, S., and Brass, J. A. (1986). Fire and chaparral management at the chaparral/urban interface. *Fremontia* **14**(3), 28–30.

Rincker, C. M. (1954). Effect of heat on impermeable seeds of alfalfa, sweet clover and red clover. *Agron. J.* **46**, 247–250.

Risch, S. J. (1977). Effect of isolation of hollyhocks (*Althaea rosea*) on seed predation by a curculionid beetle. *J. Kans. Entomol. Soc.* **50**, 149–156.

Risch, S. J., and Carroll, C. R. (1986). Effects of seed predation by a tropical ant on competition among weeds. *Ecology* **67**, 1319–1327.

Risser, P. G., Birney, E. C., Blocker, H. D., May, S. W., and Wiens, J. A. (1981). "The True Prairie Ecosystem," US/IBP Synth. Ser. 16. Hutchinson Ross, New York.

Rissing, S. W. (1986). Indirect effects of granivory by harvester ants: Plant species composition and reproductive increase near ant nests. *Oecologia* **68**, 231–234.

Riswan, S. (1982). Ecological studies on primary, secondary and experimentally cleared mixed dipterocarp forest and kerangas forest in East Kalimantan, Indonesia. Ph. D. Thesis, Aberdeen University, Aberdeen, Scotland.

Rizzini, C. T. (1977). Nota sobre um embriao dormente em Leguminosa esclerodermica. *Rodriguesia* **42**, 33–39.

Roach, D. A. (1983). Buried seed and standing vegetation in two adjacent tundra habitats, northern Alaska. *Oecologia* **60**, 359–364.

Roberts, E. H. (1960). The viability of cereal seed in relation to temperature and moisture. *Ann. Bot. (London)* [N.S.] **24**, 12–30.

Roberts, E. H., ed. (1972a). "Viability of Seeds." Chapman & Hall, London.

Roberts, E. H. (1972b). Dormancy: A factor affecting seed survival in the soil. *In* "Viability of Seeds" (E. H. Roberts, ed.), pp. 321–359. Chapman & Hall, London.

Roberts, H. A. (1958). Studies on the weeds of vegetable crops. I. Initial effects on cropping the weed seeds in the soil. *J. Ecol.* **46**, 759–768.

Roberts, H. A. (1962). Studies on the weeds of vegetable crops. II. Effect of six years of cropping on the weed seeds in the soil. *J. Ecol.* **50**, 803–813.

Roberts, H. A. (1963a). Studies on the weeds of vegetable crops. III. Effect of different primary cultivations on the weed seeds in the soil. *J. Ecol.* **51**, 83–95.

Roberts, H. A. (1963b). Studies on the weeds of vegetable crops. IV. Further observations on the effects of different primary cultivations. *J. Ecol.* **51**, 323–332.

Roberts, H. A. (1966). The seed population of the soil and its implications for weed control. *Proc. Ir. Crop Protection Conf., 1966* pp. 14–22.

Roberts, H. A. (1968). The changing population of viable weed seeds in an arable soil. *Weed Res.* **8**, 253–256.

Roberts, H. A. (1970). Viable weed seeds in cultivated soils. *Rep. Natl. Veg. Res. Stn., 1969* pp. 25–38.

Roberts, H. A. (1981). Seed banks in soil. *Adv. Appl. Biol.* **6**, 1–55.

Roberts, H. A. (1983). Weed seeds in horticultural soils. *Sci. Hortic. (Canterbury, Engl.)* **34**, 1–11.

Roberts, H. A., and Boddrell, J. E. (1983). Field emergence and temperature requirements for germination in *Solanum sarrachoides* Sendt. *Weed Res.* **23**, 247–252.

Roberts, H. A., and Dawkins, P. A. (1967). Effect of cultivation on the numbers of viable weed seeds in soil. *Weed Res.* **7**, 290–301.

Roberts, H. A., and Feast, P. M. (1972). Fate of seeds of some annual weeds in different depths of cultivated and undisturbed soil. *Weed Res.* **12**, 316–324.

Roberts, H. A., and Feast, P. M. (1973a). Emergence and longevity of seeds of annual weeds in cultivated and undisturbed soil. *J. Appl. Ecol.* **10**, 133–143.

Roberts, H. A., and Feast, P. M. (1973b). Changes in the numbers of viable weed seeds in soil under different regimes. *Weed Res.* **13**, 298–303.

Roberts, H. A., and Neilson, J. E. (1981). Changes in the soil seed bank of four long-term crop/herbicide experiments. *J. Appl. Ecol.* **18**, 661–668.

Roberts, H. A., and Neilson, J. E. (1982a). Seasonal changes in the temperature requirements for germination of buried seeds of *Aphanes arvensis* L. *New Phytol.* **92**, 159–166.

Roberts, H. A., and Neilson, J. E. (1982b). Seed banks of soils under vegetable cropping in England. *Weed Res.* **22**, 13–16.

Roberts, H. A., and Ricketts, M. E. (1979). Quantitative relationships between the weed flora after cultivation and the seed population in the soil. *Weed Res.* **19**, 269–275.

Roberts, H. A., and Stokes, F. G. (1965). Studies on the weeds of vegetable crops. V. Final observations on an experiment with different primary cultivations. *J. Appl. Ecol.* **2**, 307–315.

Roberts, J. T., and Heithaus, E. R. (1986). Ants rearrange the vertebrate-generated seed shadow of a neotropical fig tree. *Ecology* **67**, 1046–1051.

Roberts, T. L., Carson, W. P., and Vankat, J. L. (1984). The seed bank and the initial revegetation of disturbance sites in Hueston Woods State Nature Preserve. *In* "Hueston Woods State Park and Nature Preserve" (G. E. Willeke, ed.), pp. 150–155. Miami University, Oxford, Ohio.

Robertson, D. J. (1985). "Freshwater Wetland Reclamation in Florida: An Overview," Publ. No. 03–000–033. Florida Inst. Phosphate Res., Bartow.

Rockwood, L. L. (1973). The effect of defoliation on seed production in six Costa Rican tree species. *Ecology* **54**, 1363–1369.

Rogers, K. H., and Breen, C. M. (1980). Growth and reproduction of *Potamogeton crispus* in a South African lake. *J. Ecol.* **68**, 561–571.

Roland, A. E., and Smith, E. C. (1969). The flora of Nova Scotia. *Proc. N. S. Inst. Sci.* **26**, 5–238, 277–743.

Rollet, B. (1983a). La régénération naturelle dans les trouées, un processus général de la dynamique des forêts tropicales humides. *Rev. Bois For. Trop.* **201**, 3–34.

Rollet, B. (1983b). *Rev. Bois For. Trop.* **202**, 19–34,

Rolston, M. P. (1978). Water impermeable seed dormancy. *Bot. Rev.* **44**, 365–396.

Rosenthal, G. A., and Bell, E. A. (1979). Naturally occurring toxic non-protein amino acids. *In* "Herbivores" (G. A. Rosenthal and D. H. Janzen, eds.), pp. 353–385. Academic Press, New York.

Rosenzweig, M. L., and Sterner, P. W. (1970). Population ecology of desert rodent communities: Body size and seed husking as bases for heteromyid coexistences. *Ecology* **51**, 217–224.

Rowe, J. S. (1972). Forest regions of Canada. *Can. For. Serv. Dep. Fish. Environ., Publ.* **1300**.

Rowe, J. S. (1983). Concepts of fire effects on plant individuals and species. *In* "The Role of Fire in Northern Circumpolar Ecosystems" (R. W. Wein and D. A. MacLean, eds.), pp. 135–154. Wiley, New York.

Rudolf, P. O. (1974). *Aesculus* L. *U.S., Dep. Agric., Agric. Handb.* **450**, 195–200.

Rundel, P. W., and Parsons, D. J. (1984). Post-fire uptake of nutrients by diverse ephemeral herbs in chamise chaparral. *Oecologia* **61**, 285–288.

Ryvarden, L. (1971). Studies in seed dispersal. I. Trapping of diaspores in the alpine zone at Finse, Norway. *Norw. J. Bot.* **18**, 215–226.

Ryvarden, L. (1975). Studies in seed dispersal. II. Winter-dispersed species at Finse, Norway. *Norw. J. Bot.* **22**, 21–24.

Sabatier, D. (1985). Saisonnalité et déterminisme du pic fructificationen forêt guyanaise. *Rev. Ecol. (Terre Vie)* **40**, 289–320.

Sabiiti, E. N., and Wein, R. W. (1987). Fire and *Acacia* seeds: A hypothesis of colonization success. *J. Ecol.* **74**, 937–946.

Sagar, G. R. (1982). An introduction to the population dynamics of weeds. *In* "Biology and Ecology of Weeds" (W. Holzner and M. Numata, eds.), pp. 161–168. Junk, The Hague.

Saini, H. S., Bassi, P. K., and Spencer, M. S. (1986). Interactions among ethephon, nitrate and after-ripening in the release of dormancy of wild oat (*Avena fatua*) seed. *Weed Sci.* **34**, 43–47.

Salisbury, E. (1970). The pioneer vegetation of exposed muds and its biological features. *Philos. Trans. R. Soc. London, Ser. B* **259**, 207–255.

Salisbury, E. (1975). The survival value of modes of dispersal. *Proc. R. Soc. London, Ser. B* **188**, 183–188.

Salisbury, E. J. (1942). "The Reproductive Capacity of Plants." Bell, London.

Salisbury, E. J. (1961). "Weeds and Aliens." Collins, London.

Salmerón, R. (1984). Germinación de semillas acumuladas en el suelo de una selva húmeda tropical Los Tuxtlas, Veracruz, México. Tesis Profesional, Universidad Nacional Autónoma de México, México, D. F.

Salter, J. (1857). On the viability of seeds after prolonged submersion in the sea. *J. Linn. Soc. London* **1**, 140–142.

Sampford, M. R. (1962). "An Introduction to Sampling Theory." Oliver & Boyd, Edinburgh.

Sampson, A. W. (1944). Plant succession on burned chaparral lands in northern California. *Univ. Calif. Agric. Exp. Stn., Bull.* **685**.

Sarmiento, G., and Monasterio, M. (1983). Life forms and phenology. *In* "Ecosystems of the World" (F. Bourlière, ed.), Vol. 13, pp. 79–108. Elsevier, Amsterdam.

Sarukhán, J. (1974). Studies on plant demography: *Ranunculus repens* L., *R. bulbosus* L. and *R. acris* L. II. Reproductive strategies and seed population dynamics. *J. Ecol.* **62**, 151–177.

Sarukhán, J. (1978). Studies of the demography of tropical trees. *In* "Tropical Trees as Living Systems" (P. B. Tomlinson and M. H. Zimmermann, eds.), pp. 163–184. Cambridge Univ. Press, London and New York.

Sarukhán, J. (1980). Demographic problems in tropical systems. *In* "Demography and Evolution in Plant Populations" (O. T. Solbrig, ed.), pp. 161–198. Univ. of California Press, Berkeley.

Sarukhán, J., and Gadgil, M. (1974). Studies on plant demography: *Ranunculus repens* L., *R. bulbosus* L. and *R. acris* L. III. A mathematical model incorporating multiple modes of reproduction. *J. Ecol.* **62**, 921–936.

Sastroutomo, S. S. (1981). Turion formation, dormancy and germination of curley pond-weed, *Potamogeton crispus* L. *Aquat. Bot.* **10**, 161–173.

Saulei, S. M. (1984). Natural regeneration following clear-fell logging operations in the Gogol Valley, Papua New Guinea. *Ambio* **13**, 351–354.

Saulei, S. M., and Swaine, M. D. (1988). Dynamics of seed rain and soil seed banks during forest clearance and subsequent regrowth in the Gogol Valley, Papua New Guinea. *J. Ecol.* **76**, 1133–1152.

Sayers, R. L. , and Ward, R. T. (1966). Germination responses in alpine species. *Bot. Gaz. (Chicago)* **127**, 11–16.

Schlising, R. A. (1980). Seed destruction of California morning glories (Convolvulaceae: *Calystegia*) by bruchid beetles. *Madroño* **27**, 1–16.

Schmidt, K. P., and Lawlor, L. R. (1983). Growth rate projection and life history sensitivity for annual plants with a seed bank. *Am. Nat.* **121**, 525–539.

Schmidt, R. H., Jr. (1979). A climatic delineation of the 'real' Chihuahuan desert. *J. Arid Environ.* **2**, 243–250.

Schneider, R. L., and Sharitz, R. R. (1986). Seed bank dynamics in a southeastern riverine swamp. *Am. J. Bot.* **73**, 1022–1030.

Schneider, R. L., and Sharitz, R. R. (1988). Hydrochory: A critical influence in regeneration of a bald cypress-water tupelo forest. *Ecology* **69**, 1055–1063.

Schoen, D. J., and Lloyd, D. G. (1983). The selection of cleistogamy and heteromorphic diaspores. *Biol. J. Linn. Soc.* **23**, 303–322.

Scholander, P. F., Flagg, W., Hock, R. J., and Irving, L. (1953). Studies on the physiology of frozen plants and animals in the Arctic. *J. Cell. Comp. Physiol.* **42**, Suppl. I, 1–56.

Schopmeyer, C. S. (1974). "Seeds of Woody Plants in the United States," Agric. Handb. No. 450. USDA For. Serv., Washington, D.C.

Schulz, J. P. (1960a). "Ecological Studies on Rain Forest in Northern Surinam." N. V. Noord-Hollandsche Uitgevers Maatschappij, Amsterdam.

Schulz, J. P. (1960b). *Verh. K. Ned. Akad. Wet., Afd. Natuur., Reeks* 2 **53**, 1–267.

Schupp, E. W. (1988). Seed and early seedling predation in the forest understory and in treefall gaps. *Oikos* **51**, 71–78.

Schupp, E. W., Howe, H. F., Augspurger, C. K., and Levy, D. J. (in press). Arrival and survival in tropical treefall gaps. *Ecology*.

Schweizer, E. E., and Zimdahl, R. L. (1984a). Weed seed decline in irrigated soil after six years of continuous corn (*Zea mays*) and herbicides. *Weed Sci.* **32**, 76–83.

Schweizer, E. E., and Zimdahl, R. L. (1984b). Weed seed decline in irrigated soil after rotation of crops and herbicides. *Weed Sci.* **32**, 84–89.

Schweizer, E. E., Lybecker, D. W., Zimdahl, R. L., and King, R. P. (1986). Weed management strategies in corn based on bioeconomic modeling. *Proc. West. Soc. Weed Sci.* **39**, 126 (abstr.).

Schwerzel, P. J. (1976). The effect of depth of burial in soil on the survival of some common Rhodesian weed seeds. *Rhod. Agric. J.* **73**, 97–99.

Schwerzel, P. J., and Thomas, P. E. L. (1979). Effects of cultivation frequency on the survival of seeds of six weeds commonly found in Zimbabwe Rhodesia. *Zimbabwe Rhod. Agric. J.* **26**, 195–199.

Scott, D. R. (1970). Feeding of *Lygus* bugs (Hemiptera: Miridae) on developing carrot and bean seed: Increased growth and yields of plants grown from seed. *Ann. Entomol. Soc. Am.* **63**, 1604–1608.

Scott, J. K. (1982). The impact of destructive insects on reproduction in six species of *Banksia* L.f. (Proteaceae). *Aust. J. Zool.* **30**, 901–921.

Scott, J. K., and Black, R. (1981). Selective predation by white-tailed Black Cockatoos on fruit of *Banksia attenuata* containing the seed-eating weevil *Alphitopis nivea*. *Aust. Wildl. Res.* **8**, 421–430.

Sculthorpe, C. D. (1967). "The Biology of Aquatic Vascular Plants." Edward Arnold, London.

Sexsmith, J. J. (1967). Varietal differences in seed dormancy of wild oats. *Weeds* **15**, 252–255.

Sharma, K. P., Khan, T. I., and Bhardwaj, N. (1984). Temperature-regulated seed germination in *Neptunia oleracea* Lour. and its ecological significance. *Aquat. Bot.* **20**, 185–188.

Shaver, G. R., and Chapin, F. S., III (1980). Response to fertilization by various plant growth forms in an Alaskan tundra: Nutrient accumulation and growth. *Ecology* **61**, 662–675.

Shaver, G. R., Gartner, B. L. , Chapin, F. S., III, and Linkins, A. E. (1983). Revegetation of arctic disturbed sites by native tundra plants. *In* "Permafrost: Fourth International Conference Proceedings," pp. 1133–1138. National Academy Press, Washington, D.C.

Shaw, M. F. (1929). A microchemical study of the fruit coat of *Nelumbo lutea*. *Am. J. Bot.* **16**, 259–276.

Shreve, F. (1942). The desert vegetation of North America. *Bot. Rev.* **8**, 195–246.

Shreve, F. (1951). Vegetation of the Sonoran Desert. *Carnegie Inst. Washington Publ.* **591**.

Shukla, R. P., and Ramakrishnan, P. S. (1982). Comparative study on field germination and establishment of early vs. late successional trees in northeastern India. *Proc. Indian Natl. Sci. Acad., Part B* **48**, 115–120.

Sickels, F. A., and Simpson, R. L. (1985). Growth and survival of giant ragweed (*Ambrosia trifida* L.) in a Delaware River freshwater tidal wetland. *Bull. Torrey Bot. Club* **112**, 368–375.

Sidhu, S. S. (1971). Some aspects of the ecology of black medick (*Medicago lupulina* L.). Ph.D. Thesis, University of Western Ontario, London, Canada.

Sidhu, S. S., and Cavers, P. B. (1977). Maturity-dormancy relationships in attached and detached seeds of *Medicago lupulina* L. (black medick). *Bot. Gaz. (Chicago)* **138**, 174–182.

Sih, A., Crowley, P., McPeek, M., Petranka, J., and Strohmeier, K. (1985). Predation, competition, and prey communities: A review of field experiments. *Annu. Rev. Ecol. Syst.* **16**, 269–312.

Silvertown, J. W. (1980a). Leaf-canopy-induced dormancy in a grassland flora. *New Phytol.* **85**, 109–118.

Silvertown, J. W. (1980b). The evolutionary ecology of mast seeding in trees. *Biol. J. Linn. Soc.* **14**, 235–250.

Silvertown, J. W. (1981). Seed size, lifespan, and germination date as co-adapted features of plant life history. *Am. Nat.* **118**, 860–864.

Silvertown, J. W. (1982). "Introduction to Plant Population Ecology." Longman, London.

Silvertown, J. (1988). The demographic and evolutionary consequences of seed dormancy. *In* "Plant Population Biology" (A. J. Davy, M. J. Hutchings, and A. R. Watkinson, eds.). Blackwell, Oxford (in press).

Simms, E. L., and Rausher, M. D. (1987). Cost and benefits of plant resistance to herbivory. *Am. Nat.* **130**, 570–581.

Simpson, B. B., Neff, J., and Moldenke, A. (1976). Flower - flower visitor - predator interactions. *In* "Convergent Evolution in Warm Deserts" (O. T. Solbrig and G. H. Orians, eds.). Dowden, Hutchinson, & Ross, New York.

Simpson, G. M. (1966). A study of germination in the seed of wild rice (*Zizania aquatica*). *Can. J. Bot.* **44**, 1–9.

Simpson, R. L., Leck, M. A., and Parker, V. T. (1985). The comparative ecology of *Impatiens capensis* (Balsaminaceae) in central New Jersey. *Bull. Torrey Bot. Club* **112**, 295–311.

Sivori, E., Nakayama, F., and Cigliano, E. (1968). Germination of achira seed (*Canna* sp.) approximately 550 years old. *Nature (London)* **219**, 1269–1270.

Slater, J. A. (1972). Lygaeid bugs (Hemiptera: Lygaeidae) as seed predators of figs. *Biotropica* **4**, 145–151.

Slingsby, P., and Bond, W. J. (1985). The influence of ants on the dispersal distance and seedling recruitment of *Leucospermum conocarpodendron* (L.) Buek (Proteaceae). *S. Afr. J. Bot.* **51**, 30–34.

Smela, S. (1987). The influence of soil cracking on radon–222 flux from a Panamanian soil. B.A. Senior Thesis, Princeton University, Princeton, New Jersey.

Smith, A. J. (1975). Invasion and ecesis of bird-disseminated woody plants in a temperate forest sere. *Ecology* **56**, 19–34.

Smith, A. P. (1987). Respuestas de hierbas del sotobosque tropical a claros ocasionados por la caído de árboles. *Rev. Biol. Trop.* **35**, Suppl. 1, 111–118.

Smith, C. C. (1970). The coevolution of pine squirrels (*Tamiaxiueua*) and conifers. *Ecol. Monogr.* **40**, 349–371.

Smith, C. C. (1975). The coevolution of plants and seed predators. *In* "Coevolution of Animals and Plants" (L. E. Gilbert and P. H. Raven, eds.), pp. 53–77. Univ. of Texas Press, Austin.

Smith, C. F. (1942). The fall food of brushfield pocket mice. *J. Mammal.* **23**, 337–339.

Smith, H. (1982). Light quality, photoperception, and plant strategy. *Annu. Rev. Plant Physiol.* **33**, 481–518.

Smith, J. H., and Clark, M. B. (1960). Growth and survival of Engelmann spruce and alpine fir on seed spots at Bolean Lake, B. C., 1954–59. *For. Chron.* **36**, 46–49.

Smith, L. M., and Kadlec, J. A. (1983). Seed banks and their role during drawdown of a North American marsh. *J. Appl. Ecol.* **20**, 673–684.

Smith, L. M., and Kadlec, J. A. (1985a). Predictions of vegetation change following fire in Great Salt Lake marsh. *Aquat. Bot.* **21**, 43–51.

Smith, L. M., and Kadlec, J. A. (1985b). The effects of disturbance on marsh seed banks. *Can. J. Bot.* **63**, 2133–2137.

Smith, L. M., and Kadlec, J. A. (1985c). Fire and herbivory in a Great Salt Lake marsh. *Ecology* **66**, 259–265.

Smith, R. H., and Bass, M. H. (1972). Relation of artificial pod removal to soybean yields. *J. Econ. Entomol.* **65**, 606–608.

Smith, T. J., III (1987a). Seed predation in relation to tree dominance and distribution in mangrove forests. *Ecology* **68**, 266–273.

Smith, T. J., III (1987b). Effects of seed predators and light level on the distribution of *Avicennia marina* (Forsk.) Vierh. in tropical tidal forests. *Estuarine, Coastal Shelf Sci.* **25**, 43–51.

Smith, T. J., III (1988). Structure and succession in tropical, tidal forests: The influence of seed predators. *Proc. Ecol. Soc. Aust.* **15** (in press).

Smythe, N. (1978). The natural history of the Central American agouti (*Dasyprocta punctata*). *Smithson. Contrib. Zool.* **257**, 1–52.

Soderstrom, E. (1986). The effect of secondary forest on the soil seed bank of primary forest. *In* "Tropical Biology: An Ecological Approach" (F. E. Putz, ed.), OTS 86-1, pp. 97–100. Organization for Tropical Studies, Durham, North Carolina.

Soholt, L. F. (1973). Consumption of primary production by a population of kangaroo rats (*Dipodomys merriami*) in the Mojave Desert. *Ecol. Monogr.* **43**, 357–376.

Solbreck, C., and Sillen-Tullberg, B. (1986a). Seed production and seed predation in a patchy and time-varying environment. *Oecologia* **71**, 51–58.

Solbreck, C., and Sillen-Tullberg, B. (1986b). The role of variable weather for the dynamics of a seed-seed predator system. *Oecologia* **71**, 59–62.

Sork, V. L. (1983). Mammalian seed dispersal of pignut hickory during three fruiting seasons. *Ecology* **64**, 1049–1056,

Sork, V. L. (1985). Germination response in a large-seeded neotropical tree species, *Gustavia superba* (Lecythidaceae). *Biotropica* **17**, 130–136.

Sork, V. L. (1987). Effects of predation and light on seedling establishment in *Gustavia superba*. *Ecology* **68**, 1341–1350.

Sork, V. L., and Boucher, D. H. (1977). Dispersal of sweet pignut hickory in a year of low fruit production, and the influence of predation by a curculionid beetle. *Oecologia* **29**, 289–299.

Sousa, W. P. (1984). The role of disturbances in natural communities. *Annu. Rev. Ecol. Syst.* **15**, 353–391.

Specht, R. L. (1981). Heathlands. *In* "Australian Vegetation" (R. H. Groves, ed.), pp. 253–275. Cambridge Univ. Press, London and New York.

Spira, T. P., and Wagner, L. K. (1983). Viability of seeds up to 211 years old extracted from adobe brick buildings of California and northern Mexico. *Am. J. Bot.* **70**, 303–307.

Stamp, N. E. (1984). Effect of defoliation by checkerspot caterpillars (*Euphydryas phaeton*) and sawfly larvae (*Macrophya nigra* and *Tenthredo grandis*) on their host plant (*Chelone* spp.). *Oecologia* **63**, 275–280.

Stanton, M. L. (1984). Seed variation in wild radish: Effect of seed size on components of seedling and adult fitness. *Ecology* **65**, 1105–1112.

Stanton, M. L. (1985). Seed size and emergence time within a stand of wild radish (*Raphanus raphanistrum* L.): The establishment of a fitness hierarchy. *Oecologia* **67**, 524–531.

Starfield, A. M., and Bleloch, A. L. (1986). "Building Models for Conservation and Wildlife Management." Macmillan, New York.

Stebbins, G. L. (1971). Adaptive radiation of reproductive characteristics in angiosperms. II. Seeds and seedlings. *Annu. Rev. Ecol. Syst.* **2**, 237–260.

Stebbins, G. L. (1974). "Flowering Plants, Evolution Above the Species Level." Harvard Univ. Press, Cambridge, Massachusetts.

Steenis, J. H., and Warren, J. (1959). Management of needlerush for improving waterfowl habitat in Maryland. *Proc. Annu. Conf. Southeast. Assoc. Game Fish Comm.* **13**, 296–298.

Stephenson, A. G. (1980). Fruit set, herbivory, fruit reduction, and the fruiting strategy of *Catalpa speciosa* (Bignoniaceae). *Ecology* **61**, 57–64.

Stephenson, A. G. (1981). Flower and fruit abortion: Proximate causes and ultimate functions. *Annu. Rev. Ecol. Syst.* **12**, 253–279.

Stevens, O. A. (1932). The number and weight of seeds produced by weeds. *Am. J. Bot.* **19**, 784–794.

Stocking, S. K. (1962). Some aspects of the influence of fire and sodium-calcium borate on chaparral vegetation. M. S. Thesis, San Francisco State University, San Francisco, California.

Stocking, S. K. (1966). Influences of fire and sodium-calcium borate on chaparral vegetation. *Madroño* **18**, 193–244.

Stoller, E. W., and Wax, L. M. (1974). Dormancy changes and fate of some annual weed seeds in the soil. *Weed Sci.* **22**, 151–155.

Stone, E. C. (1951). The stimulative effect of fire on the flowering of the golden Brodiaea (*Brodiaea ixiodes* Wats. var. *lugens* Jepson). *Ecology* **32**, 534–537.

Stone, E. C., and Juhren, G. (1951). The effect of fire on the germination of the seed of *Rhus ovata*. *Am. J. Bot.* **38**, 368–372.

Stone, E. C., and Juhren, G. (1953). Fire stimulated germination. *Calif. Agric.* **9**, 13–14.

Strickler, G. S., and Edgerton, P. J. (1976). Emergent seedlings from coniferous litter and soil in eastern Oregon. *Ecology* **57**, 801–807.

Sugden, A. M., Tanner, E. V. J., and Kapos, V. (1985). Regeneration following clearing in a Jamaican montane forest: Results of a ten-year study. *J. Trop. Ecol.* **1**, 329–351.

Sunderland, N. (1960). The production of the *Striga* and *Orobanche* germination stimulant by maize roots. *J. Exp. Bot.* **11**, 236–245.

Swaine, M. D., and Hall, J. B. (1983). Early succession on cleared forest land in Ghana. *J. Ecol.* **71**, 601–627.

Swaine, M. D., and Whitmore, T. C. (in press). On the definition of ecological species groups in tropical rain forests. *Vegetatio.*

Sweeney, J. R. (1956). Responses of vegetation to fire: A study of the herbaceous vegetation following chaparral fires. *Univ. Calif., Berkeley, Publ. Bot.* **28**, 143–249.

Sweeney, J. R. (1967). Ecology of some "fire type" vegetations in northern California. *Proc. Tall Timbers Fire Ecol. Conf.* **7**, 111–125.

Symington, C. F. (1933). The study of secondary growth on rain forest sites in Malaya. *Malay. For.* **2**, 107–117.

Tacey, W. H., and Glossop, B. L. (1980). Assessment of topsoil handling techniques for rehabilitation of sites mined for bauxite within the Jarrah forest of Western Australia. *J. Appl. Ecol.* **17**, 195–201.

Tamm, C. O. (1956). Further observations on the survival and flowering of some perennial herbs. I. *Oikos* **7**, 273–292.

Taylor, C. E. (1982). Reproductive ecology of some tropical pioneer trees. Ph.D. Thesis, University of Aberdeen, Aberdeen, Scotland.

Taylor, C. J. (1960). "Synecology and Silviculture in Ghana." Thomas Nelson, Edinburgh.

Taylor, F. (1979). Convergence to the stable age distribution in populations of insects. *Am. Nat.* **113**, 511–530.

Taylor, G. B. (1981). Effect of constant temperature treatments followed by fluctuating temperatures on the softening of hard seeds of *Trifolium subterraneum* L. *Aust. J. Plant Physiol.* **8**, 547–558.

Taylorson, R. B. (1970). Changes in dormancy and viability of weed seeds in soils. *Weed Sci.* **18**, 265–269.

Taylorson, R. B. (1972). Phytochrome controlled changes in dormancy and germination of buried weed seeds. *Weed Sci.* **20**, 417–422.

Templeton, A. R., and Levin, D. A. (1979). Evolutionary consequences of seed pools. *Am. Nat.* **114**, 232–249.

Terborgh, J. (1986). Keystone plant resources in the tropical forest. *In* "Conservation Biology: Science of Scarcity and Diversity" (M. E. Soulé, ed.), pp. 330–344. Sinauer, Sunderland, Massachusetts.

Terpstra, R. (1986). Behavior of weed seed in soil clods. *Weed Sci.* **34**, 889–895.

Tevis, L., Jr. (1958a). A population of desert ephemerals germinated by less than one inch of rain. *Ecology* **39**, 688–695.

Tevis, L., Jr. (1958b). Interrelations between the harvester ant *Veromessor pergandei* (Mayr) and some desert ephemerals. *Ecology* **39**, 695–704.

Thomas, A. G. (1979a). Weed survey of cultivated land in Saskatchewan. *Agric. Can. Weed Surv. Ser. Publ.* **79–2.**

Thomas, A. G. (1979b). Weed survey of cultivated land in Manitoba. *Agric. Can. Weed Surv. Ser. Publ.* **79–3.**

Thomas, P. A., and Wein, R. W. (1985a). The influence of shelter and the hypothetical effect of fire severity on the postfire establishment of conifers from seed. *Can. J. For. Res.* **15**, 148–155.

Thomas, P. A., and Wein, R. W. (1985b). Delayed emergence of four conifer species on postfire seedbeds in eastern Canada. *Can. J. For. Res.* **15**, 727–729.

Thompson, J. N. (1978). Within-patch structure and dynamics in *Pastinaca sativa* and resource availability to a specialized herbivore. *Ecology* **59**, 443–448.

Thompson, J. N. (1982). "Interaction and Coevolution." Wiley, New York.

Thompson, J. N. (1983). The use of ephemeral plant parts on small host plants: How *Depressaria leptotaeniae* (Lepidoptera: Oecophoridae) feeds on *Lomatium dissectum* (Umbelliferae). *J. Anim. Ecol.* **52**, 281–291.

Thompson, J. N. (1985). Post-dispersal seed predation in *Lomatium* spp. (Umbelliferae): Variation among individuals and species. *Ecology* **66**, 1608–1616.

Thompson, J. N., and Willson, M. F. (1978). Disturbance and dispersal of fleshy fruits. *Science* **200**, 1161–1163.

Thompson, K. (1978). The occurrence of buried viable seeds in relation to environmental gradients. *J. Biogeogr.* **5**, 425–430.

Thompson, K. (1987). Seeds and seed banks. *New Phytol.* **106**, Suppl., 23–34.

Thompson, K., and Grime, J. P. (1979). Seasonal variation in the seed banks of herbaceous species in ten contrasting habitats. *J. Ecol.* **67**, 893–921.

Thompson, K., Grime, J. P., and Mason, G. (1977). Seed germination in response to diurnal fluctuations in temperature. *Nature (London)* **267**, 147–148.

Tiner, R. W., Jr. (1984). "Wetlands of the United States: Current Status and Recent Trends." U.S. Fish Wildl. Serv., National Wetlands Inventory, Newton Corner, Massachusetts.

Titus, J. (1988). Soil seed bank of a hardwood floodplain swamp in Florida. *Castanea* (in press).

Toole, E. H., and Brown, E. (1946). Final results of the Duvel buried seed experiment. *J. Agric. Res.* **72**, 201–210.

Troumbis, A., and Larigauderie, A. (1987). Why does *Cistus* dominated Greek maquis not senesce? *Bull. Ecol. Soc. Am.* **68**, 433 (abstr.).

Troumbis, A., and Trabaud, L. (1986). Comparison of reproductive biological attributes of two *Cistus* species. *Oecol. Plant.* **7**, 235–250.

Turkington, R., and Burdon, J. J. (1983). The biology of Canadian weeds. 57. *Trifolium repens* L. *Can. J. Plant Sci.* **63**, 243–266.

Turkington, R., and Franko, G. D. (1980). The biology of Canadian weeds. 41. *Lotus corniculatus* L. *Can. J. Plant Sci.* **60**, 965–979.

Turnbull, C. L., and Culver, D. C. (1983). The timing of seed dispersal in *Viola nuttallii*: Attraction of dispersers and avoidance of predators. *Oecologia* **59**, 360–365.

Turnbull, C. L., Beattie, A. J., and Hanzawa, F. M. (1983). Seed dispersal by ants in the Rocky Mountains. *Southwest. Nat.* **28**, 289–293.

Turner, J. H. (1933). The viability of seeds. *Bull. Misc. Inf., R. Bot. Gard.* pp. 257–268.

Turner, R. M. (1982a). Great Basin desert scrub. *Desert Plants* **4**, 145–155.

Turner, R. M. (1982b). Mohave desert scrub. *Desert Plants* **4**, 157–168.

Turner, R. M., and Brown, D. E. (1982). Sonoran desert scrub. *Desert Plants* **4**, 181–221.

Tyndall, R. W., Teramura, A. H., Mulchi, C. L., and Douglass, L. W. (1986). Seed burial effect on species presence along a mid-Atlantic beach. *Can. J. Bot.* **64**, 2168–2170.

Ueckert, D. N. (1973). Effects of leaf-footed bugs on mesquite reproduction. *J. Range Manage.* **26**, 227–229.

Uhl, C., and Clark, K. (1983). Seed ecology of selected Amazon Basin successional species. *Bot. Gaz. (Chicago)* **144**, 419–425.

Uhl, C., Clark, K., Clark, H., and Murphy, P. (1981). Early plant succession after cutting and burning in the upper Rio Negro region of the Amazon Basin. *J. Ecol.* **69**, 631–649.

Uhl, C., Clark, H., Clark, K., and Maquirino, P. (1982a). Successional patterns and processes associated with slash-and-burn agriculture in the upper Rio Negro region of the Amazon Basin. *Biotropica* **14**, 249–254.

Uhl, C., Jordan, C., Clark, K., Clark, H., and Herrera, R. (1982b). Ecosystem recovery in Amazon caatinga forest after cutting, cutting and burning, and bulldozer clearing treatments. *Oikos* **38**, 313–320.

Ungar, I. A. (1978). Halophyte seed germination. *Bot. Rev.* **44**, 233–264.

Ungar, I. A. (1984). Autecological studies with *Atriplex triangularis* Willdenow. *USDA For. Serv. Gen. Tech. Rep. INT* **INT-172**, 40–52.

Ungar, I. A. (1987a). Population biology of halophyte seeds. *Bot. Rev.* **53**, 301–334.

Ungar, I. A. (1987b). Population characteristics, growth, and survival of the halophyte *Salicornia europaea*. *Ecology* **68**, 71–102.

Ungar, I. A., and Riehl, T. E. (1980). The effect of seed reserves on species composition in zonal halophyte communities. *Bot. Gaz. (Chicago)* **141**, 447–452.

Upadhyaya, M. K., Turkington, R., and McIlvride, D. (1986). The biology of Canadian weeds. 75. *Bromus tectorum* L. *Can. J. Plant Sci.* **66**, 689–709.

Urbanska, K. M., and Schutz, M. (1986). Reproduction by seed in alpine plants and revegetation research above timberline. *Bot. Helv.* **96**, 43–60.

U. S. Department of Agriculture (1948). "Woody-Plant Seed Manual," USDA For. Serv. Misc. Publ. 654. USDA, Washington, D.C.

Vallance, K. B. (1950). Studies on the germination of the seeds of *Striga hermonthica*. I. The influence of moisture treatment, stimulation, dilution and after-ripening on germination. *Ann. Bot. (London)* [N.S.] **14**, 347–363.

Vanden Born, W. H. (1971). Green foxtail: Seed dormancy, germination and growth. *Can. J. Plant Sci.* **51**, 53–59.

Vandermeer, J. H. (1974). Relative isolation and seed predation in *Calliandra grandiflora*, a mimosaceous legume from the highlands of Guatemala. *Biotropica* **6**, 267–268.

Vandermeer, J. H. (1975). A graphical model of insect seed predation. *Am. Nat.* **109**, 147–160.

van der Pijl, L. (1969). "Principles of Dispersal in Higher Plants." Springer-Verlag, Berlin.

van der Valk, A. G. (1981). Succession in wetlands: A Gleasonian approach. *Ecology* **62**, 688–696.

van der Valk, A. G. (1986). The impact of litter and annual plants on recruitment from the seed bank of a lacustrine wetland. *Aquat. Bot.* **24**, 13–26.

van der Valk, A. G., and Davis, C. B. (1976). The seed banks of prairie glacial marshes. *Can. J. Bot.* **54**, 1832–1838.

van der Valk, A. G., and Davis, C. B. (1978). The role of seed banks in the vegetation dynamics of prairie glacial marshes. *Ecology* **59**, 322–335.

van der Valk, A. G., and Davis, C. B. (1979). A reconstruction of the recent vegetational history of a prairie marsh, Eagle Lake, Iowa, from its seed bank. *Aquat. Bot.* **6**, 29–51.

van der Valk, A. G., and Verhoeven, J. T. A. (1988). The restoration of quaking fens from floating forests: The role of the seed bank and understory species. *Vegetatio* **76**, 3–13.

van der Valk, A. G., and Welling, C. H. (1988). The development of zonation in freshwater wetlands: An experimental approach. *Proc. Int. Symp. Veg. Struct., 1987* (in press).

van der Valk, A. G., Welling, C. H., and Pederson, R. L. (1988). Predicting vegetation change in a freshwater wetland: A test of a priori predictions. *In* "Freshwater Wetlands and Wildlife" (R. R Sharitz and J. W. Gibbons, eds.) DOE-CONS 860326. Off. Tech. Info. USDOE. (in press).

van der Vegte, F. W. (1978). Population differentiation and germination ecology in *Stellaria media* (L.) Vill. *Oecologia* **37**, 231–245.

VanDevender, T. R., and Spaulding, W. G. (1983). Development of vegetation and climate in the Southwestern United States. *In* "Origin and Evolution of Deserts" (S. G. Wells and D. R. Haragan, eds.), pp. 131–156. Univ. of New Mexico Press, Albuquerque.

Van Esso, M. L., Ghersa, C. M., and Soriano, A. (1986). Cultivation effects on the dynamics of a Johnson grass seed population in the soil profile. *Soil Tillage Res.* **6**, 325–335.

Van Wagner, C. E. (1978). Age class distribution and the forest fire cycle. *Can. J. For. Res.* **8**, 220–227.

Van Wagner, C. E. (1983). Fire behaviour in northern conifer forests and shrublands. *In* "The Role of Fire in Northern Circumpolar Ecosystems" (R. W. Wein and D. A. MacLean, eds.), pp. 65–80. Wiley, New York.

Vasek, F. C., and Barbour, M. G. (1977). Mojave desert scrub vegetation. *In* "Terrestrial Vegetation of California" (M. G. Barbour and J. Major, eds.), pp. 835–867. Wiley, New York.

Vázquez-Yanes, C. (1974). Studies on the germination of seeds of *Ochroma lagopus* Swartz. *Turrialba* **24**, 176–179.

Vázquez-Yanes, C. (1976a). Estudios sobre ecofisiología de la germinación en una zona calido-húmeda. *In* "Investigaciones sobre la Regeneración de Selvas Altas en Veracruz, México" (A. Gómez-Pompa, C. Vázquez-Yanes, S. del Amo Rodríguez, and A. Butanda Cervana, eds.), pp. 279–387. Compañia Editorial Continental, México, D. F.

Vázquez-Yanes, C. (1976b). Seed dormancy and germination in secondary vegetation tropical plants: The role of light. *Comp. Physiol. Ecol.* **1**, 30–34.

Vázquez-Yanes, C. (1981). Germinación de dos especies de Tiliaceas arbóreas de la vegetación secundaria tropical: *Belotia campbellii* y *Heliocarpus donnell-smithii*. *Turrialba* **31**, 81–82.

Vázquez-Yanes, C., and Orozco Segovia, A. (1982a). Seed germination of a tropical rain forest pioneer tree (*Heliocarpus donnell-smithii*) in response to diurnal fluctuation of temperature. *Physiol. Plant.* **56**, 295–298.

Vázquez-Yanes, C., and Orozco Segovia, A. (1982b). Germination of the seeds of a tropical rain forest shrub, *Piper hispidum* SW. (Piperaceae) under different light qualities. *Phyton* **42**, 143–149.

Vázquez-Yanes, C., and Orozco Segovia, A. (1984). Ecophysiology of seed germination in the tropical humid forests of the world: A review. *In* "Physiological Ecology of Plants in the Wet Tropics" (E. Medina, H. A. Mooney, and C. Vázquez-Yanes, eds.), pp. 37–50. Junk, The Hague.

Vázquez-Yanes, C., and Orozco Segovia, A. (1986). Dispersal of seeds by animals: Effect on light-controlled dormancy in *Cecropia obtusifolia*. *In* "Frugivores and Seed Dispersal" (A. Estrada and T. H. Fleming, eds.), pp. 71–77. Junk, Dordrecht.

Vázquez-Yanes, C., and Orozco Segovia, A. (1987). Fisiología ecológica de semillas en la Estación de Biología Tropical "Los Tuxtlas," Veracruz, México. *Rev. Biol. Trop.* **35**, 85–96.

Vázquez-Yanes, C., and Smith, H. (1982). Phytochrome control of seed germination in the tropical rain forest pioneer trees *Cecropia obtusifolia* and *Piper auritum* and its ecological significance. *New Phytol.* **92**, 477–485.

Vegis, A. (1964). Dormancy in higher plants. *Annu. Rev. Plant Physiol.* **15**, 185–215.

Venable, D. L. (1985). The evolutionary ecology of seed heteromorphism. *Am. Nat.* **126**, 577–595.

Venable, D. L., and Brown, J. S. (1988). The selective interactions of dispersal, dormancy, and seed size as adaptations for reducing risk in variable environments. *Am. Nat.* **131**, 360–384.

Venable, D. L., and Lawlor, L. (1980). Delayed germination and dispersal in desert annuals: Escape in space and time. *Oecologia* **46**, 272–282.

Venable, D. L., Burquez, A., Corral, G., Morales, E., and Espinosa, F. (1987). The ecology of seed heteromorphism in *Heterosperma pinnatum* in central Mexico. *Ecology* **68**, 65–76.

Viereck, L. A. (1973). Wildfire in the taiga of Alaska. *Quat. Res.* **3**, 465–495.

Villiers, T. A. (1973). Ageing and the longevity of seeds in field conditions. *In* "Seed Ecology" (W. Heydecker, ed.), pp. 265–287. Pennsylvania State Univ. Press, University Park.

Vincent, E. M., and Cavers, P. B. (1978). The effects of wetting and drying on the subsequent germination of *Rumex crispus* L. *Can. J. Bot.* **56**, 2207–2217.

Vitousek, P. M. (1983). Nitrogen turnover in a ragweed-dominated first-year old field in southern Indiana. *Am. Midl. Nat.* **110**, 46–53.

Vlahos, S., and Bell, D. T. (1986). Soil seed bank components of the northern jarrah forest of Western Australia. *Aust. J. Ecol.* **11**, 171–179.

Vogl, R. J. (1973). Ecology of knobcone pine in the Santa Ana Mountains, California. *Ecol. Monogr.* **43**, 125–143.

Vogl, R. J. (1980). The ecological factors that produce perturbation dependent ecosystems. *In* "The Recovery Process in Damaged Ecosystems" (J. Cairns, ed.), pp. 63–94. Ann Arbor Sci. Publ., Ann Arbor, Michigan.

Walker, L. R., Zasada, J. C., and Chapin, F. S., III (1986). The role of life history processes in primary succession on an Alaskan floodplain. *Ecology* **67**, 1243–1253.

Waloff, N., and Richards, O. W. (1977). The effect of insect fauna on growth, mortality and natality of broom, *Sarothamnus scoparius*. *J. Appl. Ecol.* **14**, 787–798.

Walter, H. (1985). "Vegetation of the Earth and Ecological Systems of the Geobiosphere," 3rd Engl. ed. Springer-Verlag, Berlin.

Walter, H., and Box, E. O. (1983). Middle Asian deserts. *In* "Ecosystems of the World" (N. E. West, ed.), Vol. 5, pp. 79–104. Am. Elsevier, New York.

Wapshere, A. J. (1974). An ecological study of an attempt at biological control of Noogoora Burr (*Xanthium strumarium*). *Aust. J. Agric. Res.* **25**, 275–292.

Warnes, D. D., and Andersen, R. N. (1984). Decline of wild mustard (*Brassica kaber*) seeds in soil under various cultural and chemical practices. *Weed Sci.* **32**, 214–217.

Washitani, I., and Saeki, T. (1984). Leaf-canopy inhibition of germination as a mechanism for the disappearance of *Amaranthus patulus* Bertol. in the second year of a secondary succession. *Jpn. J. Ecol.* **34**, 55–61.

Washitani, I., and Takenaka, A. (1986). 'Safe sites' for the seed germination of *Rhus javanica*: A characterisation by responses to temperature and light. *Ecol. Res.* **1**, 71–82.

Watson, A. K., and Renney, A. J. (1974). The biology of Canadian weeds. 6. *Centaurea diffusa* and *C. maculosa*. *Can. J. Plant Sci.* **54**, 687–701.

Watt, A. S. (1947). Pattern and process in the plant community. *J. Ecol.* **35**, 1–22.

Watt, A. S. (1974). Senescence and rejuvenation in an ungrazed chalk grassland in Breckland: The significance of litter and of moles. *J. Appl. Ecol.* **11**, 1157–1171.

Weaver, S. E., and Lechowicz, M. J. (1982). The biology of Canadian weeds. 56. *Xanthium strumarium* L. *Can. J. Plant Sci.* **63**, 211–225.

Weaver, S. E., and Riley, W. R. (1982). The biology of Canadian weeds. 53. *Convolvulus arvensis* L. *Can. J. Plant Sci.* **62**, 461–472.

Webb, L. J., Tracey, J. G., and Williams, W. T. (1972). Regeneration and pattern in the subtropical rain forest. *J. Ecol.* **60**, 675–695.

Wee, Y. C. (1974). Viable seeds and spores of weed species in peat soil under pineapple cultivation. *Weed Res.* **14**, 193–196.

Wein, R. W. (1974). "Recovery of Vegetation in Arctic Regions after Burning," Inf. Can.

Rep. No. 74–6. Environmental-Social Committee, Northern Pipelines Task Force on Northern Oil Development.

Weiner, J., and Solbrig, O. T. (1984). The meaning and measurement of size hierarchies in plant populations. *Oecologia* **61**, 334–336.

Weis, I. M. (1982). The effects of propagule size on germination and seedling growth in *Mirabilis hirsuta*. *Can. J. Bot.* **60**, 1868–1874.

Weller, M. W., and Fredrickson, L. H. (1974). Avian ecology of a managed glacial marsh. *Living Bird* **12**, 269–291.

Weller, M. W., and Spatcher, C. E. (1965). Role of habitat in the distribution and abundance of marsh birds. *Spec. Rep., Iowa State Agric. Home Econ. Exp. Stn.* **43**.

Welling, C. H. (1987). Reestablishment of perennial emergent macrophytes during a drawdown in a lacustrine marsh. M.S. Thesis, Iowa State University, Ames.

Welling, C. H., Pederson, R. L., and van der Valk, A. G. (1988). Recruitment from the seed bank and the development of emergent zonation during a drawdown in a prairie wetland. *J. Ecol.* **76**, 483–496.

Wellington, A. B., and Noble, I. R. (1985). Seed dynamics and factors limiting recruitment of the mallee *Eucalyptus incrassata* in semi-arid south-eastern Australia. *J. Ecol.* **73**, 657–666.

Wells, P. V. (1962). Vegetation in relation to geological substratum and fire in the San Luis Obispo Quadrangle, California. *Ecol. Monogr.* **32**, 79–103.

Wells, P. V. (1969). The relation between mode of reproduction and extent of speciation in woody genera of the California chaparral. *Evolution (Lawrence, Kans.)* **23**, 264–267.

Wells, P. V. (1977). Post-glacial origin of the present Chihuahuan Desert less than 11,500 years ago. *In* "Transactions of the Symposium of the Biological Resources of the Chihuahuan Desert Region, U.S. and Mexico" (R. H. Wauer and D. H. Riskind, eds.), Trans. Proc., Ser. 3, pp. 67–83. U.S. Dept. Int., Natl. Park Serv., Washington, D.C.

Went, F. W. (1942). The dependence of certain annual plants on shrubs in southern California deserts. *Bull. Torrey Bot. Club* **69**, 100–114.

Went, F. W. (1949). Ecology of desert plants. II. The effect of rain and temperature on germination and growth. *Ecology* **30**, 1–13.

Went, F. W. (1957). "The Experimental Control of Plant Growth." Chronica Botanica, Waltham, Massachusetts.

Went, F. W. (1969). A long-term test of seed longevity. *Aliso* **7**, 1–12.

Went, F. W. (1973). Competition among plants. *Proc. Natl. Acad. Sci. U.S.A.* **70**, 585–590.

Wentworth, T. R. (1981). Vegetation on limestone and granite in the Mule Mountains, Arizona. *Ecology* **62**, 469–482.

Werner, P. A. (1977). Colonization success of a "biennial" plant species: Experimental field studies of species cohabitation and replacement. *Ecology* **58**, 840–849.

Werner, P. A. (1979). Competition and coexistence of similar species. *In* "Topics in Plant Population Biology" (O. T. Solbrig, S. Jain, G. B. Johnson, and P. H. Raven, eds.), pp. 287–310. Columbia Univ. Press, New York.

Werner, P. A., and Rioux, R. (1977). The biology of Canadian weeds. 24. *Agropyron repens* (L.) Beauv. *Can. J. Plant Sci.* **57**, 905–919.

Wertis, B. A., and Ungar, I. A. (1986). Seed demography and seedling survival in a population of *Atriplex triangularis* Willd. *Am. Midl. Nat.* **116**, 152–162.

Wesley, S. L., Hill, N. M., and van der Kloet, S. P. (1986). Seed banks of *Vaccinium angustifolium* Aiton on managed and unmanaged barrens in Nova Scotia. *Rev. Ecol.* **113**, 309–316.

Wesolowski, M. (1979a). [Species composition and number of weed seeds in the soils of southeast Poland. Part I. Soils of the lowland region.] *Ann. Univ. Mariae Curie-Sklodowska, Sect. E* **34**, 23–36.

Wesolowski, M. (1979b). [Species composition and number of weed seeds in the soils of southeast Poland. Part II. Soils of the mountainous region.] *Ann. Univ. Mariae Curie-Sklodowska, Sect. E* **34**, 37–47.

Wesson, G., and Wareing, P. F. (1967). Light requirements of buried seeds. *Nature (London)* **213**, 600–601.

Wesson, G., and Wareing, P. F. (1969a). The role of light in the germination of naturally occurring populations of buried weed seeds. *J. Exp. Bot.* **20**, 401–413.

Wesson, G., and Wareing, P. F. (1969b). The induction of light sensitivity in weed seeds by burial. *J. Exp. Bot.* **20**, 414–425.

West, D., and Whigham, D. F. (1975–1976). Seed germination of arrow arum (*Peltandra virginica* L.). *Bartonia* **44**, 44–49.

West, N. E. (1979). Survival patterns of major perennials in salt desert shrub communities of southwestern Utah. *J. Range Manage.* **32**, 442–445.

West, N. E. (1983a). Great Basin-Colorado Plateau sagebrush semi-desert. *In* "Ecosystems of the World" (N. E. West, ed.), Vol. 5, pp. 331–349. Am. Elsevier, New York.

West, N. E. (1983b). Intermountain salt-desert shrubland. *In* "Ecosystems of the World" (N. E. West, ed.), Vol. 5, pp. 375–397. Am. Elsevier, New York.

West, N. E. (1983c). Colorado Plateau-Mohavian blackbrush semi-desert. *In* "Ecosystems of the World" (N. E. West, ed.), Vol. 5, pp. 399–411. Am. Elsevier, New York.

West, N. E., and Ibrahim, K. L. (1968). Soil-vegetation relationships in the shadscale zone of southeastern Utah. *Ecology* **49**, 445–456.

West, N. E., Rea, K. H., and Harniss, R. O. (1979). Plant demographic studies in sagebrush-grass communities of southeastern Idaho. *Ecology* **60**, 376–388.

Westoby, M., Cousins, J. M., and Grice, A. C. (1982). Rate of decline of some soil seed populations during drought in western New South Wales. *In* "Ant–plant Interactions in Australia" (R. C. Buckley, ed.), pp. 7–16. Junk, The Hague.

Whigham, D. F., Simpson, R. L., and Leck, M. A. (1979). The distribution of seeds, seedlings, and established plants of arrow arum (*Peltandra virginica* (L.) Kunth) in a freshwater tidal wetland. *Bull. Torrey Bot. Club* **106**, 193–199.

Whipple, S. A. (1978). The relationship of buried, germinating seeds to vegetation in an old-growth Colorado subalpine forest. *Can. J. Bot.* **20**, 403–413.

White, A. S. (1983). The effects of thirteen years of annual prescribed burning on a *Quercus ellipsoides* community in Minnesota. *Ecology* **64**, 1081–1085.

Whitford, W. G. (1978). Foraging in seed harvester ants *Pogonomyrmex* spp. *Ecology* **59**, 185–189.

Whitmore, T. C. (1978). Gaps in the forest canopy. *In* "Tropical Trees as Living Systems" (P. B. Tomlinson and M. H. Zimmermann, eds.), pp. 639–655. Cambridge Univ. Press, London and New York.

Whitmore, T. C. (1983). Secondary succession from seed in tropical rain forests. *For. Abstr.* **44**, 767–779.

Whitmore, T. C. (1984). "Tropical Rain Forests of the Far East," 2nd ed. Oxford Univ. Press (Clarendon), London and New York.

Whittaker, J. B. (1979). Invertebrate grazing, competition and plant dynamics. *In* "Population Dynamics" (R. M. Anderson, B. D. Turner, and L. R. Taylor, eds.), pp. 207–222. Blackwell, Oxford.

Wicklow, D. T. (1977). Germination responses in *Emmenanthe penduliflora* (Hydrophyllaceae). *Ecology* **58**, 201–205.

Wilhelm, G., Mengler, J., and Squires, L. (1987). "Des Plaines River Wetlands Demonstration Project: Seed Bank Study," preliminary results. Morton Arboretum, Lisle, Illinois.

Williams, G. C. (1966). "Adaptations and Natural Selection." Princeton Univ. Press, Princeton, New Jersey.

Williams, K. S. (1985). Climatic influences on weeds and their herbivores: Biological control of St. John's Wort in British Columbia. *Proc. Int. Symp. Biol. Control Weeds, 6th, 1984* pp. 127–132.

Williams, W. A., and Elliott, J. R. (1960). Ecological significance of seed coat impermeability to moisture in Crimson, Subterranean and Rose clovers in a mediterranean-type climate. *Ecology* **41**, 733–742.

Williams, W. A., Cook, O. D., and Kay, B. L. (1974). Germination of native desert shrubs. *Calif. Agric.* **28**, 13.

Williams-Linera, G., and Ewel, J. J. (1984). Effect of autoclave sterilization of a tropical andept on seed germination and seedling growth. *Plant Soil* **82**, 263–268.

Willson, M. F. (1973). Evolutionary ecology of plants, a review. Part IV. Plant/animal interactions. *Biologist* **55**, 89–105.

Willson, M. F. (1983). "Plant Reproductive Ecology." Wiley, New York.

Wilson, D. E., and Janzen, D. H. (1972). Predation on *Scheelea* palm seeds by bruchid beetles: Seed density and distance from the parent palm. *Ecology* **53**, 954–959.

Wilson, R. G., Kerr, E. D., and Nelson, L. A. (1985). Potential for using weed seed content in the soil to predict future weed problems. *Weed Sci.* **33**, 171–175.

Wilson, S. D., and Keddy, P. A. (1985). Plant zonation on a shoreline gradient: Physiological response curves of component species. *J. Ecol.* **73**, 851–860.

Wilson, S. D., Keddy, P. A. and Randall, D. L. (1985). The distribution of *Xyris difformis* along a gradient of exposure to waves: An experimental study. *Can. J. Bot.* **63**, 1226–1230.

Windell, J. T., Willard, B. E., Cooper, D. J., Foster, S. Q., Knud-Hanson, C. F., Rink, L. P., and Kiladis, G. N. (1986). An ecological characterization of Rocky Mountain montane and subalpine wetlands. *U.S. Fish Wildl. Serv., Biol. Rep.* **86**(11).

Windus, J. L. (1987). Seed predation on *Gentiana saponaria* (Ohio). *Restoration Manage. Notes* **4**, 81.

Winter, D. M. (1960). The development of the seeds of *Abutilon theophrasti*. II. Seed coat. *Am. J. Bot.* **47**, 157–162.

Wisheu, I. C. (1987). Species richness-standing crop relationships on an infertile shoreline in Nova Scotia. M.S. Thesis, University of Ottawa, Ontario.

Wood, D. M., and Andersen, M. C. (1988). The effect of predispersal seed predators on colonization of *Aster ledophyllus* on Mount St. Helen. *Am. Midl. Nat.* (in press).

Woodell, S. R. J., Mooney, H. A., and Hill, A. J. (1969). The behavior of *Larrea divaricata* (creosote bush) in response to rainfall in California. *J. Ecol.* **57**, 37–44.

Worsham, K. D., Klingman, A. C., and Moreland, D. E. (1962). Promotion of germination of *Striga asiatica* seed by coumarin derivatives and effects on seedling development. *Nature (London)* **145**, 199–201.

Wourms, J. P. (1972). The developmental biology of annual fishes. III. Pre-embryonic and embryonic diapause of variable duration in the eggs of annual fishes. *J. Exp. Zool.* **182**, 389–414.

Wright, E. (1931). The effect of high temperature on seed germination. *J. For.* **29**, 679–687.

Wright, H. A., and Bailey, A. W. (1982). "Fire Ecology: United States and Southern Canada." Wiley, New York.

Wright, H. A., and Klemmedson, J. O. (1965). Effects of fire on bunch grass of the sage brush-grass region in southern Idaho. *Ecology* **46**, 680–688.

Wright, S. L. (1986). Prescribed burning as a technique to manage insect pests of oak regeneration. *In* "Proceedings of Prescribed Burning in the Midwest—State of the Art". University of Wisconsin-Stevens Point, and USDA For. Serv., Northeast. For. Exp. Stn.

Wulff, R. D. (1986a). Seed size variation in *Desmodium paniculatum*. I. Factors affecting seed size. *J. Ecol.* **74**, 87–96.

Wulff, R. D. (1986b). Seed size variation in *Desmodium paniculatum*. II. Effects on seedling growth and physiological performance. *J. Ecol.* **74**, 97–114.

Yeaton, R. I., and Cody, M. L. (1976). Competition and spacing in plant communities: The northern Mohave desert. *J. Ecol.* **64**, 689–696.

Young, J. A., and Evans, R. A. (1975). Germinability of seed reserves in a big sagebrush community. *Weed Sci.* **23**, 358–364.

Young, J. A., Evans, R. A, Raguse, C. A., and Larson, J.R. (1981). Germinable seeds and periodicity of germination in annual grasslands. *Hilgardia* **49**, 1–37.

Young, K. R. (1985). Deeply buried seeds in a tropical wet forest in Costa Rica. *Biotropica* **17**, 336–338.

Young, K. R., Ewel, J. J., and Brown, B. J. (1987). Seed dynamics during forest succession in Costa Rica. *Vegetatio* **71**, 157–173.

Youngman, B. J. (1951). Germination of old seeds. *Kew Bull.* pp. 423–426.

Zackrisson, O. (1977). Influence of forest fires on the north Swedish boreal forest. *Oikos* **29**, 22–32.

Zammit, C., and Hood, C. W. (1986). Impact of flower and seed predators on seed-set in two *Banksia* shrubs. *Aust. J. Ecol.* **11**, 187–193.

Zammit, C., and Westoby, M. (1987a). Population structure and reproductive status of two *Banksia* shrubs at various times after fire. *Vegetatio* **70**, 11–20.

Zammit, C., and Westoby, M. (1987b). Seedling recruitment strategies in obligate seeding and resprouting *Banksia* shrubs. *Ecology* **68**, 1984–1992.

Zammit, C., and Westoby, M. (1988a). Predispersal seed losses and the survival of seed and seedlings of two serotinous *Banksia* shrubs in burnt and unburnt heath. *J. Ecol.* **76**, 200–214.

Zammit, C., and Westoby, M. (1988b). Seedling recruitment strategies of obligate-seeding and resprouting *Banksia* shrubs from a fire prone heath. *J. Ecol.* (in press).

Zammit, C. A., and Zedler, P. H. (1988). The influence of dominant shrubs, fire, and time since fire on soil seed banks in mixed chaparral. *Vegetatio* **75**, 175–187.

Zavitkovski, J., and Newton, M. (1968). Ecological importance of snowbrush *Ceanothus velutinus* in the Oregon Cascades. *Ecology* **49**, 1134–1145.

Zedler, P. H. (1987). The ecology of southern California vernal pools: a community profile. *U.S., Fish Wildl. Serv., Biol. Rep.* **85**(7.11).

Zedler, P. H., Gautier, C. R., and McMaster, G. S. (1983a). Vegetation change in response to extreme events: The effect of a short interval between fires in California chaparral and coastal scrub. *Ecology* **64**, 809–818.

Zedler, P. H., Guehlstorff, K., Scheidlinger, C., and Gautier, C. R. (1983b). The population ecology of a dune thistle, *Circium rhothophilum* (Asteraceae). *Am. J. Bot.* **70**, 1516–1527.

Zimmerman, M. (1980a). Reproduction in *Polemonium*: Competition for pollinators. *Ecology* **61**, 497–501.

Zimmerman, M. (1980b). Reproduction in *Polemonium*: Predispersal seed predation. *Ecology* **61**, 502–506.

Zimmerman, M., and Gross, R. S. (1982). The relationship between flowering phenology and seed set in an herbaceous perennial plant, *Polemonium foliosissimum* Gray. *Am. Midl. Nat.* **111**, 185–191.

Zohary, M. (1937). Die verbreitungsökologischen Verhaltnisse der Pflanzen Palastinas. *Beih. Bot. Zentral bl., Abt. 1* **56**, 1–155.

Zohary, M. (1962). "Plant Life of Palestine." Ronald Press, New York.

Index